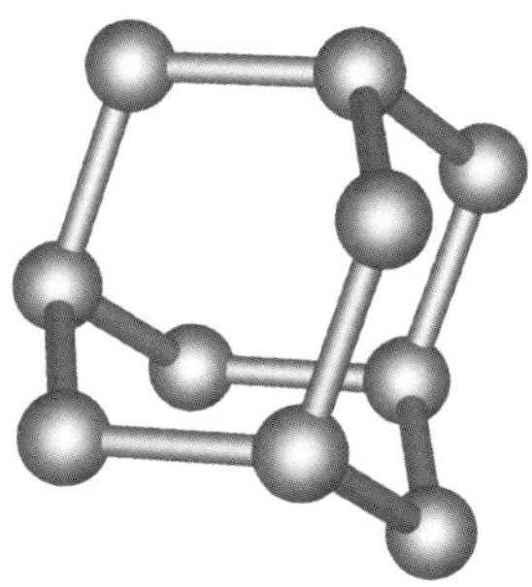

Diamondoid Molecules

With Applications in Biomedicine, Materials Science,
Nanotechnology & Petroleum Science

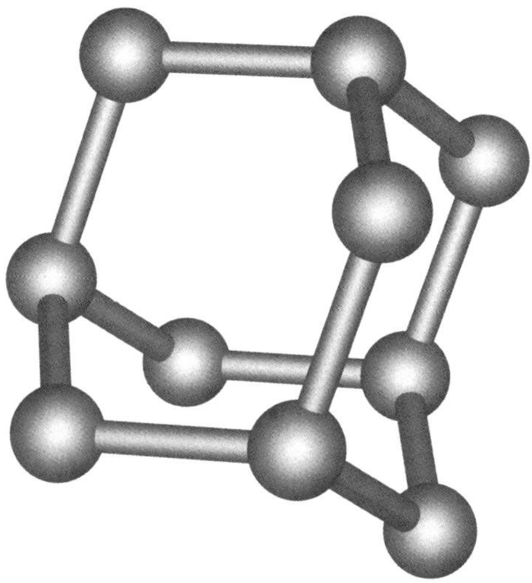

Diamondoid Molecules

With Applications in Biomedicine, Materials Science, Nanotechnology & Petroleum Science

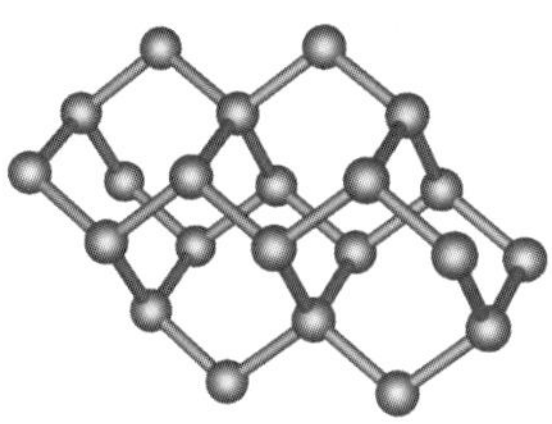

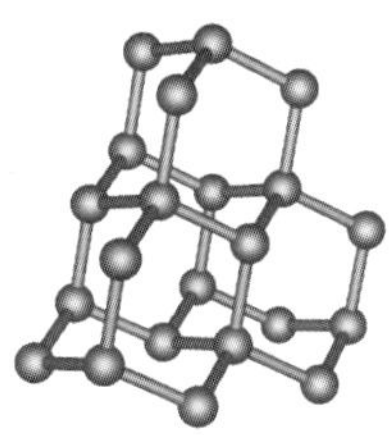

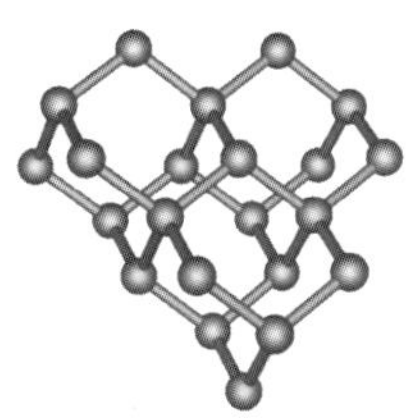

G Ali Mansoori

University of Illinois at Chicago, USA

Patricia Lopes Barros de Araujo

Elmo Silvano de Araujo

Universidade Federal Rural de Pernambuco, Brazil

World Scientific

NEW JERSEY • LONDON • SINGAPORE • BEIJING • SHANGHAI • HONG KONG • TAIPEI • CHENNAI

Published by

World Scientific Publishing Co. Pte. Ltd.
5 Toh Tuck Link, Singapore 596224
USA office: 27 Warren Street, Suite 401-402, Hackensack, NJ 07601
UK office: 57 Shelton Street, Covent Garden, London WC2H 9HE

British Library Cataloguing-in-Publication Data
A catalogue record for this book is available from the British Library.

DIAMONDOID MOLECULES
With Applications in Biomedicine, Materials Science, Nanotechnology & Petroleum Science

ISBN 978-981-4291-60-6

Typeset by Stallion Press
Email: enquiries@stallionpress.com

Printed in Singapore by Mainland Press Pte Ltd.

Contents

Preface

Diamondoids are a peculiar class of organic molecules (hydrocarbons) with unique structures and properties. They possess typical characteristics of diamond face-fused cages with hydrogen terminated dangling bonds. Diamondoids and derivatives have been of great interest since their discovery, due to their important and diverse roles in biomedicine, materials science, petroleum science, and more recently in nanotechnology. Due to their six or more linking groups, they have found major applications as templates and as molecular building blocks in nanotechnology, polymer synthesis, drug design and delivery, DNA-directed assembly, just to name a few.

This book is primarily designed to be a reference source on the fundamentals and applications of diamondoid molecules. The reader is introduced to various aspects of diamondoids including their molecular structure, chemistry, physical properties, natural sources, and their applications.

In Chapter 1, the molecular structure and chemistry of diamondoids is presented. That includes classification and crystalline structure of diamondoids, distinction between diamondoids which are well-defined molecules and nanodiamonds which are, in principle, nanoparticles. We also present the synthesis and functionalization of lower diamondoid cages including adamantane, diamantane, and triamantane.

In Chapter 2, we present a detailed information and data about the existence, genesis and role of diamondoids in petroleum and other fossil fuels. This includes the use of diamondoids as geochemical tools for petroleum characterization and their role in petroleum and natural gas flow fouling. Also methods for separation, detection and measurements of petroleum diamondoids are discussed.

In Chapter 3, we report on the available data and correlations on physical properties of diamondoids including their spectrometric, optical and thermodynamic properties.

Chapter 4 of this book is about diamondoids as molecular building blocks for nanotechnology. This includes the futuristic mechanosynthesis based on diamondoids, applications of diamondoids as molecular components of nanosystems, and their use for host–guest chemistry and in inclusion compounds.

Quantum calculation methods and results regarding prediction of the properties of diamondoids are reported in Chapter 5. That includes electronic, structural, and intermolecular interactions properties of diamondoids and derivatives.

Chapter 6 highlights the diverse and growing biomedical applications of diamondoids. Hence, subjects such as drug design and delivery in fighting infectious diseases, cancer, hypoglycemia, and diabetes with diamondoids derivatives are discussed.

Applications of diamondoids in materials science in macro and nano scales are discussed in Chapter 7. For macroscopic systems, this includes applications of diamondoids in polymer synthesis, in polymer nanocomposites and in crystal engineering. For nanosystems, we discuss applications of diamondoids in the design of diamondoids–DNA nanostructures and self-assembly of diamondoid molecules and derivatives towards NEMS and MEMS productions.

We are thankful to many family members, colleagues, collaborators and students who have helped and encouraged us to write and complete this book. We are specifically thankful to the following whose assistance and associations have been quite valuable in completing this book: L. Assoufid, R. Bagherian, L.A. Curry, A. Esche, J. Escobedo, F. Feumba, T. W. George, K. Kiani-Nassab, M.M. Moayeri, M. A. Moradi, A. Nazem, H. Ramezani, M. R. Saberi, G. R. Vakili-Nezhaad, D. Vazquez, Y. Xue, and G.P. Zhang.

The readers should bear in mind that, while writing this book, we studied as many publications as possible and available to us. However, even though we have many references for each chapter of the book, there are possibly other important publications which should have been quoted in this book, but they are omitted. The authors sincerely apologize to all others whose works, though relevant to the topics discussed in this book, but they are not quoted.

G. Ali Mansoori, Patricia Lopes Barros de Araujo
and Elmo Silvano de Araujo.

October 8, 2011

1

Molecular Structure and Chemistry of Diamondoids

1.1. Introduction

Diamondoid molecules are cage-like, ultra stable and saturated hydrocarbons. The basic repetitive unit of the diamondoids is a ten-carbon tetracyclic cage system called "adamantane" (Fig. 1.1). They are called "diamondoid" because they have at least one adamantane unit and their carbon–carbon framework is completely or largely superimposable on the diamond lattice (Balaban and Schleyer, 1978; Mansoori, 2007). The diamond lattices structure was first determined in 1913 by Bragg and Bragg using X-ray diffraction analysis (Bragg and Brag, 1913).

Diamondoids show unique properties due to their exceptional atomic arrangements. Adamantane consists of cyclohexane rings in "chair" conformation. The name adamantane is derived from the Greek language word for diamond since its chemical structure is like the three-dimensional diamond subunit as it is shown in Fig. 1.2.

1.2. Classification and Crystalline Structure of Diamondoids

The first and simplest member of the diamondoids group, adamantane, is a tricyclic saturated hydrocarbon (tricyclo[3.3.1.1(3.7)]decane, according to the von Bayer systematic nomenclature). Adamantane is followed by its

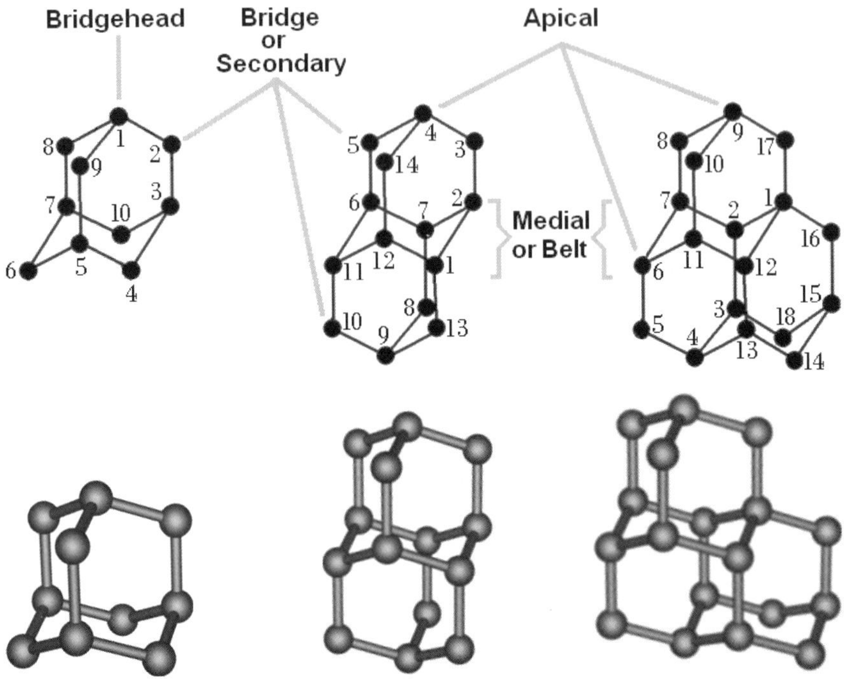

Fig. 1.1. Three-dimensional molecular structures of (from left to right) adamantane, diamantane and triamantane, the smallest diamondoids, with chemical formulas $C_{10}H_{16}$, $C_{14}H_{20}$, and $C_{18}H_{24}$, respectively. Note that the bridgehead position 1 in adamantane is equivalent to positions 3, 5, and 7, while all four secondary or bridge positions are equivalent to each other. Note that diamantane and triamantane have two types of bridgehead carbons: atoms at positions 4 and 9 of diamantane and 9 and 15 in triamantane are in equivalent *apical* positions. Carbon atoms 1, 2, 6, 7, 11, and 12 in diamantane are medial or belt; as are atoms 2 (equivalent to 12), 6 (equivalent to 4), 7 (equivalent to 11), 3 (equivalent to 13) in triamantane (Fort, 1976; Olah, 1990a; Ramezani *et al.*, 2007).

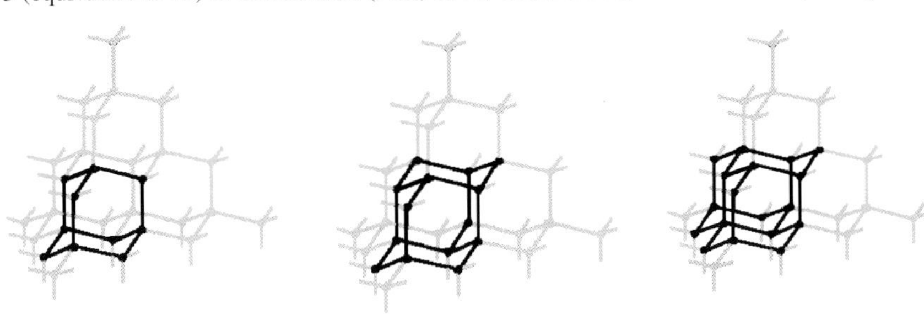

(i) **(ii)** **(iii)**

Fig. 1.2. The relation between lattice diamond structure and (i) adamantane, (ii) diamantane, and (iii) triamantane structures (Mansoori, 2007).

polymantane homologs (adamantologs): diamantane, tria-, tetra-, penta-, and hexamantane. These molecules are particularly denominated polymantanes because they are completely superimposable on the diamond lattice (see Fig. 1.2) and all carbon atoms belong in a whole adamantane unit, in addition, each pair of units is face-fused (Balaban and Schleyer, 1978). In other words, each adamantane shares six carbon atoms with an adjacent unit. Figure 1.1 illustrates the smaller diamondoid molecules, with the general chemical formula $C_{4n+6}H_{4n+12}$: adamantane ($C_{10}H_{16}$), diamantane ($C_{14}H_{20}$), and triamantane ($C_{18}H_{24}$). Each of these three lower adamantologs has only one isomer.

In 1978, Balaban and Schleyer created a systematic enumeration of polymantanes (Balaban and Schleyer, 1978). At that time, the larger known adamantolog was tetramantane, nevertheless the authors proposed a new code, based on graph-theory approach using dual-graphs. This was in anticipation of eventual preparation of higher diamondoids and to avoid use of trivial names which would probably be invented to distinguish isomeric polymantanes. Dual-graphs or dualist graphs of polymantanes are built by joining the centers of fused adamantane units. These graphs are coded using digits 1–4 as symbols of the four possible orientations in space of adamantane units in a polymantane structure (Fig. 1.3a).

Using these concepts, one could represent adamantane as a point and diamantane as an edge. Figure 1.3b shows a representation of the dualist graph for one of the isomers of tetramantane called *anti*-tetramantane (the same as [121]tetramantanes shown on Fig. 1.4). Note that this graph is acyclic linear, while Fig. 1.3c exhibits an acyclic branched graph for *iso*-tetramantane (the same as [1(2)3]tetramantane shown on Fig. 1.4).

Both isomers are classified as *cata*-condensed isomers or catamantanes as their dual graphs are open. Molecules like [12312]hexamantane (Fig. 1.4), in their turn, are *peri*-condensed polymantanes or perimantanes, as their dual-graphs are cyclic. To code polymantanes according to Balaban and Schleyer, it is necessary to number each different direction taken by the dual graph, starting from one endpoint of the longest chain. The final set of digits has to form the smallest number possible. For example, diamantane has code 1, because its dual graph is a straight line (one possible direction, smallest possible number), and triamantane has code 12 (two lines forming a 109°28′ angle, first direction is 1, second is 2). *Anti*-tetramantane is [121]tetramantane, and *iso*-tetramantane, with its tree-like dualist graph, is coded [1(2)3] (parentheses indicate a branch).

(a)

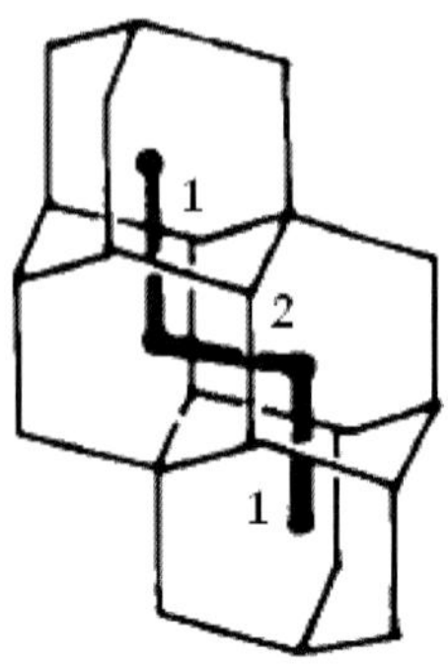

(b)

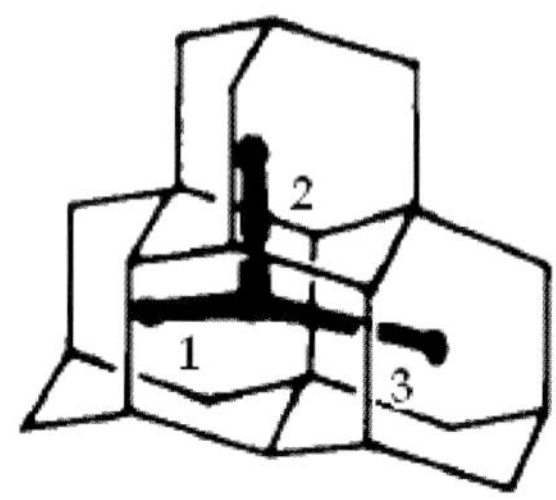

(c)

Fig. 1.3. Dualist graphs of polymantanes. (a) Four possible relative directions for face-fused adamantane units, (b) *Anti*-tetramantane and (c) *Iso*-tetramantane with their dual-graph showing the numbering of each direction (Balaban and Schleyer, 1978).

Balaban and Schleyer nomenclature is widely used for diamondoids and will be preferably used in this book.

Depending on the spatial arrangement of the adamantane units, higher polymantanes ($n \geq 4$) can have several isomers and nonisomeric equivalents. There are three possible tetramantanes, all of which are isomeric. They are depicted in Fig. 1.4 as [1(2)3] (or *iso*-), [121] (or *anti*-), and [123] (or *skew*) tetramantane (only one enantiomer is shown). [121] and [123]tetramantanes each possess two quaternary carbon atoms, whereas [1(2)3]tetramantane has three. The number of diamondoid isomers increases appreciably after tetramantane.

With regard to the remaining members of the diamondoid group, there are ten possible pentamantanes, nine of which are isomeric ($C_{26}H_{32}$) and follow the molecular formula ($C_{4n+6}H_{4n+12}$) of the homologous series and one is nonisomeric ($C_{25}H_{30}$). For hexamantane, there are 39 possible structures: 28 are regular *cata*-condensed isomers with the chemical formula $C_{30}H_{36}$, 10 are

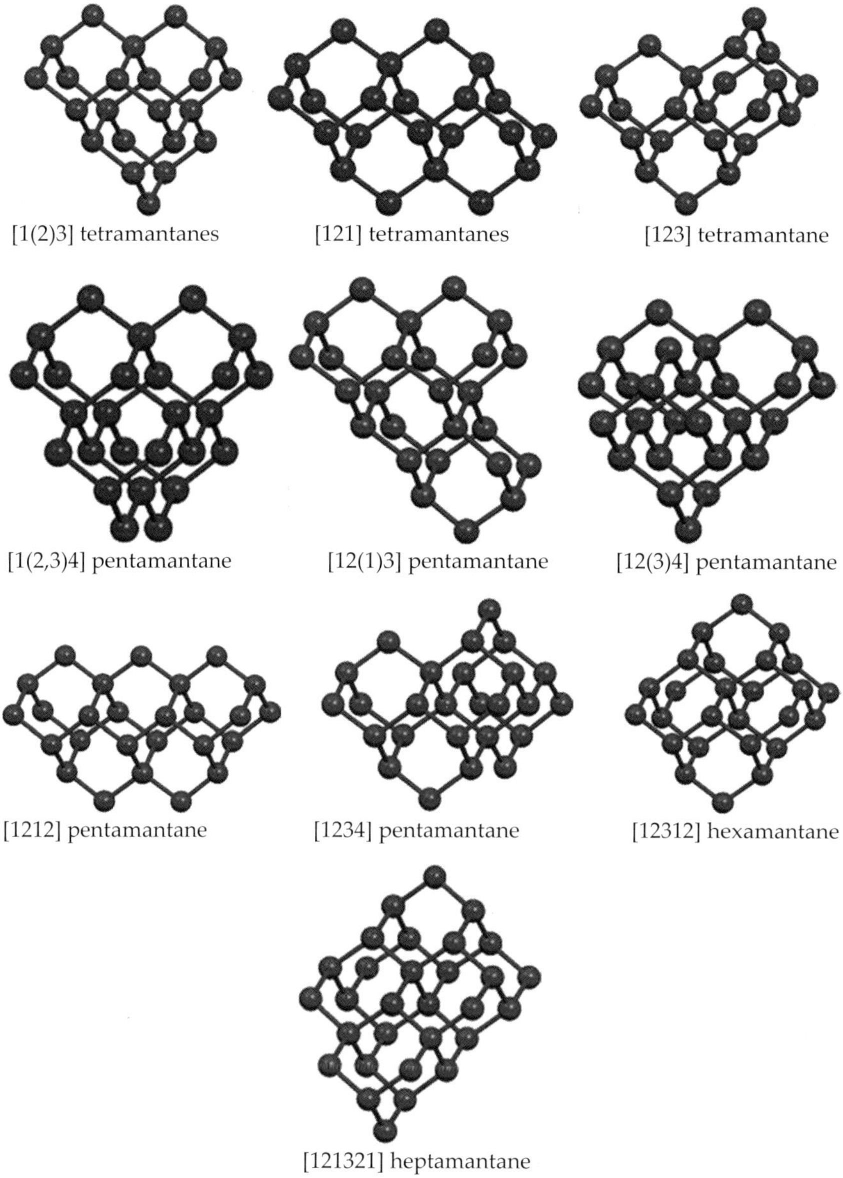

Fig. 1.4. Some of the higher diamondoids up to heptamantane. All tetramantanes are regular *cata*-condensed polymantanes. Shown pentamantanes are also regular catamantanes. *Peri*-condensed [12312] hexamantane ($C_{26}H_{30}$) and [121321] heptamantane ($C_{30}H_{34}$) and other higher diamondoids may become fundamental components in nanometric devices (see also Fig. 4.13). Adapted from CDV Diamond Group {www.chm.bris.ac.uk/pt/diamond/diamondoids.htm} 2010.

irregular *cata*-condensed isomers with the chemical formula $C_{29}H_{34}$, and one is *peri*-condensed with the chemical formula $C_{26}H_{30}$ (Carlson *et al.*, 2007). Irregular catamantanes do not follow the general formula $C_{4n+6}H_{4n+12.}$ Their codes have at least two identical digits separated by any two other digits (excluding parentheses with their contents). Figure 1.4 shows structures of some higher diamondoids. Among the diamondoids of this figure, only [12312] hexamantane and [121321] heptamantane are irregular. Table 1.1 lists some physical properties of diamondoids including their molecular weights, melting points, apparent boiling points, and normal densities.

Diamondoids melt at much higher temperatures than other hydrocarbon molecules with the same number of carbon atoms in their structures. The melting point of adamantane (269°C) is probably the highest among all organic molecules with the same molecular weights. Since diamondoids also possess low strain energy, they are more stable and stiff, resembling diamond in a broad sense. They possess superior strength-to-weight ratio.

It has been found that adamantane crystallizes in a face-centered cubic lattice, which is extremely unusual for an organic compound. The molecule therefore should be completely free from both angle strain (since all carbon atoms are perfectly tetrahedral) and torsional strain (since all C–C bonds are perfectly staggered), making it a very stable compound and an excellent candidate for various applications, as will be discussed later in this book.

At the initial growth stage, crystals of adamantane show only cubic and octahedral faces. The effects of this unusual structure on physical properties are interesting (Kabo *et al.*, 2000).

Many of the diamondoids can be brought to macroscopic crystalline forms with some special properties. For example, in its crystalline lattice, the pyramidal-shaped [1(2,3)4]pentamantane has a large void in comparison to similar crystals (Table 1.1). Although it has a diamond-like macroscopic structure, it possesses weak, noncovalent, intermolecular van der Waals attractive energies that are involved in forming a crystalline lattice (Dahl *et al.*, 2003; Mansoori *et al.*, 2003). Another example is the crystalline structure of 1,3,5,7-tetracarboxyadamantane, which is formed via carboxyl hydrogen bonds of each molecule with four tetrahedral nearest neighbors. The similar structure in 1,3,5,7-tetraiodoadamantane crystal would be formed by I–I interactions. In 1,3,5,7-tetrahydroxyadamantane, the hydrogen bonds of hydroxyl groups produce a crystalline structure

Table 1.1. Some physical properties of diamondoids (Mansoori, 2007).

Diamondoid Chemical Formula	Molecular Structure	Mw	MP [°C]	aBP [°C]	ρ[g/cc]	Crystal Structures
Adamantane $C_{10}H_{16}$		136.240	269	135.5 @ *10 nm Hg*	1.07	Cubic, fee
Diamantane $C_{14}H_{24}$		188.314	236.5	272	1.21	Cubic, Pa_3
Triamantane $C_{18}H_{24}$	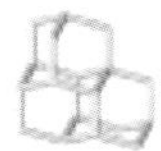	240.390	221.5	330	1.24	Orithorhombic, Fddd
Tetramantanes $C_{22}H_{28}$	[1(2)3] 	292.466	NA	NA	NA	NA
	 [121]		174	NA	1.27	Monoclinic, $P2_1/n$
	[123] 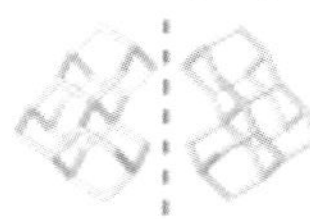		NA	NA	1.32	Triclinia, P1

(*Continued*)

Table 1.1. (*Continued*)

Diamondoid Chemical Formula	Molecular Structure	Mw	MP [°C]	aBP [°C]	ρ[g/cc]	Crystal Structures
Pentamantanes $C_{26}H_{32}$	 [1212]	344.543	NA	NA	1.26	Orthorhombic, $P2_12_12_1$
	[12(3)4] 		NA	NA	NA	Monoclinic, $P2_1/n$
	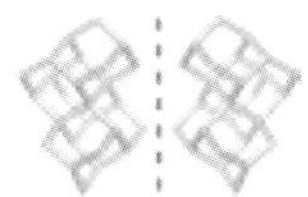 [1234]		NA	NA	1.30	NA
	[1(2,3)4] 		NA	NA	1.33	Orthorhombic, Pnma
	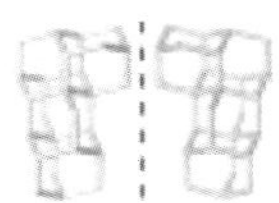 [12131]		NA	NA	1.36	Triclinic, P-l
	[12(1)3] 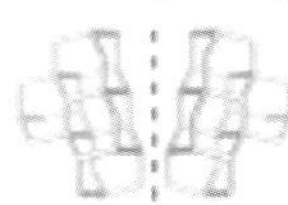		NA	NA	NA	NA

(*Continued*)

Table 1.1. (*Continued*)

Diamondoid Chemical Formula	**Molecular Structure**	**Mw**	**MP [°C]**	**aBP [°C]**	**ρ[g/cc]**	**Crystal Structures**
Cyclohexamantane (*peri-condensed*)	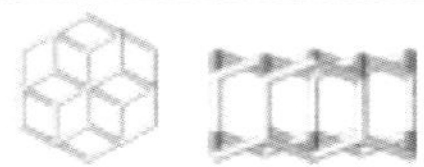 Top [12312] .Side	342.528	>314	NA	1.38	Orthorhombic, Pnma
Heptamantanes $C_{30}H_{34}$	[121321] 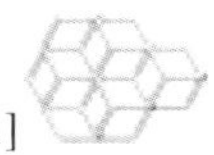	394.602	NA	NA	1.35	Monoclinic, C2/m (#12)
	 [123124]	394.602	NA	NA	NA	NA

aBP = apparent boiling point; MP = melting point; Mw = molecular weight; ρ = normal density. Adapted from ChevronTexaco (*www.moleculardiamond.com*).

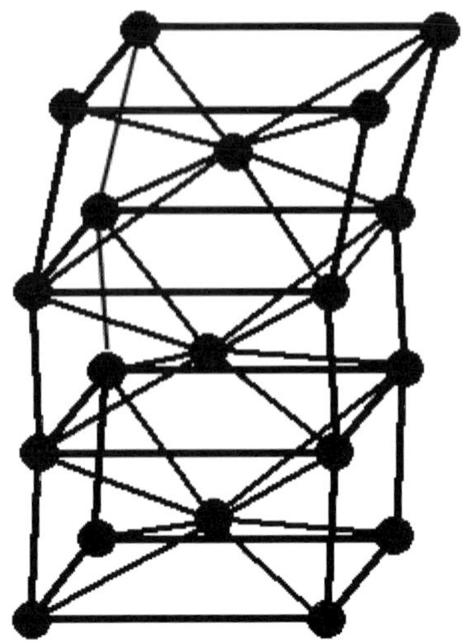

Fig. 1.5. The quasi-cubic units of crystalline network for 1,3,5,7-tetrahydroxyadamantane. Molecules are shown as spheres and hydrogen bonds as solid linking lines. This crystalline structure is similar to that of CsCl (Desiraju, 1996).

similar to inorganic compounds, like cesium chloride (CsCl) lattice (Desiraju, 1996) (Fig. 1.5).

The presence of chirality is another important feature in many diamondoids. It should be pointed out that [123]tetramantane is the smallest of the higher diamondoids to possess chirality (Balaban and Schleyer, 1978).

The vast number of structural isomers and stereoisomers is another property of diamondoids. For instance, octamantane possesses hundreds of isomers in five molecular weight classes. The octamantane class with formula $C_{34}H_{38}$ and molecular weight 446 has 18 chiral and achiral isomeric structures. Furthermore, there is unique and great geometric diversity within these isomers. For example, rod-shaped diamondoids (with the shortest one being 1.0 nm long) and disc-shaped and screw-shaped diamondoids (with different helical pitches and diameters) have been recognized (Dahl *et al.*, 2003), as shown in Table 1.1. Very interesting nanotechnology applications are proposed for these molecules, such as structural components of nanosystems (see Sec. 4.4.2 for details).

1.3. Distinction between Diamondoids and Nanodiamonds

At this point, it is important to make a distinction between diamondoid molecules and nanodiamonds (NDs). Adamantane and other diamondoids are cage-shaped hydrocarbons, constituents of petroleum, gas condensate (also called NGL or natural-gas-liquid) and natural gas reservoirs (King, 1988; Alexander and Knight, 1990; Alexander *et al.*, 1990a, 1990b; 1991; Mansoori, 2007). Diamondoid molecules can also be synthesized from a variety of organic precursor (see Sec. 1.4). ND is generally referred to diamond films with nanoscale thickness, diamond nanoparticles which may be produced as a result of crushing diamond crystals, and various other diamond-based materials at the nanoscale (see Fig. 1.6).

In comparison, there are no natural sources of NDs, except for samples found in meteorites (Grady *et al.*, 1995; Daulton *et al.*, 1996; Huss, 2005). Nevertheless, synthetic ND material is readily available from large-scale production methods of detonation synthesis (Vityaz, 2004; Dolmatov *et al.*, 2004). Nanosized diamonds can also be prepared as thin pure or doped films by microwave plasma enhanced chemical vapor deposition (MPECVD). Several substrates like glass, silicon, or metals are suitable for supporting ND crystal growth (Williams and Nesladek, 2006). Many

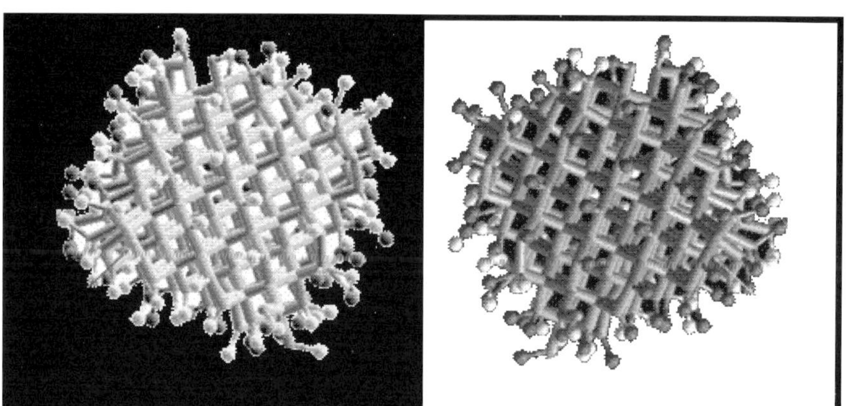

Fig. 1.6. An optimized (PM3) molecular model of nanodiamond [Redrawn for this book from the presentation made by Mochalin *et al.* (2007)] surrounded by oxygen and hydrogen atoms and purified by oxidation in air.

precursor gas mixtures, e.g., C_{60}/argon (Gruen *et al.*, 1994; Qin *et al.*, 1998), methane/hydrogen (Sung *et al.*, 1998; Zhang *et al.*, 2004) and methane/argon (Sung *et al.*, 1998) are used for carbon deposition to form ND. Diamond nanograins obtained by chemical vapor deposition (CVD) are frequently called nanocrystalline diamonds (NCD) and ultrananocrystalline diamonds (UNCD) (Williams and Nesladek, 2006).

The first publication related to diamond synthesis was by DeCarli and Jamieson (1961). They submitted samples of graphite to explosive shocks of high intensity (about 300,000 atm) and were able to separate and identify small particles of diamond having 10 μm, or less, diameter from the shocked graphite. Nanometer-sized diamonds or NDs were first produced by Danilenko and collaborators in the 60's, through detonation as it is reported in the history of the discovery of ND synthesis by Danilenko (2004). This method is nowadays used for industrial production of NDs and consists of submitting mixtures of organic explosives, such as TNT, hexogen, and octogen to an incomplete combustion reaction, inside of a detonation chamber, in vacuum or inert gas atmosphere. Hydrogen, argon, nitrogen, or carbon dioxide can be used to fill the detonation chamber when performing this dry synthesis technique. Alternatively, a wet synthesis can be carried out when ice-water or another coolant is placed into the chamber with the detonation charge (Dolmatov *et al.*, 2004) which could cool the detonation products. Carbon atoms are supplied by the mixture of explosives, thus, graphite is not added. After the detonation, the products of the condensed carbon are grinded, averaged, and purified by mechanical and magnetic methods. Then, they are treated by thermooxidation into pressurized equipments and washed to remove acids and water-soluble impurities, suspended in distilled water or dried to give homogeneous powder. The resulting product is called ultrafine-dispersed diamond (Dolmatov, 2001; Dolmatov *et al.*, 2004), ultradispersed diamond (Aleksenskii *et al.*, 1999; Danilenko, 2004a) (UDD) or nanocrystalline diamond powder (Holt, 2007).

A model for the structure of ND powder obtained through detonation synthesis is shown in Fig. 1.7.

The primary particle is made up of a ND core which is coated with an onion-like carbon shell. An external nondiamond carbon phase containing graphite inclusions and metal-oxide impurities is distributed on top of the

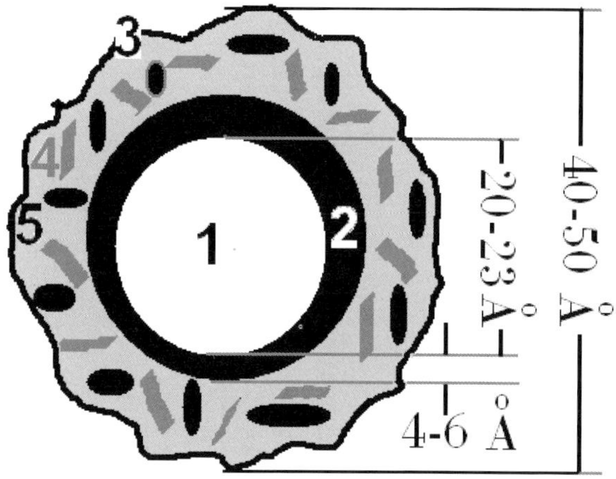

Fig. 1.7. A model structure of detonation — produced carbon: 1 — nanodiamond core, 2 — onion-like carbon shell, 3 — nanosized graphite platelets, 4 — graphite particles, 5 — metal-oxide inclusions. Depending on the oxidation processing, components 3, 4, and 5 might be eliminated, resulting in a cluster made up of a nanodiamond core and an onion-like shell. Cleaning process might destroy the onion-like shell as well (Redrawn from Aleksenskii *et al.*, 1999).

onion-like shell. Cleaning procedures consecutively etch off the bulky graphite phase, impurities, nanosized carbon platelets and may remove the onion-like shell, exposing the diamond core (Aleksenskii *et al.*, 1999).

Despites the possibility of the extraction of primary diamond nanoparticles presented in this model, transmission electron microscopy (TEM) images of detonation ND powders both before and after thermooxidation treatment show large aggregates (Fig. 1.8a–c).

These aggregates have firmly tight cores, designed as "agglutinates" and are not greatly affected by thermooxidation treatment. Nevertheless, they can be disintegrated through milling with micron-sized ceramic beads followed by sonication treatment to yield a clear colloidal solution of de-aggregated ND, as illustrated in Fig. 1.9 (Kruger *et al.*, 2005; Osawa, 2007).

While diamondoids are structurally well-characterized chemical compounds, NDs are frequently a mixture of several carbonaceous compounds. Their characteristics are governed by many factors, such as starting materials, methods of fabrication and processing (Aleksenskii *et al.*, 1999; Shenderova *et al.*, 2002; Chiganov, 2004; Dolmatov *et al.*, 2004; Larionova *et al.*, 2006; Krueger, 2008a; Nekhaev *et al.*, 2011). In

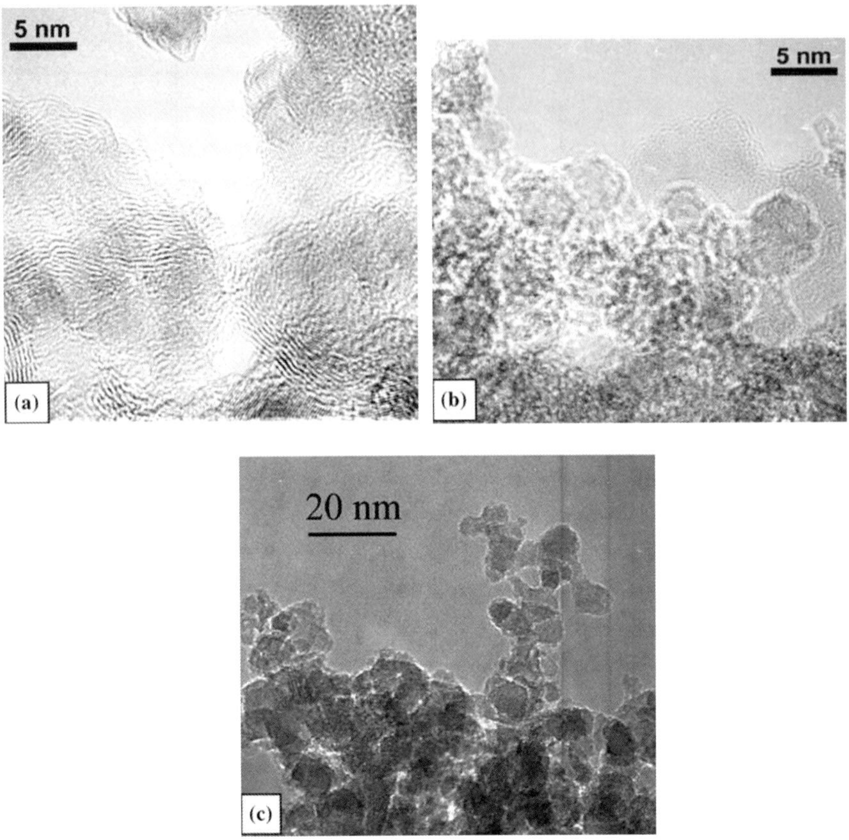

Fig. 1.8. TEM photographs (200 keV) of nanodiamond agglutinates (a) ground pristine soot prepared by wet synthesis after mechanical and magnetic cleaning, (b) pristine sample prepared by dry synthesis after thermo-oxidative treatment (HNO_3 70%, 250°C), (c) Commercial ND [Gansu Lingyun Nano — Material Co. Ltd., Lanzhou, China (Kruger *et al.*, 2005)].

addition, oxygenated groups, such as –COOH, –C=O and –OH (Kusnetsov *et al.*, 1991) and NO_3^- (Aleksenskii *et al.*, 2001) are frequently present at the surface of the ND particle aggregates, either formed during synthesis reactions or processing. The presence of reactive functional groups is important for easing surface modification in ND, but brings uncertainties concerning chemical composition of these materials. Table 1.2 shows a comparison between diamondoids and NDs.

Scientific interest in NDs has increased over the last decade primarily due to the discovery of their promising physicochemical (Aleksenskii

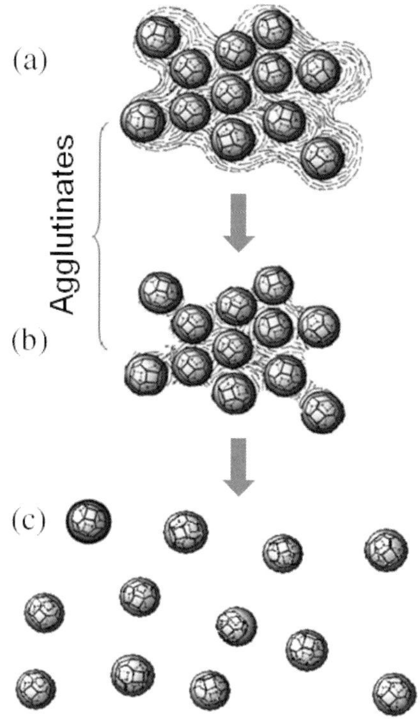

Fig. 1.9. Illustration of how a nanodiamond agglutinate is disintegrated by oxidation with hot nitric acid (a–b) followed by beads-milling (b–c). A: detonation soot containing ND particles at the cores of highly defective carbon — onions. B: high impact of beads during stirred — media milling destroys amorphous structures inside agglutinate and disperses the nanodiamond particles (Osawa, 2007).

et al., 2001; Shenderova *et al.*, 2007) and biological properties (Krueger *et al.*, 2006; Kruger, 2008b). For example, sub-micron diamond particles (800–1000 nm in diameter) fabricated by dual radiofrequency/microwave plasma activated chemical vapor deposition (RF/MW PACVD) are able to suppress the expression of cancer-related human genes while graphite powder has no such capability (Barkowicz-Mitura *et al.*, 2007). Detonation NDs are able to absorb UV light due to the presence of sp^2 carbons and impurities at the aggregates surface while are practically transparent to the visible light. Therefore, they can be included in sunscreen formulations for a number of applications (Shenderova *et al.*, 2007). Functional groups at the NDs surface allow adsorption of complex biomolecules such as

Table 1.2. Comparison between diamondoids and nanodiamonds.

Property	Diamondoids	Nanodiamonds
Chemical composition	Well-defined: Caged saturated hydrocarbons.	Not defined: Samples are mixtures of compounds.
Natural occurrence	Oil, Sediments, Natural gas condensates.[a]	Extraterrestrial sources: Meteorites.[b,c]
Size	Lower diamondoids: <1 nm. Higher diamondoids: 1–2 nm.	A broad range, depending on the method of production and purification.[d,e]
Solubility	Lower diamondoids[f]: Adamantane, 10–11.6% in C_5–C_6 alkanes; diamantane, 6.3% (cyclohexane), 5.0% (CCl_4); Triamantane: CH_2Cl_2[g] and $CHCl_3$[h] are used as solvents in synthesis procedures.	Almost insoluble in most common solvents.[i] Aqueous colloids can be obtained after breaking down of primary particles aggregates.[e]
Methods of obtention	Extraction from crude oils or gas condensates.[j] Organic synthesis.[k]	Detonation synthesis, chemical vapor deposition.[c,l,m,n]
Morphology	Crystalline solids in normal conditions.[o]	Firm aggregates of nanoparticles.[e,p]

[a]See Section 2.3.1; [b]Grady *et al.*, 1995; [c]Daulton *et al.*, 1996; [d]Larionova *et al.*, 2006;. [e]Krueger, 2005; [f]See Table 3.10; [g]Fokina *et al.*, 2007; [h]Schreiner *et al.*, 2006; [i]Liu *et al.*, 2004; [j]Vazquez and Mansoori, 2000; [k]See Chapter 5 Synthesis and Chemical Modification; [l]Gruen *et al.*, 1994; [m]Qin *et al.*,1998; [n]Zhang *et al.*, 2004. [o]See Table I for crystal structures; [p]Aleksenski *et al.*, 2000.

proteins, on diamond nanoparticles for use in separation methods and analysis (Bondar *et al.*, 2004; Kong *et al.*, 2005).

In this book, our aim is to present the state-of-the-art in diamondoid research. Although NDs are a fascinating R&D field, they will not be studied in further details.

1.4. Synthesis and Functionalization of Diamondoid Cages

In 1990, Professor George A. Olah (Nobel Prize Laureate in Chemistry in 1994) edited a book entitled: *Cage Hydrocarbons* dedicated to "*Paul Schleyer for Thirty Years of Adamantane Chemistry and His Sixtieth*

Birthday" (Olah, 1990a). In that book, Professor Olah and coauthors highlighted many aspects of diamondoids as well as nondiamondoid cage compounds, with an emphasis on synthesis, and offered a comprehensive review of the, then, past, present, and future of the chemistry of these substances. Thus, the reader interested in details on diamondoids synthetic routes will find a great deal of information in Olah's book, as well as in a number of original articles on adamantane (Prelog and Seiwerth, 1941; Stetter *et al.*, 1956; Schleyer, 1957; Schleyer and Donaldson, 1960; Schleyer *et al.*, 1962; Olah *et al.*, 1984), on diamantane (Cupas *et al.*, 1965; Gund *et al.*, 1970; Gund *et al.*, 1973) and triamantane (Williams *et al.*, 1966; Hollowood *et al.*, 1980) and on [121]tetramantane (Burns *et al.*, 1976, 1978). There are also review papers published on the chemistry of adamantane (Sevostyanova *et al.*, 1970; Moiseev *et al.*, 1999; Schwertfeger *et al.*, 2008) including the first review on adamantane, published in 1964, by Fort and Schleyer (1964a). In the next sections, we will mainly focus our discussion on new methods for the introduction of functional groups to the diamondoid cages. In addition, a general overview of established methods for cage moiety synthesis will be presented.

1.4.1. *Adamantane syntheses: from nearly trace amounts to semi-quantitative yields*

Originally diamondoids were considered as hypothetical molecules since they could neither be isolated from a natural repository nor made through rational organic synthesis until 1933 when adamantane was discovered and isolated from petroleum (Landa *et al.*, 1933). The natural source of diamondoids was the only effective source until 1941 when adamantane was synthesized for the first time, though the yield was very low (McKervey, 1980; Prelog and Seiwerth, 1941; Navratilova and Sporka, 2000).

In 1941, Prelog and Seiwerth published the first route to adamantane cage synthesis (Prelog and Seiwerth, 1941) but they achieved practically trace amounts of the target compound. Nevertheless, adamantane crystalline and its high-melting point served to confirm its structure to be the same as the natural adamantane found by Landa and Machacek in petroleum samples from Hodonin Oil field, in Moravia, Czechoslovakia, in 1933 (Landa *et al.*, 1933). Later, the 1956 Stetter's synthesis (Stetter *et al.*, 1956) represented an important increase in the overall yield of adamantane to 6%, although their proposed relatively complex four-step process still left room for improvement.

In 1957, Schleyer introduced a Lewis acid-catalyzed rearrangement to produce adamantane (Schleyer, 1957). A Lewis acid is any acid that can accept a pair of electrons and form a coordinate covalent bond [45]. According to Schleyer endo-trimethylenenorbornane, which could be produced readily, rearranged to adamantane when refluxed overnight with aluminum chloride ($AlCl_3$) as shown in Fig. 1.10.

This reaction allowed a simple preparation of adamantane with 10% overall yield, which was unprecedented at that time, and opened up the field of diamondoids synthesis and application for further studies worldwide. For example, the number of patents related to adamantane increased rapidly after the publication of Schleyer's synthesis of adamantane (see Fig. 1.20).

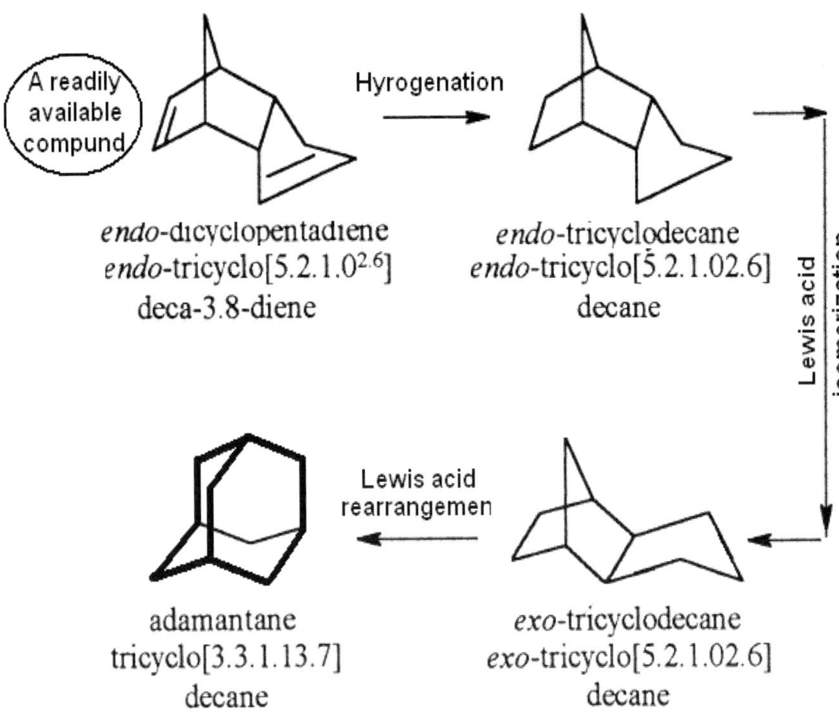

Fig. 1.10. Various stages in synthesis of adamantane (Mansoori, 2007) based on the work reported by Schleyer (1957) and Schleyer and Donaldson (1960).

Table 1.3. Superacid isomerization of *endo*- and *exo*-Trimethylenenorbonane to produce adamantane (Olah, 1990b).

Acid System	Acid:Hydrocarbon Molar Ratio	Temperature [°C]	Time [h]	*Exo*-isomer [% yield]	Adamantane Isolated [% yield]
$HF\text{-}SbF_5$, 1:1	1:3 (*endo*-)	100	18	—	47.0
CF_3SO_3H + SbF_5, 1:1	1:3 (*exo*-)	0	16	28.3	58.2
CF_3SO_3H + SbF_5, 1:1	1:3 (*exo*-)	0	50	27.0	68.0
CF_3SO_3H + SbF_5, 1:1	1:1 (*endo*-)	0	10	0.4	95.4
CF_3SO_3H + SbF_5, 1:1	1:1 (*endo*-)	0	7	0.1	98.0
CF_3SO_3H + $B(OSO_2)CF_3)_3$, 1:1	1:1 (*endo*-)	0	12	—	98.0

Schleyer's $AlCl_3$ catalyzed isomerization and subsequent improved synthetic routes involving replacement of $AlCl_3$ by $AlBr_3/HBr$ sludge (Schleyer and Donaldson, 1960; Schleyer *et al.*, 1962) were responsible for "the breaking of the cage molecule bottleneck", as referred to by Schleyer (Schleyer, 1990).

Further studies on superacids applied to isomerization of hydrocarbons, performed by Olah and co-workers, allowed up to 98% yields. These *semi*-quantitative yields can be explained by the higher stability of the 1-adamantanyl cation formed during the isomerization process; thus, all intermediate ions generated in the cationic reaction of C_{10} hydrocarbon are converted into this species. Table 1.3 shows superacid systems used for the isomerization of *endo* and *exo* trimethylenenorbonane to adamantane. In Fig. 1.11, we report a comparison of adamantane synthesis by various investigators as discussed above.

1.4.2. *Diamantane and triamantane syntheses*

In 1963, a then unknown C_{14} cage compound was used as the official emblem of the London IUPAC conference. "Congressane" was the

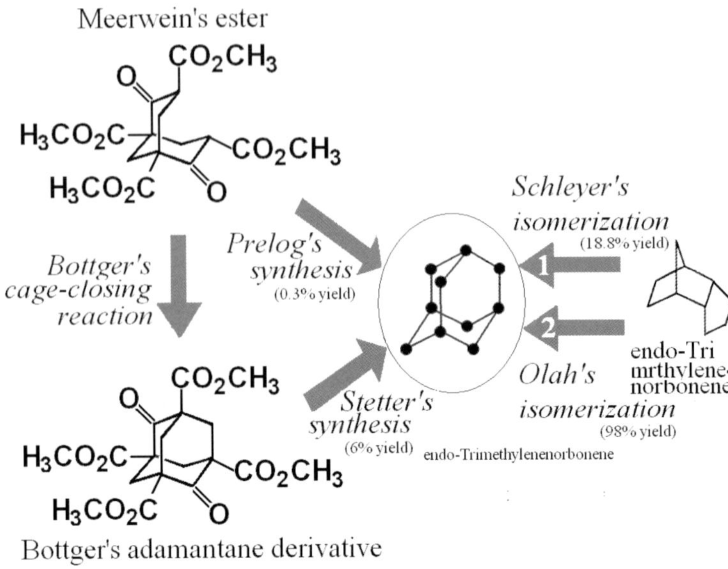

Fig. 1.11. Adamantane syntheses performed by different researchers: Bottger build a synthetic hexa-substituted adamantane cage from the Meerwein's ester (Bottger, 1937). Starting from the same reagent, Prelog prepared adamantane in 0.3% overall yield in 1941 (Prelog and Seiwerth, 1941). In 1956, Stetter *et al.* produced adamantane from the Bottger's product through a four-step process having 6% of overall yield. Simple, affordable, $AlCl_3$-catalyzed, single-step syntheses were achieved by Schleyer in 1957 (Schleyer, 1957), using as catalyst, followed by further improved synthetic routes having $AlBr_3$/HBr as catalyst (Schleyer and Donaldson, 1960; Schleyer *et al.*, 1962), and Olah and colleagues *semi*-quantitative reaction (data from Olah *et al.*, 1984).

suggested name for this compound and a challenge for its synthesis was posed to the chemical community of that time (Schleyer, 1990). Two years later Cupas and co-workers published the first synthesis of congressane (later renamed as diamantane as shown by Fig. 1.1b) using the same Lewis acid-catalyzed rearrangement synthesis approach as was used for adamantane synthesis (Cupas *et al.*, 1965). Figure 1.12 shows this route for diamantane preparation, with a yield of only 1%, and two additional $AlBr_3$-catalyzed syntheses, also published by Gund and co-workers in 1970 (Gund *et al.*, 1970) and 1973 (Gund *et al.*, 1973), in which yields were improved to 30% and 60%, respectively.

The synthesis of the next member of the lower diamondoid, triamantane, was accomplished shortly after the first diamantane synthesis. Williams and

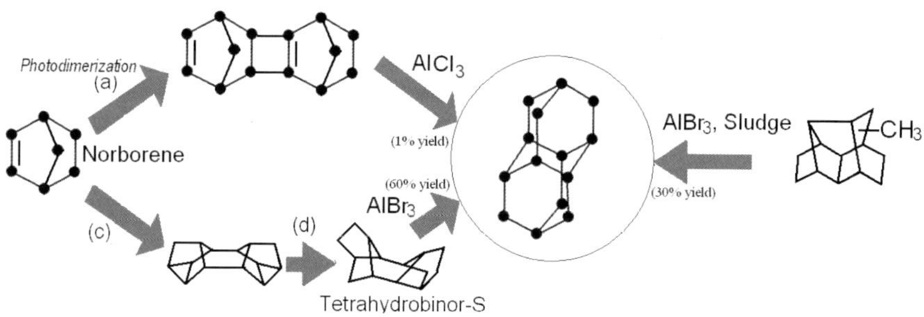

Fig. 1.12. Syntheses of diamantane from various C_{14}-precursors. [a]Cupas *et al.*, 1965; [b]Gund *et al.*, 1970; [c]Gund *et al.*, 1973, conditions: $CoBr_2$, $2Ph_3P$; BF_3; [d]H_2, PtO_2, HCl.

colleagues found an appropriate C_{18} candidate to undergo bromide halide-catalyzed isomerization reaction (compound III, Fig. 1.13) (Williams *et al.*, 1966). The starting material for their synthesis of triamantane is cyclooctatetraene. This compound undergoes dimerization and thermal rearrangements to give the C_{16} diene I. The two extra carbon atoms needed to build the target diamondoid were added through cyclopropanation of the double bonds, resulting in compound II. The next step was the hydrogenolysis of the highly strained cyclopropane rings to give C_{18} precursor III. The yield of the isomerization step from III toward triamantane was 2–5%.

Hollowood *et al.* (1980) improved triamantane synthesis by starting it from norbonadiene, in five stages (Fig. 1.13). Binor-S, the norbonadiene [4 + 4] dimer gives a mixture of precursors IV and V in three steps, which is isomerized into triamantane with 60% yield.

Near quantitative yield of triamantane was achieved through superacid-catalyzed isomerization, similar to what was obtained for adamantane and diamantane. Farooq *et al.* (1988) prepared triamantane from C_{18} precursors and a variety of superacidic catalysts, including mixtures of triflic acid and SbF_3 ($CF_3SO_3H:SbF_3$) and boron tris-triflate ($B(OSO_2CF_3)_3$) to obtain triamantane with 70–74% yield. When bromoadamantane is added in small quantities, it acts as a carbocation initiator, and increases the yield of the isomerization reaction to 95% (Fig. 1.13).

It should be pointed out that the Schleyer's Lewis acid-catalyzed rearrangement method, which is based on diamondoids thermodynamic stability during carbocation rearrangements, has had little or no success to synthesize higher diamondoids beyond triamantane. While attempts to

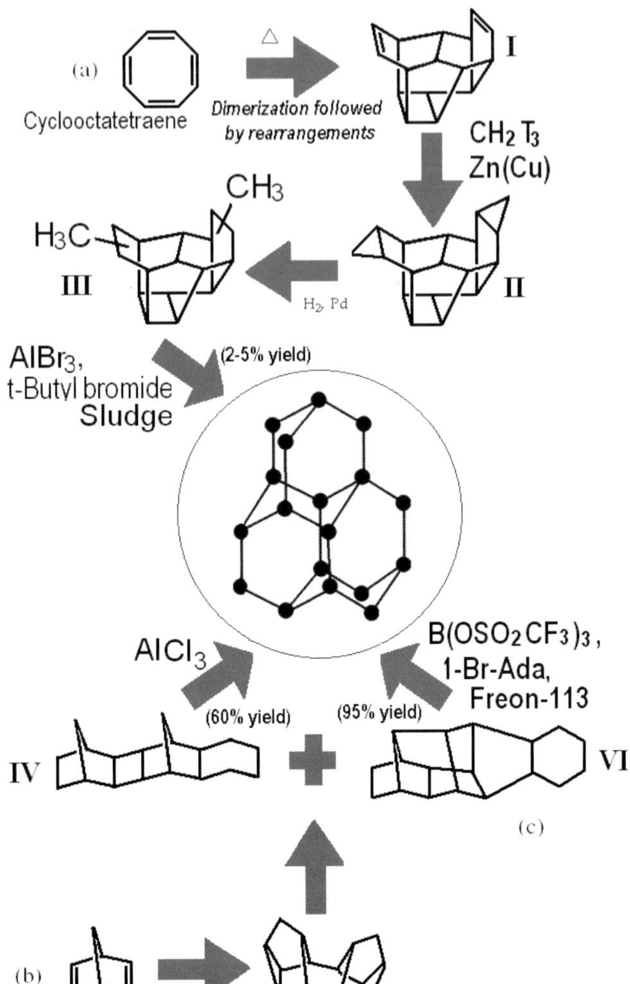

Fig. 1.13. Syntheses of triamantane from heptacyclooctadecanes precursors. 1-Br-Ada = 1-bromoadamantane. [a]Williams *et al.*, 1966. [b]Hollowood *et al.*, 1980. [c]Olah, 1990b.

synthesize lower diamondoids (adamantane, diamantane, triamantane) have been successful through Lewis acid-catalyzed rearrangement, it is no longer a method of choice for synthesis of higher polymantanes (McKervey, 1980). Producing these heavier diamondoids via the above-mentioned rearrangement synthesis method has not been convenient because of their complex structural properties and their lower thermal stability.

The synthesis of higher diamondoids is a challenging and difficult process. For example, after numerous efforts, the anti isomer of tetramantane was synthesized with only 10% yield (McKervey, 1980). The usage of zeolites as the catalyst in synthesis of adamantane has been investigated and different types of zeolites have been tested for achieving better catalyst activity and selectivity in adamantane formation reactions (Navratilova *et al.*, 2000). In the early 2000's, Shibuya *et al.* (2002) represented two convenient methods for synthesis of enantiomeric adamantane derivatives. Recently a short report on the synthesis of hexamantane is reported using low-temperature plasmas generated in supercritical xenon with limited success (Stauss *et al.*, 2010). Otherwise, high molecular weight diamondoids have not been synthesized as of this date. Present day efforts are towards more economical separation of high molecular weight diamondoids from petroleum fluids and synthesis of lower molecular weight diamondoids (Dahl *et al.*, 2003; Vakili-Nezhaad *et al.*, 2005; Mansoori, 2007).

We envision that with advances in our knowledge about inter- and intra-molecular interactions through *ab initio* methods, dynamic combinatorial chemistry, molecular design and nano-catalysts new routes for synthesis of higher diamondoids would be discovered (Mansoori, 2005).

1.4.3. *Functionalization of diamondoids*

Diamondoid derivatives are becoming of increased importance as nanoscale building blocks (Chapter 4), as carriers for biologically active substances (Chapter 6), as well as new material for polymer applications, among others (Chapter 7). Most of the diamondoid applications are based on adamantane chemistry; hence, methods for the substitution of adaman tane H^+ ions for a variety of functional groups have been intensively studied over the last five decades (Fort and Schleyer, 1964a, 1964b; Sevostyanova *et al.*, 1970; Moiseev *et al.*, 1999; Mansoori, 2007; Xue and Mansoori, 2008; Bagrii *et al.*, 2010). Some of the first reported direct substitution reactions were adamantanes halogenation (Stetter *et al.*, 1968; Peterson and Wahl, 1968), carboxylation (Koch and Haaf, 1960), cyanation followed by hydrolysis to produce amides (Haaf, 1961), and nitrates (Smith and Williamson, 1961). From simple 1-substituted adamantanes, such as bromoadamantanes, adamantyl carboxylic acid and amides, a great variety of derivatives were synthesized using well-established methods (Fort and Schleyer, 1964a).

Bromination, nitration, and carboxylation of adamantane are nucleophylic substitution reactions of hydrogen leading to bridgehead-substituted derivatives. Such mechanisms probably involve the formation of 1-adamantanyl cation, for example, as shown in Fig. 1.14 for adamantane bromination reaction substitution (Stetter *et al.*, 1959; Stetter and Wulff, 1960).

In contrast, bridge 2-substituted derivatives are more difficult to obtain, since the position 2 (bridge) in adamantane (see Fig. 1.1) is less reactive than the bridgehead position (Fort and Schleyer, 1964b; Schleyer *et al.*, 1964).

Substitution at the bridge positions frequently involves 1,2-hydride transfer. As an example, the direct oxidation of adamantane to adamantanone in concentrated sulfuric acid (Fig. 1.15) occurs in multiple stages (Geluk and Schlatmann, 1967, 1968).

First, adamantane is oxidized to 1-adamantanol, then hydroxyl elimination and 1,2-hydride transfer leads to the equilibrium between 1- and 2-adamantanyl cations. Then 2-adamantanol is formed from 2-adamantanyl cation reaction with water. Intermolecular hydride transfer occurs between adamantanyl cation and 2-adamantanol, in which the cation acts as a hydride ion acceptor. This disproportionation reaction yields adamantanone and adamantane (Geluk and Schlatmann, 1967, 1968).

Functionalization of diamantane, triamantane, and tetramantane are also gaining importance in diamondoids chemistry. These functionalized molecules are explored through combinatorial chemistries for their expected biological activity and their possible nanotechnology applications. Moreover, natural occurrence of diamantane, triamantane, and higher diamondoids in petroleum and other fossil fuels of different origins is allowing

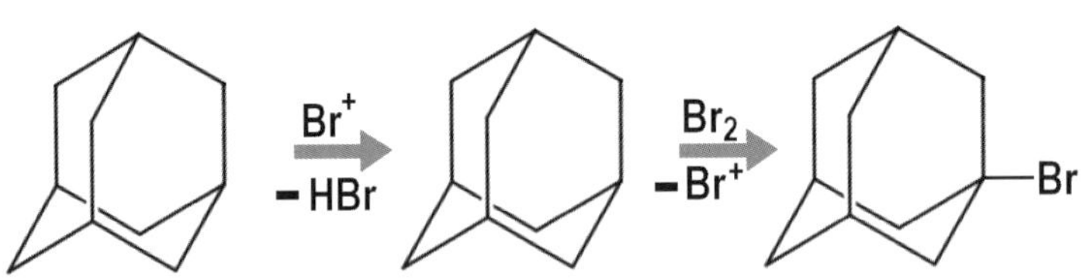

Fig. 1.14. Bromination of adamantane with Br_2.

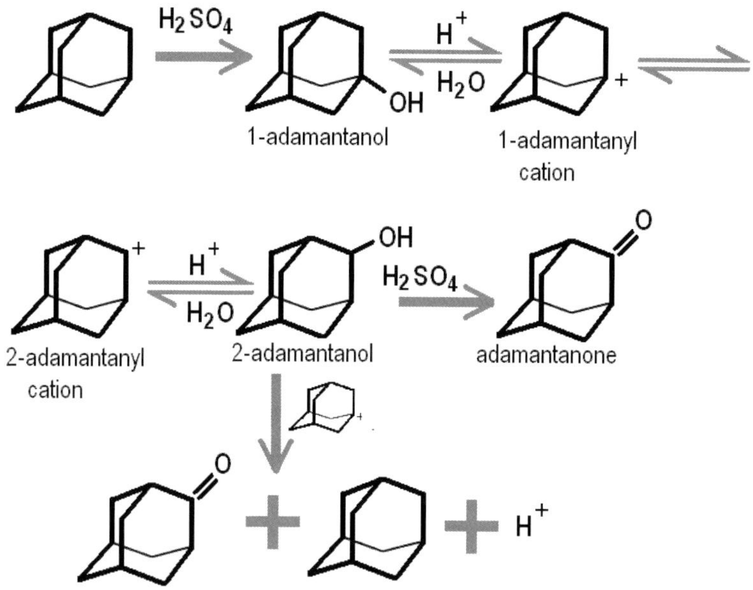

Fig. 1.15. Oxidation of adamantane to adamantanone in concentrated H_2SO_4. Direct oxidation of adamantane produces 1-adamantanol, which is dehydrated to form 1-adamantanyl cation. An equilibrium between 1- and 2-adamantanyl cation is established by 1,2-hydride transfer. Reaction of 2-adamantanyl cation and water forms 2-adamantanol, which is oxidized to adamantanone. Then 2-adamantanol also reacts by disproportionation, transferring hydride to an adamantanyl cation, to give adamantanone and adamantane (data from Geluk and Schlatmann, 1967).

the access to considerable amounts of these cage compounds (see Sec. 2.2). Thus, methods to chemically modify diamondoids other than adamantane have been published over the past years (Liu *et al.*, 2002; Davis and Liu, 2006; Tkatchenko *et al.*, 2006; Schreiner *et al.*, 2006; Fokin *et al.*, 2007; Fokina *et al.*, 2007; Davis *et al.*, 2008; Schwertfeger *et al.*, 2008, 2009). Since 2002, many patents on functionalized higher diamondoids have been filed. These patents show broad spectra of the utility of these molecules for biomedical, chemical, nanotechnology, pharmaceutical, and other applications (Dahl *et al.*, 2002a, 2002b, 2004; Liu *et al.*, 2002, 2003, 2009; Dammel, 2006; Sciamanna *et al.*, 2007; Mansoori, 2007; Yang *et al.*, 2008) including higher heterodiamondoids, as well (Liu *et al.*, 2010).

Figure 1.16 shows an example of selective bromination of diamantane to give *bis-apical* derivative 4,9-dibromodiamantane. For numbering and

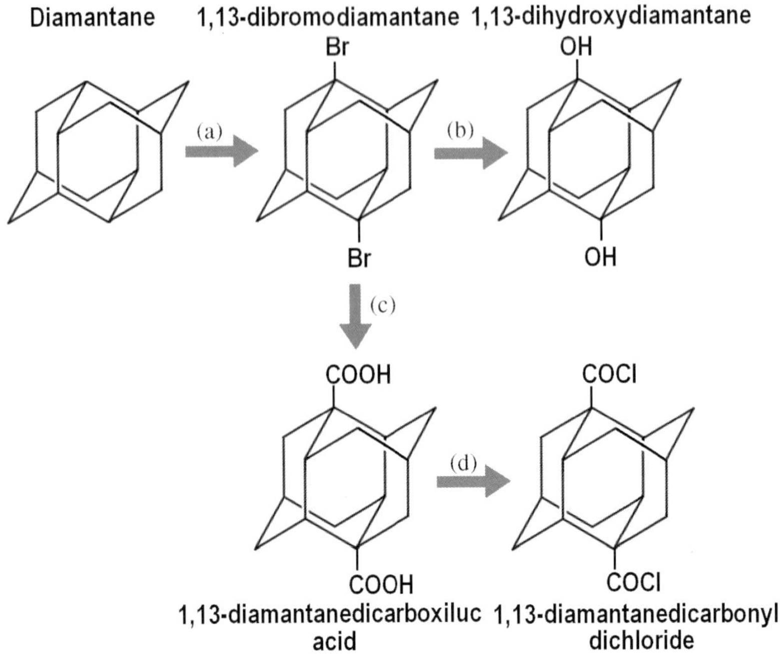

Fig. 1.16. *Bis-apical* bromination of diamantane and derivatization to 4,9-dihydroxy and 4,9-diamantanedicarboxylic acid derivative compounds. {Reagents; conditions; yields}: (a) {Br_2, Freon 113, Fe; room temperature; 62%}. (b) {DMF, H_2O, HCl; reflux; 70%}. (c) {Ag_2SO_4, H_2SO_4, formic acid; 100%}; and (d) {$SOCl_2$; reflux; 100%} (data from Davis and Liu, 2006).

positions of "4 and 9" see diamantane in the top row in Fig. 1.1. Reaction conditions include Freon-113 as solvent and iron as catalyst to give 62% yield (Davis and Liu, 2006).

This method differs from classical *apical* bromination of diamantane, in which aluminum bromide is used as catalyst to give a mixture of *apical* 4-bromodiamantane, *bis-apical* 4,9- and medial-*apical* 1,4-dibromodiamantane obtained along with traces of both medial 1-, and *bis*-medial 1,6-bromodiamantane; then products are separated by column chromatography over alumina (Gund *et al.*, 1971). Diamantane medial positions are more reactive than *apical* positions (Gund *et al.*, 1970; Olah *et al.*, 1985), hence, bromination at room temperature in the absence of Lewis acid catalysts leads to 1-bromodiamantane with 80% yields, which can be further brominated to 1,4- and 1,6-derivatives with bromine under reflux (Gund *et al.*, 1971). Dibromine derivative can be conveniently converted into dihydroxy or dicarboxy diamantane, as shown in Fig. 1.16.

Bifunctional diamondoid derivatives, especially highly symmetrical *bis-apical* 4,9-dibromodiamantane are useful as starting materials for the synthesis of polymers containing cage moieties in the backbone. Such polymers are reported to be thermoset resins of high thermal and thermo-oxidative stability (Malik *et al.*, 1991). *Apically*-substituted diamondoids are promising molecular building blocks for nanotechnology applications; for example, as improved AFM tips (see Figs. 4.6e and 4.6f) (Carlson *et al.*, 2007), or components of self-assembled monolayers (Dameron *et al.*, 2005).

Triamantane substitution reactions are even more challenging when it comes to selectivity, due to their increased number of tertiary C–H bonds. Again, a mixture of several monobromo-derivatives is obtained by reaction with neat bromine, which gives *medial* 2-bromotriamantane as main product (37% in preparative yield) after recrystalization from *n*-hexane (Schreiner *et al.*, 2006). Reaction of triamantane with 100% nitric acid followed by hydrolysis also resulted in a mixture of *medial* 2-hydroxytriamantane (20%), and 3-hydroxy (38%), and *apical* 9-hydroxytriamantane (21%) derivatives (Schreiner *et al.*, 2006).

Extended reaction times of triamantane with excess of nitric acid yielded a mixture of triamantanediols after hydrolysis (Fig. 1.17) (Fokina *et al.*, 2007), in which 3,9-dihydroxytriamantane (Fig. 1.16b) was the main product, while the *bis-apical* 9,15-derivative (Fig. 1.17f) had only 4% in yield. Thus, in such conditions, where disubstitution reaction is

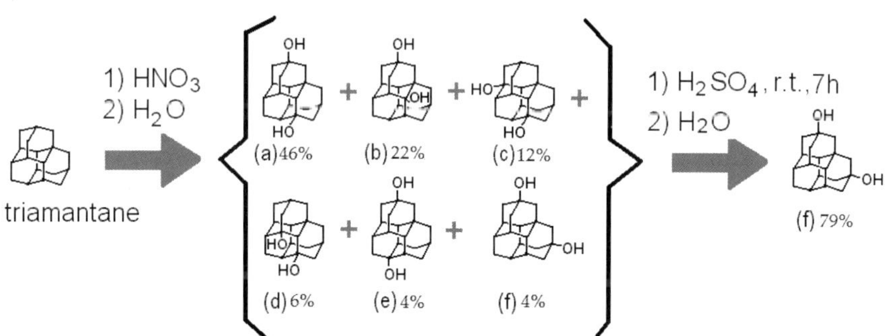

Fig. 1.17. *Bis-apical* nitroxylation of triamantane (data from Fokina *et al.*, 2007).

kinetically controlled, there is poor selectivity and low percentage of apically-substituted compounds.

Conversely, selective formation of 9,15-dyhydroxytriamantane was obtained after thermodynamically controlled isomerization of the mixture of triamantanediols (**a**)–(**f**) (Fig. 1.17) with 98% H_2SO_4. This result is explained by the higher calculated relative stability of *bis-apical* derivative (**f**) in comparison with diols (**a**)–(**e**). For example, *bis-medial* compound (**d**) is the least stable of the mixture and its formation ΔG_{298} is 6.0 kg/mol higher than ΔG_{298} of compound **f** (Fokina *et al.*, 2007). It is worthwhile to mention that *bis-apical* nitroxylation may also be performed as a one-pot reaction, starting from triamantane and nitric acid. After reaction first step, acid excess is removed by evaporation, then, isomerization with H_2SO_4, followed by hydrolysis, gives compound (**f**) with 80% yield.

Apical derivatives of tetramantane are more difficult to produce. Their reactions with electrophiles present high selectivity towards *medial* 2-derivatives of [121]tetramantane, possibly due to the high stability of correspondent 2-tetramantanyl cations relative to other cations to be formed from the same structure. The 6-tetramantanyl cation, for example, is the least stable of the family and its calculated formation ΔG_{298} is *c.a.* 5 kcal/mol higher than formation ΔG_{298} of 2-tetramantanyl cation. Thus, a simple functionalization reaction, such as mono-bromination with bromine in chloroform gives *medial* 2-bromo[121]tetramantane in high yields (Schreiner *et al.*, 2006). Figure 1.18 summarizes the results of bis-functionalization reactions in [121]tetramantane. The data reported in this figure verify how 2-derivatives predominate in dibromination: (**b**) 2,17%, (**c**) 2,18%, (**d**) 2,6%, and (**e**) 2,13% diols represent 75% of the product of bromination/hydrolysis reaction.

In addition, *bis-medial* 2,17-dibromo[121]tetramantane (compound **f**) could still be recovered with 17% yield even after hydrolysis procedure, while only traces of *bis-apical* 6,13-dihydroxy[121]tetramantane (compound **a**) is produced. Nitroxylation reaction exhibited lower selectivity and predominance of *apical*-derivatives (**d**) and (**a**). Nevertheless, the same isomeric tertiary diols were also present. Similarly to what was obtained with triamantane, these mixtures of tetramantanediols yielded *bis-apical* derivative (**a**) when subjected to thermodynamically-controlled isomerization reaction allowing yields of up to 85% for *bis-apical* derivative (**a**). Another evidence

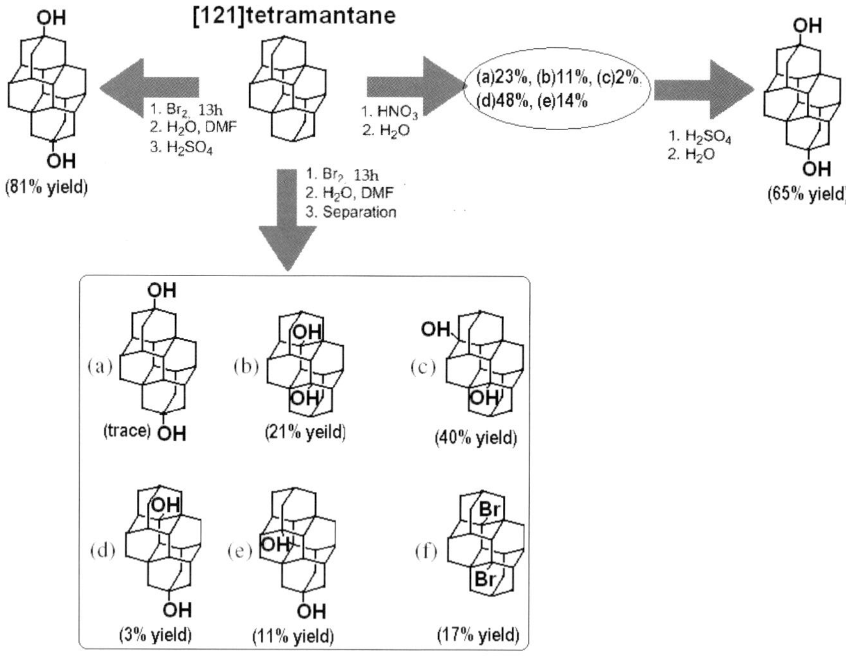

Fig. 1.18. Bifunctionalization reactions of [121]tetramantane (data Fokina *et al.*, 2007).

of compound (**a**) higher stability, when compared to *medial* derivatives, relies on difference in formation free energies, ΔG_{298}, between (**a**) and (**b**) which is *c.a.* 3 kcal/mol in favor of the *bis-apical* isomer (**a**) (Fokina *et al.*, 2007).

Thermodynamically-controlled functionalization reactions show a favorable outlook for large-scale production of functionalized diamondoids, especially for preparation of rather attractive and amenable *bis-apical* diamondoid diols, as these reactions present high selectivity, good yields and give bifunctional products easily convertible into a number of different compounds.

Other approaches for *apical* diamondoids involve oxidation under UV irradiation and are described in the following paragraphs. Such procedures demand strong oxidant agents and are more efficient in producing monofunctional diamondoids.

Good yields for preparation of *mono-apical* 9-derivatives of triamantane were achieved with photooxidation with diacetyl to produce 9-acetyltriamantane (43% yield). As this reagent (diacetyl) preferentially chooses

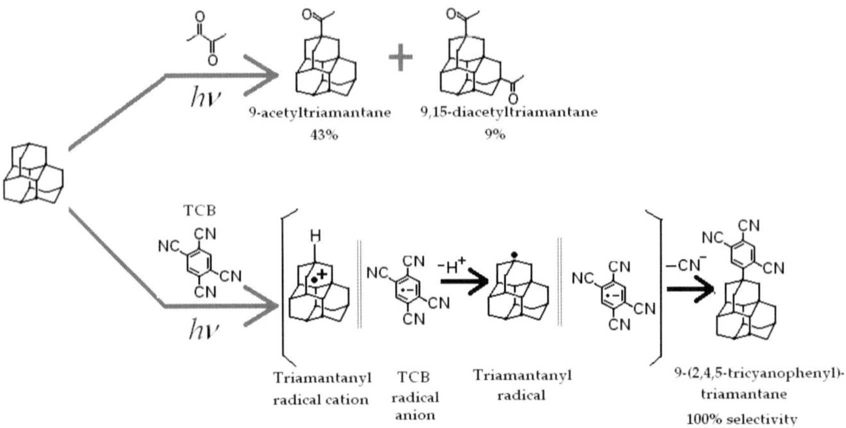

Fig. 1.19. Methods for selective apical substitution in triamantane (Schreiner *et al.*, 2006).

less sterically hindered positions, the second substitution occurs at the *apical* position 15, to give the *bis-apical* diacetyl derivative as a minor product (Fig. 1.19) (Schreiner *et al.*, 2006).

Single-electron-transference (SET) oxidation conditions also have rendered selective *apical* substitution in triamantane. In SET oxidation, triamantanyl radical cation is formed under UV photoirradiation in the presence of 1,2,4,5-tetracyanobenzene (TCB) and excess of triamantane, to yield 9-(2,4,5-tricyanophenyl)triamantane exclusively, in 43% preparative yield (Fig. 1.19) (Schreiner *et al.*, 2006).

Both photooxidation with biacetyl and SET oxidation with TCB were used to selectively produce *apical* derivatives of [121]tetramantane. *Apical* 6-acetyl[121tetramantane] was the main product of photoacetylation (57%) along with the *bis apical* product (11%), while oxidation with TCB resulted in the *apical* aromatic compound with 100% selectivity, in 40% yield (Schreiner *et al.*, 2006).

1.5. Concluding Remarks

In this chapter, we have presented the diamondoid molecules which are cage-like organic (saturated hydrocarbons) with unique structures and properties. These molecules are ringed compounds, which have a diamond-like structure consisting of a number of six-member carbon rings fused together. More explicitly, they consist of repeating units of ten carbon atoms forming a tetra-cyclic cage system. They are called "diamondoid" because their carbon–carbon framework constitutes the fundamental repeating unit in the diamond lattice structure.

Also we have introduced, the general aspects of how to define and classify diamondoids. Difference between diamondoids and NDs are pointed out. The Lewis acid-catalyzed rearrangement of hydrocarbons to synthesize lower diamondoids is presented and its limitations for the synthesis of higher diamondoids are discussed. Classical methods of diamondoid cage chemical syntheses and functionalization, along with recent methods for higher diamondoid chemical modification set the tone for organic chemistry of diamondoids.

The discovery of adamantane cage in 1933 and its direct synthesis in 1957 has turned this diamondoid and its derivatives into readily available compounds. Since 1960, the interest in practical applications of diamondoid molecules has steadily increased. A consequence of these findings can be seen through the number of adamantane-related issued US Patents in the past 55+ years (Fig. 1.20).

The first issued patent, in May 1960, was filed by Richard E. Ludwig (1960) (du Pont & Co., Wilmington, DE) a few years after the publication

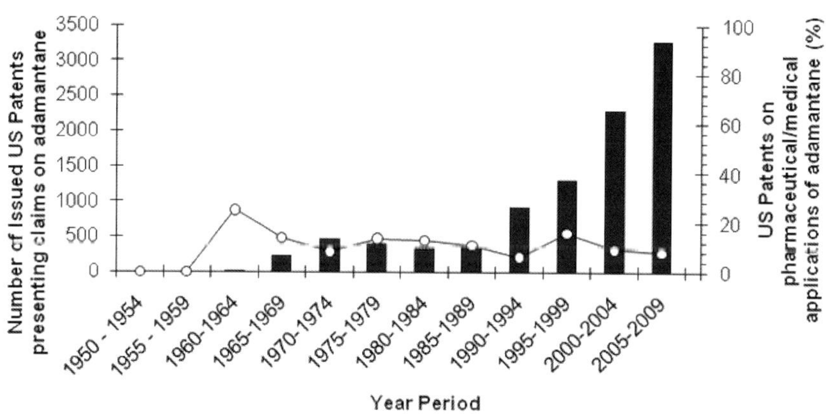

Fig. 1.20. US issued patents related to adamantane (black columns, left *Y*-axes) and percentage of patents on pharmaceutical/medical application of adamantane and its derivatives (empty circles/black line, right *Y*-axes), from January 1950 to December 2009.

of Schleyer's 1957 work, and presented an improved synthesis of adamantane. Since the 1990's, the rate of claims of products and processes related to diamondoids has rapidly increased. By the second half of the 2010 decade, total patents issued on adamantane alone reached more than 3,200 (Fig. 1.20, black columns).

Diamondoids family of compounds and their derivatives (with over 20,000 variants) is one of the best candidates for templates and molecular building blocks for synthesis of high temperature polymers, nanotechnology, drug design, drug delivery, drug targeting, DNA-directed assembly, DNA-protein nanostructure formation, and in host-guest chemistry. In the rest of this book, we present various aspects of diamondoids and their applications.

Bibliography

Aleksenskii, A. E., Baidakova, M. V., Vul', A. Ya. and Siklitskii, V. I. (1999). *Phys. Solid State*, **41**: 668–671.

Aleksenskii, A. E., Osipov, V. Y., Vul', A. Y., Ber, B. Y., Smirnov, A. B., Melekhin, V. G., Adriaenssens, G. J. and Iakoubovskii, K. (2001). *Phys. Solid State*, **43**: 145–150.

Alexander, R. A., Knight, C. E. and Whitehurst, D. D. (1990a). US Patent No. 4952747.

Alexander, R. A., Knight, C. E. and Whitehurst, D. D. (1990b). US Patent No. 4952749.

Alexander, R. A. and Knight, C. E. (1990). US Patent No. 4952748.

Alexander, R. A., Knight, C. E. and Whitehurst, D. D. (1991). US Patent No. 4982049.

Bagrii, E. I., Safir, R. E. and Arinicheva, Yu. A. (2010). *Petroleum Chem.*, **50,** 1–16.

Bakowicz-Mitura, K., Bartosz, G., Mitura, S. (2007). *Surface Coatings Tech.*, **201**: 6131–6135.

Balaban, A. T. and Schleyer, P. v. R. (1978). *Tetrahedron* **34**: 3599–3609.

Bondar', V. S., Pozdnyakova, I. O. and Puzyr', A. P. (2004). *Phys. Solid State*, **46**: 758–760.

Bragg, W. H. and Bragg, W. L. (1913). *Nature* **91**: 557–557.

Burns, W., McKervey, M. A., Mitchell, T. R. B. and Rooney, J. J. (1978). *J. Am. Chem. Soc.*, **100**: 906–911.

Burns, W., Mitchell, T. R. B., McKervey, M. A., Rooney, J. J., Ferguson, G. and Roberts, P. (1976). *J. Chem. Soc., Chem. Commun.*, 893–895.

Carlson, R. M., Dahl, J. E. and Liu, S. (2007). US Patent No. US 7309476 B2.

CDV Diamond Group. (2010). www.chm.bris.ac.uk/pt/diamond/diamondoids.htm.

Chiganov, A. S. (2004). *Phys. Solid State*, **46**: 605–606.

Cupas, C., Schleyer P. v. R. and David J. Trecker, D. J. (1965). *J. Am. Chem. Soc.*, **87**: 917–918.

Dahl, J. E., Carlson, R. M. and Liu, S. (2002a). US Patent n. 20020130407 A1.

Dahl, J. E., Carlson, R. M. and Liu, S. (2002b). US Patent n. 20020177743 A1.

Dahl, J. E., Liu, S.G. and Carlson, R. M. (2003). *Science* **299**: 96–99.

Dahl, J. E., Carlson, R. M. and Liu, S. (2004). US Patent n. 20040262744 A1.

Dameron, A. A., Charles, L. F. and Weiss, P. S. (2005). *J. Am. Chem. Soc.*, **127**: 8697–8704.

Dammel, R. R. (2006). US Patent n. US 7033728 B2.

Danilenko, V. V. (2004). *Phys. Solid State*, **46**: 595–599.

Daulton, T. l., Eisenhour, D. D., Bernatowicz, T. J., Lewis, R. S., Buseck, P. R. (1996). *Geochim. Cosmochim. Acta.*, **60**: 4853–4872.

Davis, M. C., Dahl, J. E. P. and Carlson, R. M. K. (2008). *Synt. Comm.*, **38**: 1153–1158.

Davis, M. C. and Liu, S. (2006). *Synt. Comm.*, **36**: 3509–3514.

DeCarli, P. S. and Jamieson, J. C. (1961). *Science* **133**: 1821–1822.

Desiraju, G. R. (1996). *J Mol Structure*, **374**: 191–198.

Dolmatov, V. Yu. (2001). *Russ. Chem. Rev.*, **70**: 607–626.

Dolmatov, V. Yu., Veretennikova, M. V., Marchukov, V. A. and Sushchev, V. G. (2004). *Phys. Solid State* **46**: 611–615.

Farooq, O., Farnia, S. M. F., Stephenson, M. and Olah, G. A. (1988). *J. Org. Chem.*, **53**: 2840–2843.

Fokin, A. A., Butova, E. D., Chernish, L. V., Fokina, N. A., Dahl, J. E. P., Carlson, R. M. K. and Schreiner, P. R. (2007). *Org. Lett.* **9**: 2541–2544.

Fokina, N. A., Tkachenko, B. A., Merz, A., Serafin, M., Dahl, J. E. P., Carlson, R. M. K., Fokin, A. A. and Peter R. Schreiner, P. R. (2007). *Eur. J. Org. Chem.*, 4738–4745.

Fort, R. C. (1976). *Adamantane, The Chemistry of Diamond Molecules.* Studies in Organic Chemistry; Vol. 5, Marcel Dekker, New York.

Fort, R. C. and Schleyer, P. v. R. (1964a). *Chem. Rev.*, **64**: 277–300.

Fort, R. C. and Schleyer, P. v. R. (1964b). *J. Am. Chem. Soc.*, **86:** 4194–4195.

Geluk, H. W., Schlatmann, J. L. M. A., Chemical Communications (1967). (London), 426a–426a.

Geluk, H. and Schlatmann, J. (1968). *Tetrahedron*, **24**: 5361–5368.

Grady, M. M., Lee, M. R., Arden, J. W. and Pillinger, C. T. (1995). *Earth Planetary Sci. Lett.*, **136**: 677–692.

Gruen, D. M., Liu, S., Krauss, A. R., Luo, J. and Pan, X. (1994). *Appl. Phys. Lett.* **64**: 1502–1504.

Gund, T. M., Nomura, M., Williams Jr., V. Z., Schleyer, P. v. R. and Hoogzand, C. (1970). *Tet. Let.*, **11:** 4875–4878.

Gund, T. M., Schleyer, P. v. R. and Hoogzand, C. (1971). Tet. Lett., 1583–1586.

Gund, T. M., Thielecke, W. and Schleyer, P. v. R. (1973). *Org. Synth.* **53**: 30–34.

Gund, T. M., Williams, Jr., V. Z., Osawa, E. and Schleyer, P. v. R. (1970). *Tet. Lett.* 3877–3880.

Haaf, W. (1961). *Angew. Chem.*, **73**: 144–144.

Hollowood, F. S., McKervey, M. A., Hamilton, R. and Rooney, J. J. (1980). *J. Org. Chem.*, **45**: 4954–4958.

Holt, K. B. (2007). *Phyl. Trans. R Soc. A* **365**: 2845–2861.

Huss, G. R. (2005). *Element* **1**: 97-100.

Kabo, G. J., Blokhin, A. V., Charapennikau, M. B., Kabo, A. G., Sevruk, V. M. (2000). *Thermochim Acta.*, **345**: 125–133.

King, W. J. (1988). SPE Paper #17761, *SPE Gas Tech. Symp. Proceed.*, 469–490, Soc. Petroleum Eng. Int'l, Richardson, TX.

Koch, H. and Haaf, W. (1960). *Angew. Chem.*, **72**: 628–628.

Kong, X. L., Huang, L. C. L., Hsu, C.-M., Chen, W.-H., Han, C.-C. and Chang, H.-C. (2005). *Anal. Chem.* **77**: 259–265.

Krueger, A. (2008a). *Chem. Eur. J.* **14**: 1382–1390.

Krueger, A. (2008b). *Chem. Eur. J.* **14**: 1382–1390.

Krueger, A., Kataoka, F., Ozawa, M., Fujino, T., Suzuki, Y., Alenksenskii, A. E., Vul, A. Ya. and Osawa, E. (2005). *Carbon* **43**: 1722–1730.

Krueger, A., Liang, Y., Jarre, G. and Stegk, J. (2006). *J. Mater. Chem.* **16**: 2322–2328.

Kuznetsov, V. L., Aleksandrov, M. N., Zagoruiko, I. V., Chuvilin, A. L., Moroz, E. M., Kolomiichuk, V. N., Likholobov, V. A., Brylyakov, P. M. and Sakovitch, G. V. (1991). *Carbon* **29**: 665–668.

Landa, S., Machacek, V., Mzourek, M. and Landa, M. (1933). *Chim. Ind.* (Publ. No. 506).

Larionova, I., Kusnetsov, V., Frolov, A., Shenderova, O., Moseenkov, S. and Mazov, I. (2006). *Diamond Relat. Mater.* **15**: 1804–1808.

Liu, S., Carlson, R. M., Dahl, J. E. and Qureshi, W. R. (2002). CIPO Patent n. CA 2469858.

Liu, S., Carlson, R. M., Dahl, J. E. and Qureshi, W. R. (2003). US Patent n. US 2003/0199710 A1.

Liu, S., Carlson, R. M. and Dahl, J. E. (2010). US Patent n. 7,649,056 B2.

Liu, S., Dahl, J. E. and Carlson, R. M. (2009). US Patent n. US 7488565 B2.

Liu, Y., Gu, Z., Margrave, J. L. and Khabashesku, V. N. (2004). *Chem. Mater.*, **16**: 3924–3930.

Malik, A. A., Archibald, T. G., Baum, K. and Unroe, M. R. (1991). *Macromolecules* 24: 5266–5268.

Mansoori, G. A. (2005). *Principles of Nanotechnology (Molecular-Based Study of Condensed Matter in Small Systems).* World Scientific. Pub. Co., Hackensack, NJ.

Mansoori, G. A. (2007). *Advances in Chemical Physics* **136**: 207–258.

Mansoori, G. A., Assoufid, L., George, T. F. and Zhang, G. (2003). Proceedings of *Nanotech 2003*; San Francisco, CA.

McKervey, M. A. (1980). *Tetrahedron* **36**: 971–992.

Mochalin, V. N., Osswald, S., Portet, C., Yushin, G., Hobson, C., Havel, M., Gogotsi, Y. (2007). Presented in Materials Research Society Fall Meeting, Boston, MA, USA, 2007; 1039, p 1039–P11-03.

Moiseev, I. K., Makarova, N. V. and Zemtsova, M. N. (1999). *Russ. Chem. Rev.*, **68**: 1001–1020.

Navratilova, M., Sporka, K. (2000). *Appl. Catalysis A, General*, 203: 127–132.

Nekhaev, A. I., Bagrii, E. I. and Maximov, A. L. (2011). *Petroleum Chemistry* **51**: 86–95.

Olah, G. A. (1990a). *Cage Hydrocarbons*, John Wiley & Sons, Hoboken, NJ.

Olah, G. A. (1990b). In *Cage Hydrocarbons*, G. A. Olah, editor, Chapter 4, pp. 103–153, John Wiley & Sons, Hoboken, NJ.

Olah, G. A., Laali, K. and Farooq, O. (1984). *J. Org. Chem.* **49**: 4591–4594.

Olah, G. A., Prakash, G. K. S., Shih, J. G., Krishnamurthy, V. V., Mateescu, G. D., Liang, G., Sipos, G., Buss, V., Gund, T. M., Schleyer, P. v. R. (1985). *J. Am. Chem. Soc.*, **107**: 2764–2772.

Osawa, E. (2007). *Diamond Relat. Mater.* **16**: 2018–2022.

Peterson, M. R. and Wahl, G. H. (1968). *Chem. Comm.* (London), 1552–1553.

Prelog, V. and Seiwerth, R. (1941). *Ber.*, **71**: 1769–1772. DOI: 10.1002/cber.19410741109.

Qin, L. C., Zhou, D., Krauss, A. R. and Gruen, D. M. (1998). *NanoStruct. Mater.* **10**: 649–660.

Ramezani, H., Mansoori, G. A. and Saberi, M. R. (2007). *J. Comput'l & Theor'l Nanoscience* **4**(1): 96–106.

Schleyer, P. v. R. (1957). *J. Am. Chem. Soc.* **79**: 3292–3292.

Schleyer, P. v. R. and Donaldson, M. M. (1960). *J. Am. Chem. Soc.*, **82**: 4645–4651.

Schleyer, P. v. R (1990). In *Cage Hydrocarbons*, G. A. Olah, editor, Chapter 1, p. 1–38, John Wiley & Sons, Hoboken, NJ.

Schleyer, P. v. R (2009). www.chem.uga.edu/schleyer/index.html.

Schleyer, P. v. R., Donaldson, M. M., Nicholas, R. D. and Cupas, C. (1962). *Org. Synth.*, **42**: 8–11.

Schreiner, P. v. R., Fokina, N. A., Tkachenko, B. A., Hausmann, H., Serafin, M., Dahl, J. E. P., Liu, S., Carlson, R. M. K. and Fokin, A. A. (2006). *J. Org. Chem.*, **71**: 6709–6720.

Schleyer, P. v. R., Fort, R. C., Watts, W. E., Comisarow, M. B. and Olah, G. A. (1964). *J. Am. Chem. Soc.*, **86:** 4195–4197.

Schwertfeger, H. Fokin and A. A. Schreiner, P. R. (2008). *Angewandte Chemie — Int'l Ed English*, **47**: 1022–1036.

Schwertfeger, H., Würtele, C., Serafin, M., Hausmann, H., Carlson, R. M. K., Dahl, J. E. P. and Schreiner, P. R. (2008). *J. Org. Chem.*, **73**: 7789–7792.

Schwertfeger, H., Würtele, C., Hausmann, H., Dahl, J. E. P., Carlson, R. M. K., Fokin, A. A., Schreiner, P. R. (2009). *Adv. Synth. Catal.*, **351**: 1041–1054.

Sevostyanova, V. V., Krayushkin, M. M. and Yurchenko, A. G. (1970). *Russ. Chem. Rev.*, **39**: 817–833.

Sciamanna, S. F., Liu, S. and Mukai, A. (2007). US Patent n. 20070037909 A1.

Shenderova, O., Grichko, V., Hens, S., Walch, J. (2007). *Diamond Relat. Mater.* **16**: 2003–2008.

Shenderova, O. A., Zhirnov, V. V. and Brenner, D. W. (2002). *Crit. Rev. Solid State Mater. Sci.*, **27**: 227–356.

Smith, G. W. and Williams, H. D. (1961). *J. Org. Chem.*, **26**: 2207–2212.

Stetter, H., Bänder, O.-E. and Neumann, W. (1956). *Chem. Ber.*, **89**: 1922–1926. DOI: 10.1002/cber.19560890821.

Stetter, H., Krause, M. and Last, W.-D. (1968). *Angew. Chem.*, **80**: 970–971.

Stetter, H., Schwarz, M. and Hirschhorn, A. (1959). *Angew. Chem.*, **71**: 429–430.

Stetter, H., Wulff, C. (1960). *Chem. Ber.* **93:** 1366–1371.

Stauss, S., Miyazoe, H., Shizuno, T., Saito, K., Sasaki, T. and Terashima, K. (2010). *Japanese J. Applied Physics*, **49**: 070213–070213-3.

Sung, S. L., Guo, X. J., Huang, K. P., Chen, F. R., Shih, H. C. (1998). *Thin Solid Films*, **315**: 345–350.

Tkachenko, B. A., Fokina, N. A., Chernish, L. V., Dahl, J. E. P., Liu, S., Carlson, R. M. K., Fokin, A. A. and Schreiner, P. R. (2006). *Org. Lett.* **8**: 1767–1770.

Vakili-Nezhaad, G. R., Mansoori, G. A. and Kiani-Nassab, K. (2005). *Int'l J. Nanoscience & Nanotechnology*, 65–74.

Vazquez, D. and Mansoori, G. A. (2000). *J. Petroleum Sci. & Eng.*, **26**: 49–55.

Vityaz, P. A. (2004). *Phys. Solid State*, **46**: 591–595.

Williams, O. A. and Nesladek, M. (2006). *Phys. Stat. Sol.* **203**: 3375–3386.

Williams, Jr., V. Z., Schleyer, P. v. R., Gleicher, G. J. and Rodewald, L. B. (1966). Triamantane. *J. Am. Chem. Soc.*, **88**: 3862–3863.

Xue, Y. and Mansoori, G. A. (2008). *International Journal of Nanoscience*, **7**: 63–72.

Xue, Y. and Mansoori, G. A. (2010). *Int'l J Molecular Sciences* **11**(1): 288–303.

Yang, W., Fabbri, J. D., Melosh, N. A., Hussain, Z. and Shen, Z.-X. (2008). US Patent no. 20080191598 A1.

Zhang, W. J., Wu, Y., Chan, C. Y., Wong, W. K., Meng, X. M., Bello, I., Lifshitz, Y., Lee, S. T. (2004). *Diamond Relat. Mater.* **13**: 1037–1043.

2

Diamondoids in Petroleum and other Fossil Fuels

2.1. Introduction

As described in Chapter 1, underground fossil fuel reservoirs including coal, petroleum, gas condensate, and natural gas are the natural resources of diamondoid molecules. The simplest representative of such compounds is the symmetrical tricyclodecane $C_{10}H_{16}$, which was isolated in 1933 from petroleum in the vicinity of Hodonin in Moravia, Czechoslovakia (Landa *et al.*, 1933). This compound was named as adamantane, a noun derived from the Greek word for diamond — αδαμαξ — "the invincible" (Seiwerth, 1996). In this chapter, we report some aspects of the origin of diamondoids (alkyladamantanes) in fossil fuels. In addition, we will present the state-of-the-art in the technology of using diamondoids content of petroleum fluids as a geochemical tool in such reservoir fluids characterization. We also report on the possible role of diamondoids in petroleum and natural gas production fouling.

As we discuss in the remaining chapters of this book, diamondoids are considered as one of the best molecular building blocks (MBBs) for biomedicine, materials science, and nanotechnology. As diamondoids are almost entirely constituted by sp^3 bonded carbons arranged in a three-dimensional network, these stiff, dense, covalent solids present many advantages over other types of MMBs (2007). Considering that the major source of diamondoids is petroleum fluids, their occurrence in, and understanding of their identification and separation from, petroleum fluids are also presented here.

2.2. Diamondoids in Fossil Fuels

Most of the literature data on naturally occurring diamondoids are about petroleum diamondoids. Naturally occurring adamantane, present in petroleum, is generally accompanied by small amounts of alkylated adamantane: 2-methyl-; 1-ethyl-; and probably 1-methyl-; 1,3-dimethyladamantane; and others (Katz *et al.*, 2008; Nekhaev *et al.*, 2011). Diamantane, triamantane and their alkyl-substituted compounds are also present in certain petroleum crude oils (Wingert, 1992). Wingert (1992) reported the formula, molar mass, and structure for diamondoids found in petroleum as shown in Fig. 2.1.

Concentration of adamantane in petroleum is the highest among all the petroleum diamondoids as reported in Table 2.1. Tetramantane, pentamantane, and hexamantane were discovered for the first time in 1995 in a gas condensate (also known as NGL standing for natural gas liquid) produced from a very deep petroleum reservoir (~6800 m below the earth's surface) located in the US Gulf Coast, along with significant amounts of alkyl-substituted diamantanes and triamantanes (Lin and Wilk, 1995). A group of investigators reported isolation of crystals of [12312]hexamantane or cyclohexamantane ($C_{26}H_{30}$) as shown in Fig. 1.4, from distilled Gulf of Mexico petroleum using reverse-phase high-performance liquid chromatography (HPLC) (Dahl *et al.*, 2003a). They determined the structure of $C_{26}H_{30}$ using X-ray diffraction, mass spectroscopy, and ^{1}H, ^{13}C-NMR spectroscopy. They also used experimental Raman spectra of crystalline diamond, adamantane, and nanophase diamond to indirectly identify frequencies in the experimental Raman spectra of this *peri*-condensed hexamantane isomer. See Secs. 3.2.3.2–3.2.3.4 for Raman spectroscopy data of representative diamondoids.

Identification and isolation of crystals of many new medium-size and higher-order diamondoids from petroleum have been reported by Dahl and co-workers (2003b). These investigators reported the isolation of multiple families of higher diamondoid molecules containing 4 to 11 diamond-crystal cages from petroleum and supplied X-ray structures for three representative families. Separations of higher diamondoids from

Formula, Name, Molecular wt	Structure	Formula, Name, Molecular wt	Structure
$C_{10}H_{16}$ Adamantane Mw=136	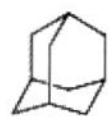	$C_{16}H_{24}$ 4, 9 - dimethyldiamantane Mw=216	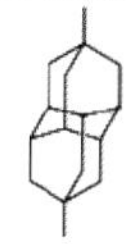
$C_{14}H_{24}$ 1, 2, 5, 7 - Tetramethyladamantane Mw=192	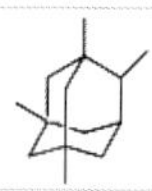	$C_{15}H_{22}$ 1 - Methyldiamantane Mw=202	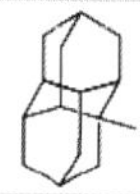
$C_{12}H_{20}$ 1 - Ethyladamantane Mw=164	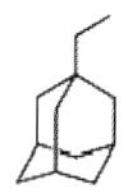	$C_{16}H_{24}$ 1, 4 - Dimethyldiamantane Mw=216	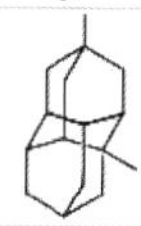
$C_{13}H_{22}$ 1 - Ethyl - 3 - methyladamantane Mw=178	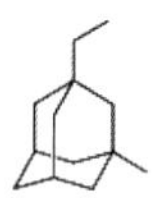	$C_{16}H_{24}$ 2, 4 - Dimethyldiamantane Mw=216	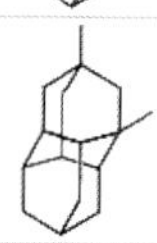
$C_{14}H_{24}$ 1 - Ethyl - 3, 5 - dimethyladamantane Mw=192	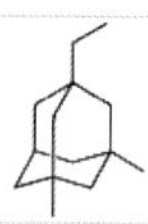	$C_{16}H_{24}$ 4, 8 - Dimethyldiamantane Mw=216	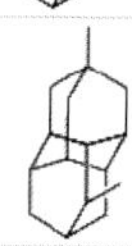
$C_{12}H_{20}$ 1 - Ethyladamantane Mw=164	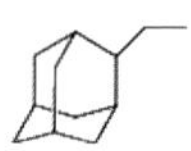	$C_{17}H_{26}$ 1, 3, 4 - Trimethyldiamantane Mw=230	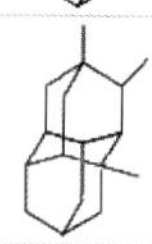
$C_{14}H_{20}$ Diamantane Mw=188	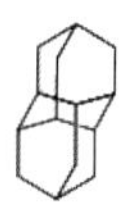	$C_{15}H_{22}$ 3 - Methyldiamantane Mw=202	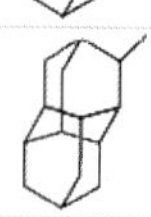
$C_{15}H_{22}$ 4 - Methyldiamantane Mw=202	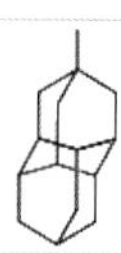	$C_{16}H_{24}$ 3, 4 - Dimethyldiamantane Mw=216	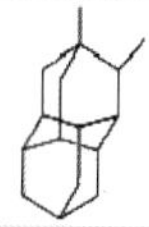

Fig. 2.1. Diamondoid molecules found in petroleum (Modified from Wingert, 1992).

selected petroleum feedstock were performed by distillation, and the removal of nondiamondoids was achieved by pyrolysis at 400–450°C. Aromatic and polar compounds were removed from pyrolysis products by argentic silica gel liquid chromatography. Higher diamondoids were then isolated by a combination of reversed-phase HPLC on octadecyl silane columns and highly shape-selective Hypercarb® HPLC columns. Individual higher diamondoids were re-crystallized to high purity (Dahl *et al.*, 2003b). For some of the diamondoid molecules reported in Fig. 2.1

Fig. 2.2. Photomicrographs of [1(2,3)4] pentamantane crystallized from HPLC fractions of gas condensates and natural gas liquids from the Norphlet Formation, Gulf Coast USA and the Western Canada Basin. Crystals show visual similarities with macroscopic diamond but different physical properties (see Table 2.3). Field of view is approximately 3.5 mm. Photograph, Marilyn Olmstead, U. C. Davis (see also Dahl *et al.*, 2003).

their crystalline structures are determined by single-crystal X-ray. For example, [1(2,3)4] pentamantane forms pyramidal crystal that share some resemblance with macroscopic diamond (Fig. 2.2).

Nevertheless, weak van der Waals intermolecular interactions between the six molecules that constitute the orthorhombic unit cell and packing of crystalline [1(2,3)4] pentamantane, Fig. 2.3 (Top), allow dissolution of this diamondoid substance in acetone (Dahl *et al.*, 2003b). Also reported in Fig. 2.3 (Bottom) is the 3-D structure of [1(2,3)4] pentamantane and its relation to diamond lattice structure.

Sulfur-containing diamondoid compounds, including thiadiamondoids and adamantanethiols (Fig. 2.4) occur naturally in petroleum, especially in oils and/or gas condensates subjected to thermochemical sulfate reduction (TSR) (Heydari and Moore, 1989; Hanin *et al.*, 2002; Wei *et al.*, 2007a). TSR is one of the most important organic–inorganic interactions in hot carbonate petroleum reservoirs and is characterized by the oxidation of petroleum by sulfate (generally anhydrite), with production of H_2S, CO_2, mercaptans, elemental sulfur, and pyrobitumen (Seewald, 2003). This phenomenon is well

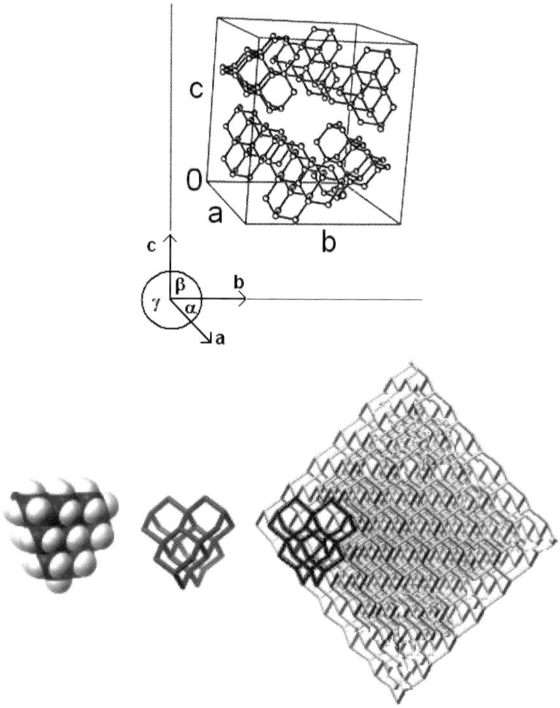

Fig. 2.3. **Top:** Stereo view of the unit cell and packing of crystalline [1(2,3)4] pentamantane (orthorhombic, Pnma space group, unit cell: $a = 12.5$ Å, $b = 11.47$ Å, $c = 12.64$ Å and $\alpha = \beta = \gamma = 90^{\circ}$, where a, b, and c are the dimension in Angstroms, and α, β, and γ are the angles of the orthorhombic unit cell. (Dahl *et al.*, 2003). **Bottom:** 3-D structure of [1(2,3)4] pentamantane and its relation to diamond lattice structure.

documented in numerous geological observations all around the world (Worden *et al.*, 1996; Machel, 2001; Cai *et al.*, 2003) and decreases commercial value of a crude oil while increases its cost of production. TSR is responsible for substantial quantities of H_2S (between 10–80%) in some high-temperature natural underground petroleum reservoirs.

The presence of alkylated 2-thiadiamantane has also been suggested by gas chromatography–mass spectrometry (GC/MS) data (Hanin *et al.*, 2002). Sulfur-containing diamondoids, particularly alkylthiaadamantanes may serve as molecular indicator of TSR occurring in deep petroleum reservoirs (Hanin *et al.*, 2002) and can be used to detect the level of risk associated with the presence of H_2S in petroleum exploration and production operations.

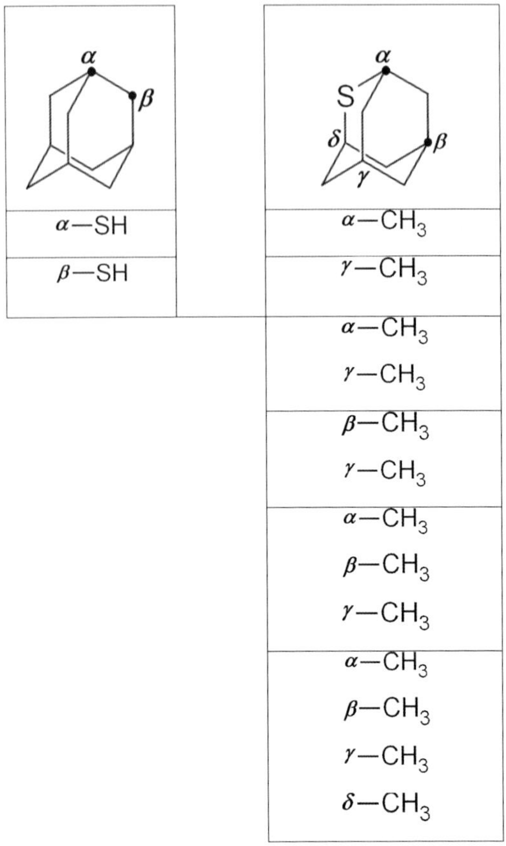

Fig. 2.4. Chemical structure of ten sulfur-containing adamantane molecules found in petroleum and/or gas condensates which have underwent thermochemical sulfate reduction (TSR) process. Notations α, β, γ, and δ refer to bonding locations on adamantane cage structure as indicated (Data from Wei *et al.*, 2007a).

In Table 2.1, we report concentration of adamantanes, diamantanes, 3-methyldiamantane +4-methyldiamantane (3-,+4-MD) and total diamondoids determined through GC/MS experiments in several petroleum and two gas condensate samples.

Variations in concentration of (3-,+4-MD) are frequently reported as a model for diamondoids behavior in crude oils. Besides being relatively abundant, these two molecules possess low volatility and chemical behavior similar to other diamondoids. Thus, they satisfactorily represent most

Table 2.1. A compilation of concentrations of diamondoids in various crude oils and condensates in [μg/g] (From Yang *et al.*, 2006; Wei *et al.*, 2007c; Wang *et al.*, 2006a).

Crude oil reservoir, Location	[A]	[D]	[3-,+4-MD]	[TDs]
Orinoco, Venezuela	31.1	4.9	1.1	NA
Cold Lake, Canada	117.0	12.0	3.8	NA
Calgary, Canada	130.0	13.6	3.9	NA
Platform Elly, US West Coast	42.1	10.4	5.5	NA
Maya, Mexico	51.6	14.6	1.7	NA
Gulf of Mexico	121.0	12.3	3.3	NA
Saudi Arabia	87.8	22.8	6.7	NA
West Delta Canyon, Gulf of Mexico	126.0	17.2	4.8	NA
Alaska North Slope, Alaska	263.0	27.5	7.3	NA
Southern Alaska	209.0	9.1	2.1	NA
South Louisiana	1880.0	195.0	51.9	NA
Race Track, California	2285.2	53.6	14.2	2346
Buena Vista, California	212.6	5.1	1.2	221
Fruitvale, California	2091.7	38.6	9.6	2137
McKittrick, California	323.4	6.3	1.7	334
South Belridge, California	1188.6	18.5	4.6	1253
Kern River, California	453.2	11.0	2.96	465
Cymric, California	713.0	14.4	4.3	728
Prudhoe Bay, Alaska	273.0	38.9	10.1	312
Gas Condensate, Location				
Hudson Canyon, US Atlantic	4038; 4266	704; 509	181; 148	NA
Tarin Basin, China	NA	NA	250	NA

[3-,+4-MD] = Concentrations of 3-methyldiamantane + 4-methyldiamantane.
[A] = Concentrations of all one-cage diamondoids c.
[D] = Concentrations of all two-cage diamondoids (Diamantane and its alkyl derivatives including (3-,+4-MD)).
NA = Not Available.
[TDs] = Concentration of Total Diamondoids.
[µg/g] = Microgram of diamondoids per gram of crude oil.

common diamondoid classes. (3-,+4-MD) concentrations in petroleum of different origins are reported in Table 2.1, along with concentrations of one-cage diamondoids (adamantane and derivatives), two-cage diamondoids (diamantane and derivatives), and total diamondoids determined through GC/MS measurements.

In Table 2.1, [A] represents concentration of all one-cage diamondoids (Adamantane and its alkyl derivatives), [D] represents concentration of all two-cage diamondoids (Diamantane and its alkyl derivatives including (3-,+4-MD)) and [TDs] represents concentration of all diamondoids (Total Diamondoids). In the case of crude oils data published by Yang *et al.* (2006), they only searched for one and two-cage diamondoids, thus, there is no [TDs] reported in their crude oils data. In some cases, when [A], [D], [3-,+4-MD], and [TDs] data are available, [TDs] is smaller than [A]+[D]+[3-,+4-MD]. This happens because many authors report [3-,+4-MD] and their variations as a model for diamondoids behavior in crude oils. As a result concentration of (3-,+4-MD) and possibly [D] and possibly [A] are counted twice, once as [D] and [A] and once in [3-,+4-MD].

Diamondoids are widespread in the bitumen of coals. In Table 2.2, we report a compilation of concentrations of (3-,+4-MD) diamondoids in various coals in microgram of diamondoids per gram of coal [μg/g] (Wang *et al.*, 2006a; Sassen and Post, 2008; Zhang *et al.*, 2005; Wei *et al.*, 2006c). By comparing Tables 2.1 and 2.2, it becomes clear that generally petroleum fluids contain significantly more [3-,+4-MD] than coals. This is also the case for adamantane and diamantane in coals compared to petroleum fluids (Imuta and Ouchi, 1973; Aczel *et al.*, 1975, 1979).

2.3. Origin of Petroleum and Genesis of Diamondoids

In order to answer the question "*How could diamondoids be present in fossil fuels?*", it is important to discuss in some details about the origin of petroleum, the mechanisms involved in its maturation and degradation in the ground and then understand the mechanisms for diamondoids

Table 2.2. A compilation of concentrations of (3-,+4-MD) diamondoids in various coals in [μg/g] = Microgram of diamondoids per gram of coal (From Wang *et al.*, 2006a; Sassen and Post, 2008; Zhang *et al.*, 2005; Wei *et al.*, 2006c).

Coal mine, location	[3-,+4-MD]
Richland, Montana	0.12
Campbell, Wyoming	0.02; 0.03
Mercer, North Dakota	0.11
Freestone, Texas	0.02
Bighorn, Montana	0.08
Rosebud mine, Montana	0.04
Jungar Basin, China	0.41; 0.58; 0.34
Colfax, Texas	0.54
Carbon, Utah	0.64
Kanawha, West Virginia	0.95
Dickenson, Virginia	1.23
Buchanan, Virginia	10.0
Greenbrier, West Virginia	23.0
Sommerset, Pennsylvania	47.0
Pennsylvania	70.0
Arkansas	206.0
Sullivan, Pennsylvania	341.0
Colombia, Pennsylvania	27.0
Luzcerne, Pennsylvania	3.18
Pappano Mine	2.64

formation and concentration in petroleum. According to one hypothesis (Hunt, 1979), when living organisms die, their organic constituents undergo physical, chemical, and biological transformations. Most of the resulting products are gaseous and water-soluble substances that are carried off by weathering agents. Subsequent oxidative processes in high-energy, oxygen-rich environments may ultimately transform organic carbon to carbon dioxide. Thus, carbon atoms present in the original organic matter would re-enter the earth's carbon cycle. However, under appropriate conditions, organic matter is preserved and temporarily withdrawn from this geochemical cycle. This preserved matter would become fossil fuels including finely dispersed buried organic matter, coal and petroleum fluids.

A good example of such process occurs in sedimentary basins because water from these basins is rich in oxygen which makes it aerobic at the surface but sediments are exposed to anaerobic environment even at a few centimeters below surface. The organic matter in these sediments, especially humic material (humus), the major organic constituent of soil which may include resins, waxes, and lipids stand a better chance of resisting degradation during long-term burial and thermal processes, becoming available for petroleum formation (Hunt, 1979; Carpentier *et al.*, 2007).

During and after sedimentation, organic matter could be submitted to three distinct transformation processes: Diagenesis, catagenesis, and metagenesis (Tissot and Welte, 1984):

Diagenesis is the earliest stage of transformation of sedimentary organic matter. During this process, biological, physical, and chemical alterations occur before extensive thermal degradation, usually in depths where temperature ranges from ambient to 50°C (Tissot and Welte, 1984). Typical biomacromolecules, such as polysaccharides, proteins, and lignins along with other biocompounds (e.g., lipids and pigments) may be consumed by burrowing (fossorial) organisms or may enter into complex formation with mineral components. Alternatively, microbial enzymatic attack is possible to convert the biomacromolecules into simpler molecules, such as amino acids, monosaccharides, fatty acids, and phenols. Part of these low molecular weight substances may react to form geopolymers that are precursors of kerogen.

The term "kerogen" is generally referred to the organic matter (oil) in oil shale released upon heating (Hunt, 1979). A more recent definition states that kerogen is a type of particulate organic matter insoluble in organic solvents that consists of various macerals originated from components of plants, animals, and bacteria, preserved in sedimentary rocks.

Macerals are microscopically recognizable constituents of a fossil fuel that can be differentiated by their physical and chemical properties. Macerals can be seen as analogs of minerals of inorganic rocks with less uniform chemical composition and a larger variety of physical characteristics (Hunt, 1979). The three main maceral groups are liptinite, vitrinite, and inertinite (Hunt, 1979; Scott, 2002).

Studies on kerogen's three main maceral groups compositions revealed that kerogen contains practically the same diversity of organic compounds as found in living organisms, including terpenoids, steroids, porphyrins, amino acids, carbohydrates, carboxylic acids, ketones, alcohols, and ether bridges (Hunt, 1979). During early burial and diagenesis, many complex organic compounds originating from former living organisms are preserved. These molecular fossils are called biomarkers or biological markers and show only slight or no alterations in structure when compared with their parent biological molecules. Biomarkers occur not only in sediments and rocks but in crude oils as well, evidencing their resistance to thermal stress and biodegradation (Peters *et al.*, 2005).

Catagenesis is the process within which thermal cracking of kerogen produces petroleum hydrocarbons (Durand, 1980; Pepper and Corvi, 1995; Behar *et al.*, 1997). Catagenesis is characterized by thermocatalytic alterations in buried organic material subjected to temperatures in the range of 50°C to about 150°C. This process lasts millions of years during which main chain scissions, decarboxilation and hydrogen disproportionation are some of the predominant chemical reactions that occur (Hunt, 1979). During catagenesis, biomarkers undergo some structural changes that are indicative of the extent of thermal stress imposed on (or suffered by) source rocks or expelled oil. As a result, biomarker occurrence can be used as an indication of the origin of expelled oil that have migrated miles away from its original formation site by comparing the key structural features and distribution of biomarkers both in migrated oil and source rocks (Peters *et al.*, 2005).

Metagenesis is the transformation stage when an organic material is exposed to temperatures in the range of 150–200°C; it undergoes extensive thermal cracking, leading to final conversion into methane and/or graphite (Landes, 1967). During this final stage, biomarkers undergo dramatic decrease in concentration or complete destruction (Peters *et al.*, 2005).

It is important to point out that the limits for the onset of each of the three processes are not rigid. In addition, many factors other than time and temperature are important in oil formation, maturation, and destruction. For example, the presence of water, high fluid pressures, and retention of products in closed systems could inhibit thermal cracking of oil, typical of metagenesis process in laboratory experiments. The absence of water, low fluid pressures, and open systems conditions appear to promote oil thermal cracking (Price, 1993). Bacteria may also play important roles in the formation of crude oil and natural gas. In well-oxygenated sedimentary environments, oxidation of organic matter by aerobic bacteria may result in its poor preservation. Conversely, lower oxygen levels may contribute to preservation of oil-prone organic matter. Biodegradation is another aspect of petroleum lifecycle that can interfere in its accumulation processes. Sometimes, such alterations are so extensive that makes geochemical evaluation of thermal maturity and source correlation almost impossible (Clayton, 1992).

The ideas of what chemical processes are involved in the formation of petroleum have evolved since 1970's into a modern vision of the "petroleum system", in which four geological components are necessary for hydrocarbon accumulation:

(i) a source rock, where petroleum is or could be generated;
(ii) a reservoir rock which is a body of rock having sufficient porosity and permeability to store and transmit fluids;
(iii) a cap rock or seal, an impermeable rock that acts as a barrier to further migration of hydrocarbon liquids, and;
(iv) a trap, a configuration of rocks suitable for containing hydrocarbons and sealed by a relatively impermeable formation through which hydrocarbons will not migrate (Magoon, 1992).

Moreover, these four geological components must be properly placed in a dynamic scenario where the source rock plays the major role as its thermal history gives the guidelines for organic matter transformations in the basin (Huc, 2003).

2.3.1. *Diamondoid genesis in petroleum*

Diamond is believed to be an abiogenic matter (not produced by the action of living organisms) since in nature it is formed from abiogenic source. However, the carbon framework for diamondoids in oil is believed to be biogenic. Diamondoids in petroleum are believed to be formed by hydrocarbon rearrangement reactions under acidic conditions, compatible with those found in clay minerals in petroleum source rocks (Katz *et al.*, 2008).

In 1964, Fort and Schleyer suggested that adamantane could have been formed during petroleum genesis from cyclic hydrocarbon precursors by Lewis acid-catalyzed isomerization and rearrangement, through a mechanism similar to that observed during the synthesis of adamantane from *endo*-tricyclodecane (Fort and Schleyer, 1964) (Fig. 1.10). Laboratory experimental data available at the time also pointed to the same conclusion. For example, when adamantane-free samples from Hodonin oil were subjected to fractioning and treatment with various catalysts, additional amounts of adamantane were formed in the fractions having boiling points in the 170–190°C range after treatment with aluminum chloride.

Likewise, diamantane and triamantane can be synthesized by superacid-catalyzed isomerization of suitable polycycloalkane precursors (McKervey, 1980; Olah, 1990) and might have their geochemical formation related to the same mechanism.

Nevertheless, higher diamondoids are unlike to be formed the same way, primarily because it is difficult to find large 22 carbons) polycyclic hydrocarbons with chemical formula Cn_6 Hn_{12} in petroleum. As higher polycyclic hydrocarbons are not reported to occur in living organisms, they cannot be structurally related biological molecules.

It is proposed that tetramantane, pentamantane, and hexamantane may be generated via homologation of the lower alkyl-substituted diamondoids at high pressures and temperatures in the natural underground fossil fuel reservoirs (Lin and Wilk, 1995).

Alkylated diamondoids are frequently found in petroleum, as discussed previously in Sec. 2 of this chapter. These substituted diamondoids may be formed by reactions between abundant methane and polymantyl cations catalytically generated from polymantanes in acidic sites. Of course, direct reaction between methane and diamondoid molecules is difficult. It is believed that the presence of clays in the underground oil reservoirs may have provided the acid sites necessary to catalytically convert polymantanes into polymantyl cations. The resulting alkyl-substituted diamondoid molecules can then be converted into higher members of the

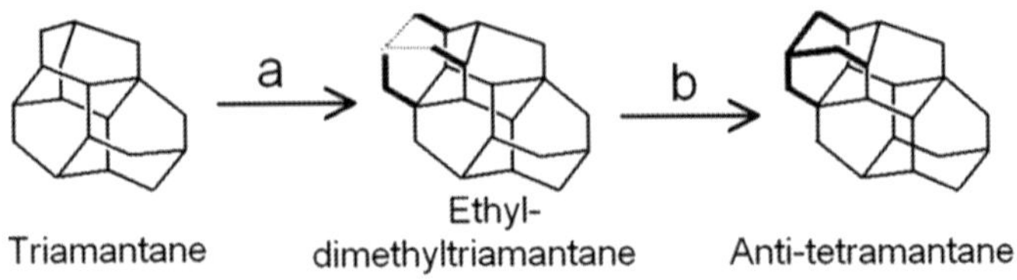

Fig. 2.5. Homologation of triamantane into *anti*-tetramantane: (a) Ethyl-dimethylation of triamantane that may occur in clay sites of rock sources, (b) Homologation of ethyl-dimethyltriamantane into *anti*-tetramantane ([121]tetramantane) under high pressure and temperature. New alkyl groups are shown in bold lines, dashed lines represent forming bonds (Replotted from Lin and Wilk, 1995).

diamondoid family (Lin and Wilk, 1995). A scheme of the homologation of triamantane into *anti*-tetramantane is shown in Fig. 2.5.

More recently efforts have been made to investigate the formation of diamondoids during thermal maturation of preserved organic matter and clarify the influence of minerals on the generation of diamondoid molecules.

Experiments involving artificial hydrothermal maturation of different types of immature sediments (sediments that have seen little migration and erosion) from organic-rich modern mud samples and carbonate sediment (Wei *et al.*, 2006a) and immature sedimentary rocks and peats (Wei *et al.*, 2007b) indicated that diamondoids could not be found in extracts of organic-rich modern sediment or immature peats prior to artificial maturation. Conversely, they were present in immature sedimentary rocks which had gone through a certain level of natural thermal maturation. These findings suggest that diamondoid generation is not related to biosynthetic pathways and has its onset during diagenesis. Diagenesis is any change in sediment, after its initial deposition and during and after its hardening.

The presence of artificially-added minerals such as montmorillonite K10 clay or acidic aluminosilicate during artificial maturation of modern mud samples and carbonate sediment revealed that they exhibit strong catalytic effect on diamondoid genesis, while $CaCO_3$ (an alkaline salt) has an inhibiting action. These minerals showed similar effects when kerogen samples of various types were subjected to artificial thermal maturation (Wei *et al.*, 2006b). When $CaCO_3$ is mixed with montmorillonite K10 clay, it tends to neutralize the catalytic activity of montmorillonite K10 clay even if only a small amount of $CaCO_3$ is present in the mixture.

Catalytic effect of such Lewis acids suggests that the natural formation of diamondoids follows a carbonium rearrangement. Carbocation rearrangement is seen for chemical synthesis of diamondoid molecules (see Sec. 1.4). Experimental results are detailed below in the later paragraphs.

Table 2.3 shows a comparison of (3-,+4-MD) yields after hydrous pyrolysis treatment of sediments and kerogen types I, II, II-S, and III (see Glossary). The combined (3-,+4-MD) group diamondoid molecules are chosen to represent all diamondoid classes because they are less volatile and relatively abundant.

Moreover, it is possible to observe similar features between (3-,+4-MD) and other diamondoid yields concerning catalytic effects of minerals, as shown in Fig. 2.6a–f. Results presented on Fig. 2.6c to Fig. 2.6e are related to experiments in which montmorillonite K10 clay (43.77% SiO_2, 18.57% Al_2O_3, 36.09% H_2O and minor oxide components), acidic aluminosilicate MS-25 (76.40% SiO_2, 23.20% Al_2O_3 and minor oxide components) or $CaCO_3$(99.99% purity) were used, respectively. This figure is very enlightening, since it is an experimental evidence to support the hypothesis of natural diamondoid formation in the very source rock bed through Lewis acid catalysis. Moreover, the negative interference of alkaline minerals is also demonstrated.

According to Table 2.3 under "Extract", column (a), various amounts of (3,+4-MD) were detected in all nonpyrolyzed extractable organic matter of mud and rock sediments. Carbonate, peat, and kerogen samples showed no detectable extract quantities of (3,+4-MD) as also reported in Table 2.3. The traces of (3,+4-MD) found in mud samples were attributed to contaminations from oil spills or oil seeps.

In addition, the presence of biomarkers, normally absent in modern sediments, served as a confirmation of the sample contamination. Mexico's Celestun Lagoon carbonate sample, nevertheless, lacked biomarkers typical of oil contaminants. German peat samples presented bulky geochemical and biomarker profiles consistent with immature modern organic-rich sediment. These results suggest that diamondoid formation is associated with processes of maturation, which these modern sediments had not gone through yet. Overall, according to Fig. 2.6 and Table 2.3, higher concentrations of diamondoids are expected to be found in samples that present lower concentrations of alkaline minerals. This

Table 2.3. Yields of 3–methyldiamantane + 4–methyldiamantane (3-,+4-MD) before and after artificial maturation pyrolysis experiments, in sediments and kerogen from various locations (From Wei *et al.*, 2006a, 2006b).

				[3-,+4-MD] Yield [ng/g]				
	Sample type	**Location (Environment or lithology)**	**TOC [wt%]**	**(a)**	**(b)**	**(c)**	**(d)**	**(e)**
Sediment	Mud	Calif., San Francisco Bay (salt marsh)	8.0	0.09	1.84	0.55	24.07	26.63
	Mud	Elkhorn Slough (marsh)	5.0	0.02	5.80	0.48	36.17	36.32
	Mud	Calif., Santa Barbara Basin (continental margin)	7.0	0.07	1.12	0.86	10.00	3.89
	Carbonate	Mexico, Celestun Lagoon (mangrove)	6.0	NA	0.14	0.11	0.44	1.47
	Rock	Brazil, Irati formation (limestone)	2.4	0.34	0.93	NT	NT	NT
	Rock	Brazil, Irati formation (shale)	7.0	0.79	1.12	NT	NT	NT
	Rock	Nevada, Vinini Creek (limestone)	1.7	1.58	3.70	NT	NT	NT
	Peat	Germany, Outcrop (fen peat)	37.9	NA	1.13	NT	NT	NT
	Peat	Germany, Outcrop (bog peat)	34.4	NA	1.53	NT	NT	NT

(Continued)

temperature range. Total pyrolysate yields were maximized at 340°C for peats, at 350°C for Irati shale and at 360°C for both limestone samples (Wei *et al.*, 2007b). Figure 2.7 shows concentration (in ng/g) of (3,+4-MD) (left Y-axis) and biomarker (right Y-axis) in various pyrolysates of Irati limestone, shale, Vinini Creek limestone (Fig. 2.7a), and peats (Fig. 2.7b).

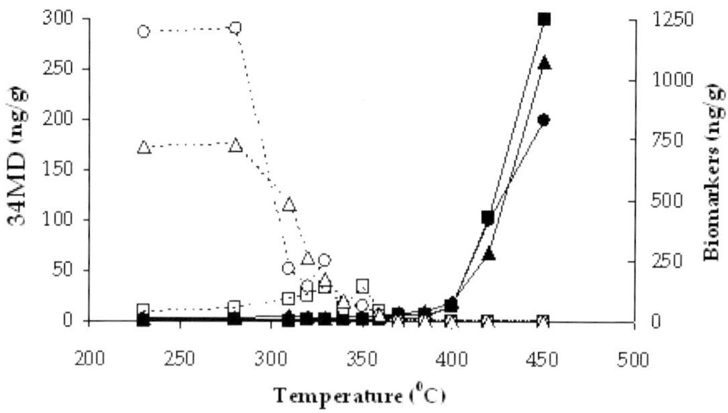

Fig. 2.7a. Concentration of (3,+4-MD) (left vertical-axis) in pyrolysates of Irati limestone(¡), Irati shale (•), and Vinini Creek (▲)*versus* Temperature. Concentration of C_{29} $\alpha\alpha\alpha$ stigmastane 20R biomarkers (right vertical-axis) in pyrolysis, Irati limestone(□), Irati shale (○), and Vinini Creek (△)*versus* Temperature. (Wei *et al.*, 2007b).

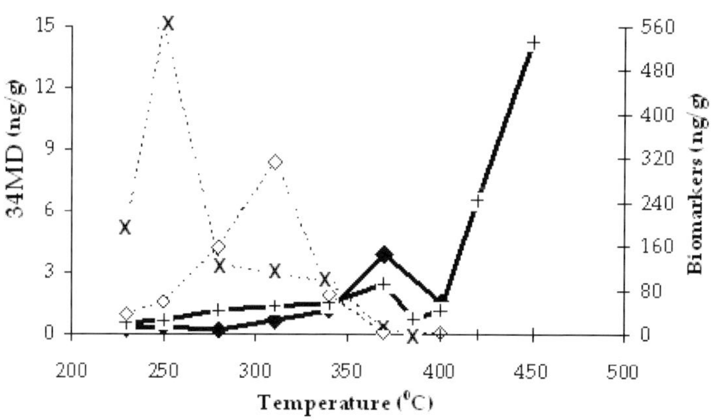

Fig. 2.7b. Concentration of (3,+4-MD) (left vertical-axis) in pyrolysates of Bog peats (+) and Fen peats (◆) *versus* Temperature. Concentration of C_{29} $\alpha\alpha\alpha$ stigmastane 20R biomarkers (right vertical-axis) in pyrolysis Bog peats (x) and Fen peats (◇) *versus* Temperature (Wei *et al.*, 2007b).

Concentration of (3,+4-MD) show little alteration in pyrolysates below 340°C (solid lines). Conversely, intense biomarker release was detected in all samples (dashed lines). In a first analysis, this may appear contradictory, because diamondoid generation is expected to increase with thermal stress. In fact, absolute diamondoid formation increased with temperature, but other petroleum hydrocarbons were simultaneously formed from kerogen cracking. Hence, the net effect results in minor variations in concentration of (3,+4-MD). The relative abundance of (3,+4-MD) markedly increases above 360°C for rocks and above 340°C for peats. This range corresponds to the beginning of oil cracking, with degradation of less stable hydrocarbons. Diamondoids, nevertheless, resist thermal stress and become more concentrated in pyrolysates obtained at higher temperatures, although intensity of diamondoid formation decreases due to loss in catalytic activity of clay minerals under such conditions. This evidence suggests that diamondoids formation and release happens not only during diagenesis but proceeds throughout catagenesis along with the liberation of other oil hydrocarbons and biomarkers.

Hydrous pyrolysis of kerogen samples produced different amounts of (3,+4-MD) depending on the type of kerogen. Naturally mature gas-prone type III kerogen, extracted from Raton Basin coal, presented the highest yield (9.68%) while type II-S kerogens produced small amounts of diamondoids (0.46 and 0.23% for Naples Beach siliceous shales and shales, respectively). Results for type III kerogen may be explained by the thermal stability of diamondoids that allows them to be preserved in higher temperatures while hydrocarbons are being cracked. Since this kerogen sample had already undergone geothermal maturation, it was diamondoid-enriched prior to hydrous pyrolysis process. Additional heating treatment promoted bond cleavage and diamondoid releasing. Type II-S kerogen is rich in sulfur, thermal decomposition of this material releases sulfur species, which may act as an inhibitor for diamondoid formation.

Figure 2.8 shows a model for adamantanes and diamantanes releasing from kerogen, after pyrolysis (Wei *et al.*, 2006b).

In this model, it is assumed that the diamondoid moieties are linked to kerogen macromolecules through polysulfide, ether or ester groups, in a similar fashion as what is observed for biomarkers (Richnow *et al.*, 1992, Schaeffer *et al.*, 1995; Hold *et al.*, 1998; Khaddor *et al.*, 2002). Thus, during

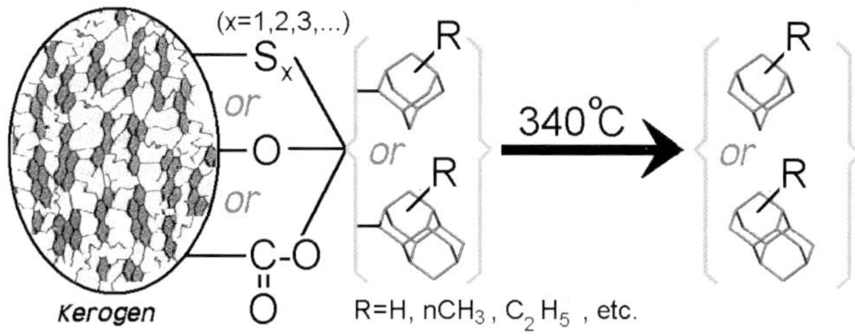

Fig. 2.8. Possible schemes for the releasing of diamondoids from kerogen pyrolysis without catalysts. Adamantanes or diamantanes bound to the kerogen skeleton with polysulfide ether or ester groups. Adapted from (Wei *et al.*, 2006b).

kerogen thermolysis, C–O and C–S bonds are cleaved and diamondoids are released along with other hydrocarbons, biomarkers, and organic sulfur compounds. Such model may be helpful in explaining differences in yields obtained from type I and type II kerogens: type II kerogens are known to bear many C-heteroatom bonds (Tissot and Welte, 1984). These weaker bonds would easily undergo scissions for diamondoid liberation. Conversely, type I kerogen present fewer ester, ether or polysulfide bonds, and thus fewer sites for diamondoid release.

The yields of (3,+4-MD) increased in pyrolysates about 3 to 14 times and about 4 to almost 110 times in kerogen samples when acidic minerals were added to sediments. $CaCO_3$ had an inhibitory effect on diamondoid yields this was particularly high on Elkhorn Slough mud sample, in which yields reduced from 5.80 ng/g to 0.48 ng/g after $CaCO_3$ addition. However, in Celestun Lagoon carbonate and type I kerogen from Brazilian limestone, very little inhibiting effect could be noticed.

An explanation of such results may be related to catalytic effects of minerals on total hydrocarbon yields and diamondoids in particular during kerogen pyrolysis experiments. Montmorillonite clays are capable of adsorbing hydrocarbons and catalyze a number of reactions under pyrolysis condition (Tennenbaum and Kaplan, 1985; Tennenbaum *et al.*, 1986; Huizinga *et al.*, 1987a, 1987b; Saxby *et al.*, 1992). For example, the presence of bentonite (a mineral which has montmorillonite as its main component) in pyrolysis experiments changed distribution of hydrocarbon classes in coal pyrolysates due to increasing aromatic content in comparison with aliphatic hydrocarbons and decreasing the overall oil yield

(Saxby *et al.*, 1992). Tennenbaum and Kaplan (1985) reported a significant catalytic effect on production of C_1 to C_6 hydrocarbons for kerogen after montmorillonite addition, with yields up to five times higher than those observed for kerogen alone. The dominance of branched hydrocarbons in the C to C_6 range of hydrocarbons was also detected. Small or no similar effects were observed with illite (an aluminosilicate) and calcite (Tennenbaum and Kaplan, 1985). Catalytic effect of montmorillonite is generally attributed to cracking via a carbonium-ion (carbocation) intermediate which forms on the acidic sites of the clay (Tannenbaum and Kaplan, 1985). Montmorillonite promotes degradation of kerogen and decreases the activation energies of the generation of gas and light oil; thus, it increases yield of low molecular weight products (Li *et al.*, 1998).

A similar carbocation mechanism is proposed for diamondoids formation during pyrolysis according to the following steps:

(i) Montmorillonite adsorbs possible organic precursors released from kerogen;
(ii) As Lewis acid sites (mainly associated with octahedral Al ions) become available with increase in temperature (Johns, 1979), protons are transferred from montmorillonite to various unsaturated organic compounds to produce carbocations;
(iii) Carbocations undergo rearrangements, deprotonations, isomerizations and subsequent reactions at Lewis acid sites to yield diamondoids (Wei *et al.*, 2006a).

Limestone and other carbonate rocks are frequently reported as petroleum source rocks (Brosse *et al.*, 1990; Pepper and Corvi, 1995; Abdullah and Connan, 2002; Lu *et al.*, 2007; Konyukhov, 2008). Oils associated with carbonates frequently present higher amounts of heavy hydrocarbons than oils derived from deposits containing active surface minerals. This fact indicates the role of clay minerals as effective hydrocarbon cracking acid catalysts in the ground (Brooks, 1948). Moreover, there are indications of a possible role of $CaCO_3$ as a catalyst in petroleum formation, especially in the presence of ammonia, during artificial maturation of organic matter (Evans and Felbeck, 1983). Nevertheless, montmorillonite clay demonstrates little or no catalytic activity at the temperature of maximum hydrocarbon yields (Evans and Felbeck, 1983), particularly with regards to heavy hydrocarbons formation (Li *et al.*, 1998).

The results discussed so far clearly indicate that both clays and carbonate minerals have strong interference in either natural or artificial thermal maturation of organic matter. Their presence can alter the course of oil generation and destruction. Results presented in Table 2.3 also confirm clays and carbonate minerals interference in diamondoids formation. In the case of hydrous pyrolysis of dried natural Celestun Lagoon carbonate and type I kerogen from Brazilian limestone, experiments indicated little difference with the addition of $CaCO_3$ since this mineral was naturally present in the samples. For the Brazilian limestone sample, despite carbonate removal prior to artificial maturation, $CaCO_3$ presence during burial may have altered organic diamondoid precursor's distribution, hampering subsequent transformation that could have ultimately created diamondoids. Thus, in the presence of added $CaCO_3$ (and absence of acidic catalysts), yields remained practically unchanged.

As discussed previously, evidence of diamondoid generation in the presence of Lewis acid sites in clay minerals has been presented over the years, but this is still a controversial subject. Hence, the role of minerals in the formation of diamondoids from modern organic-rich sediments, kerogen, or petroleum remains unsettled.

The experiments reported so far depict the complexity of diamondoid generation in petroleum. Pyrolysis data are useful for the understanding of oil formation in general and diamondoid formation in particular. However, the large number of variables involved in the process makes it difficult to model geological generation of diamondoid based on controlled laboratory experiments. For example, Wei and co-workers (2006a, 2006b) suggested that almost every organic compound could create diamondoids in the presence of suitable catalysts, but nature and distribution of the products is dependent on the original organic matter composition.

Lithology of the source rock can alter diamondoid formation process in unexpected ways, such as in condensates from Hudson Canyon, US Atlantic (Sassen and Post, 2008) reported in Table 2.1, where extremely high concentration of diamondoids could be found in oils from sandstones. The authors suggested that the similarities between condensates from Hudson Canyon and Smakover Formation of the US Golf Cost

(Wingert, 1992) indicate that laminated lime mudstones is a favorable environment to diamondoid formation because carbonate rocks may form complex natural systems containing siliciclastic minerals, including clay catalysts. Other factors, such as level of maturity of the source rock and the presence of water or inorganic gases in the generation and/or accumulation sites are believed to interfere with creation and releasing of diamondoids, making the understanding of diamondoid formation even more difficult. In fact, much work remains to be done in order to understand diamondoid genesis in fossil fuels.

2.4. Diamondoids as Geochemical Tools for Petroleum Characterization

Diamondoid research has many implications in petroleum geochemistry. Used solely or in conjunction with biomarkers investigation (Chen *et al.*, 1996; Dahl *et al.*, 1999; Grice *et al.*, 2000; Schulz *et al.*, 2001; Peters *et al.*, 2005; Azevedo *et al.*, 2008; Springer *et al.*, 2010), diamondoid molecules can be an important tool for advanced geochemical analysis, after initial screening for selection of sediments, rock and crude oil samples made through well-known geochemical methods, such as Rock-Eval pyrolysis, vitrinite reflectance, scanning fluorescence, gas chromatography, mass spectrometry, and stable isotope analysis (Peters *et al.*, 2005).

The isolation and identification of biomarkers is very helpful in petroleum exploration technology. The most important investigated biomarkers are hydrocarbons possessing intact steroid, terpenoid, or isoprenoid carbon skeleton (Seifert and Moldovan, 1986). Many biomarkers are used in petroleum exploration for correlation, biodegradation and maturity assessments (Mackenzie *et al.*, 1980; Seifert and Moldovan, 1986; vanGraas, 1990; Ekweozor and Telnaes, 1990; Requejo, 1994; Farrimond *et al.*, 1998; Obermajer *et al.*, 2000; Bennett and Olsen, 2007; Azevedo *et al.*, 2008; George *et al.*, 2008). Similarly, diamondoid molecules are proposed to play an important role in petroleum evaluation, and may be used to assess origin, extend of biodegradation and thermal maturity, as well as to identify the occurrence of petroleum. It is important to point out that, in spite of their biogenic origin, diamondoids are not biomarkers, because they do not share structural similarities with their putative biomolecular precursors.

Petroleum hydrocarbons are degradable by certain microorganisms (Atlas, 1975; Ahsan *et al.*, 1997; Trolio *et al.*, 1999). In this process, hydrocarbons are transformed into CO_2 and partially oxidized organic compounds, such as carboxylic acids. Petroleum volume and its commercial value decrease with increase in biodegradation. Thus, information on the extent of biological alterations is important in petroleum prospects. Since diamondoids present high thermal stability and are resistant to biodegradation, they may be used to characterize heavily biodegraded (Williams *et al.*, 1986; Grice *et al.*, 2000; Schulz *et al.*, 2001; Wei *et al.*, 2007c) or highly mature oils (Chen *et al.*, 1996, Jinggui *et al.*, 2000; Sassen and Post, 2008), which frequently lack or have few biomarkers.

The use of diamondoid identification and quantification in petroleum geochemistry is mainly based on GC/MS data (see Sec. 2.5). Analyses are more conveniently performed on isolated fractions of saturated hydrocarbons (known as saturates) rather than on the whole oil, because of aromatic compounds interference (Wingert, 1992; Vazquez and Mansoori, 1998; Vakili-Nezhaad *et al.*, 2005). Saturates samples are spiked with known amounts of deuterated diamondoids or other suitable substances as internal standards for quantification of diamondoids by peak areas comparisons. Reconstructed ion chromatograms of the base-peak of adamantane (with mass to charge ratio, m/z, equal to 136 and alkyl-substituted $C_nH_{n-5)}$ series); diamantane (base-peak m/z = 188 and $C_nH_{(2n-9)}$ series) and triamantane (base-peak m/z = 240 and $C_nH_{(2n-13)}$ series) are obtained by selective ion-recording (SIR), or selective ion monitoring (SIM) modes, as reported in several related literature. Peak assignments are based on elution order and are confirmed by full mass spectral data acquisition (Wang *et al.*, 2006a; Yang *et al.*, 2006; Azevedo *et al.*, 2008; Schulz *et al.*, 2001; Chen *et al.*, 1996; Zhang *et al.*, 2005, Jinggui *et al.*, 2000). Figure 2.9 shows a reconstructed ion chromatogram for the base-peak m/z = 187 obtained from GC/MS/SIR analysis of a coal extract sample. This peak belongs to the alkylated $C_nH_{(2n-9)}$ series of diamantane. Two deutero-alkyldiamantanes were used as internal standards in this particular coal extract sample (Wei *et al.*, 2006c).

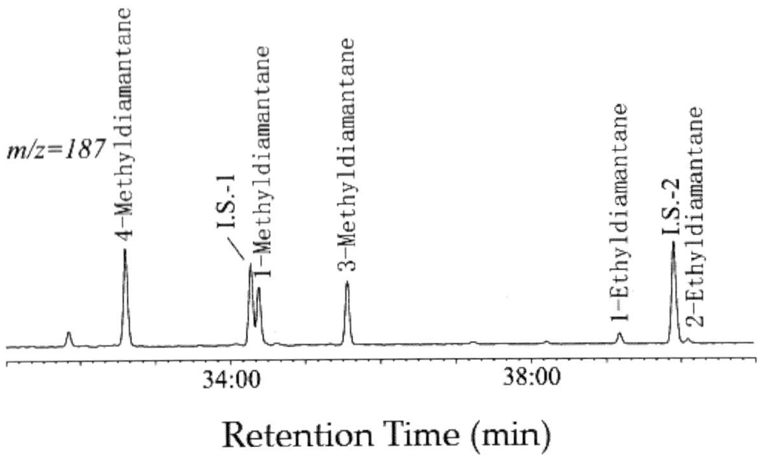

Fig. 2.9. Reconstructed ion chromatogram for the base-peak $m/z = 187$ ($C_{14}H_{19}^{+}$) of diamondoids showing distribution of methyldiamantanes and ethyldiamantanes in coal extracts (Wei *et al.*, 2006c). I.S.-1 = D_3-1-methyldiamantane; I.S.-2 = D_5-2–ethyldiamantane. I.S.-1 and I.S.-2 are synthetic internal standards (Wei *et al.*, 2006c).

2.4.1. *Diamondoid-based geological correlations for petroleum characterization*

In petroleum geology, it is customary to look for correlations which could be based on geochemical comparisons among crude oils, refined products and/or extracts from prospective source rocks. From these comparisons, it may be possible to determine whether a relationship of origin exists (Peters *et al.*, 1993, 2005; Waples and Curiale, 1999). Frequently referred correlations are: (a) **oil-source rock correlation** which is based on similarities between compositions of migrated oils and the remained bitumen in the source rocks; and (b) **oil–oil correlation** which is a comparison of chemical composition to describe relations between oils origins without necessarily identifying rock sources (Peters *et al.*, 2005). Availability of correlations is a valuable tool in locating new petroleum sites and extending existing productions.

Among all the correlations there are a number of diamondoid-based correlation ratios (such as facies parameter, biodegradation and Oil-source correlation, and differentiation assessment) which were suggested by various investigators as maturity parameter. These diamondoid-based correlation ratios are reported in Tables 2.4 and 2.5.

Table 2.4. Some diamondoid-based diagnostic correlation ratios used for petroleum evaluation. All maturity parameters are suggested for thermal maturity assessment.

Diamondoid-based ratios	Mathematical expression and main features	Applications	References
Methyladamantane Index (MAI)	$MAI = \frac{[1-MA]}{[1-MA]+[2-MA]}$	Maturity parameter	Chen *et al.*, 1996; Grice *et al.*, 2000; Sassen and Post, 2008; Wei *et al.*, 2007b; Zhang *et al.*, 2005
Methyldiamantane Index (MDI)	$MDI = \frac{[4-MD]}{[1-+3-+4-MD]}$ This may be affected by source rock lithology		Chen *et al.*, 1996; Grice *et al.*, 2000; Jinggui *et al.*, 2000; Sassen and Post, 2008; Schulz *et al.*, 2001; Wei *et al.*, 2007b; Zhang *et al.*, 2005
Ethyladamantane Index (EAI-1)	$EAI-1 = \frac{[1-EA]}{[1-+2-EA]}$	Facies parameter:	Zhang *et al.*, 2005; Wei *et al.*, 2007b
Ethyladamantane Index (EAI-2)	$EAI-2 = \frac{[2-EA]}{[1-+2-EA]}$ useful within the oil window, affected by maturity in the late gas-window range		Schulz *et al.*, 2001
Dimethyladamantane Index (DMAI-1)	$\mathrm{DMAI}-1 = \frac{[1,3-\mathrm{DMA}]}{[1,2-+1,3-\mathrm{DMA}]}$	Maturity parameter	Zhang *et al.*, 2005; Wei *et al.*, 2007b
Dimethyladamantane Index (DMAI-2)	$\mathrm{DMAI}-2 = \frac{[1,3-\mathrm{DMA}]}{[1,3-+1,4-\mathrm{DMA}]}$		Zhang *et al.*, 2005

(*Continued*)

Table 2.4. (*Continued*)

Diamondoid-based ratios	Mathematical expression and main features	Applications	References
Trimethyladamantane Indices (TMAI)	$TMAI-1=\frac{[1,3,5-TMA]}{[1,3,5-TMA+1,3,6-TMA]}$ $TMAI-1=\frac{[1,3,5-TMA]}{[1,3,5-TMA+1,3,4-TMA]}$		Zhang *et al.*, 2005
Dimethyldiamantane Indices (DMDI)	$DMDI-1=\frac{[4,9-DMD]}{[4,9-+3,4-DMD]}$ $DMDI-2=\frac{[4,9-DMD]}{[4,9-+4,8-DMD]}$	Maturity parameter	Zhang *et al.*, 2005; Wei *et al.*, 2007b
Methyltriamantane Index (MTI)	$MTI=\frac{[9-MT]}{[5-+8-+9+16-MT]}$		Wei *et al.*, 2007b

Table 2.5. Some diamondoid-based diagnostic correlation ratios used for petroleum evaluation.

Diamondoid-based ratios	Mathematical expression and main features	Applications	References
Methyladamantane / adamantane	$\frac{[1-+2-\text{MA}]}{[\text{Ada}]}$ This ratio rises with bidegradation increase	Biodegradation assessment	Grice *et al.*, 2000
Methyladamantane / diamantane	$\frac{[1-+3=+4-\text{MD}]}{[\text{Dia}]}$ This ratio rises only at extreme levels of biodegradation		
Methyladamantane / ethyladamantane	$\frac{[1-\text{MA}]}{[2-\text{MA}]}$	Oil-source correlation and differentiation	Yang *et al.*, 2006
Methyladamantane / dimethyladamantane	$\frac{[1-\text{MA}]}{[1{,}2-D\text{MA}]}$		

(*Continued*)

Table 2.5. *(Continued)*

Diamondoid-based ratios	Mathematical expression and main features	Applications	References
Methyladamantane/ trimethyladamantane	$\frac{[1-MA]}{[1,3,4-TMA]}$		
Methyladamantane/ tetramethyladamantane	$\frac{[1-MA]}{[1,2,5,7-TeMA]}$		
Dimethyladamantane/ trimethyladamantane	$\frac{[1,4-DMA]}{[1,3,4\text{-}TMA]}$		
Trimethyladamantane/ trimethyladamantane	$\frac{[1,3,5-TA]}{[1,2,5,7-TeMA]}$	Oil-source	
Trimethyladamantane/ tetramethyladamantane	$\frac{[1,3,5-TMA]}{[1,3,6-TMA]}$		
Tetramethyladamantane/ tetramethyladamantane	$\frac{[1,3,5,7-TeMA]}{[1,2,5,7-TeMA]}$		

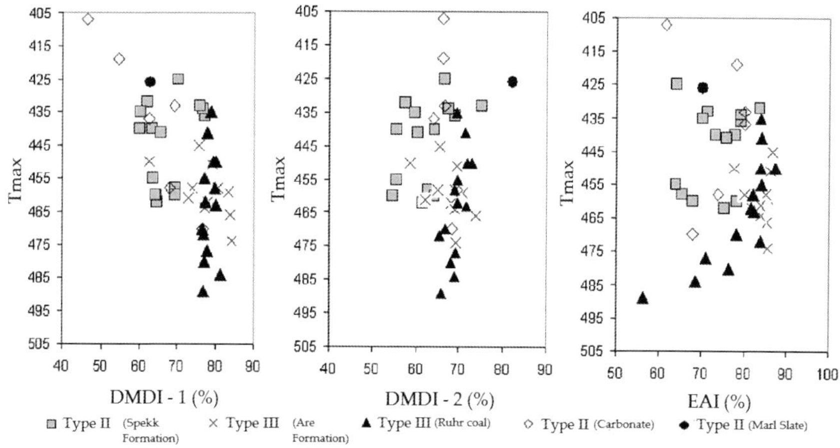

Fig. 2.10. DMDI–1, DMDI–2, and EAI parameters (%) *versus* the temperature at which the maximum release of hydrocarbons from cracking of organic matter occurs during pyrolysis (*Tmax*), for five different rock sources (Schulz *et al.*, 2001).

As an example of the use of these correlations Fig. 2.10 is reported by Schulz *et al.* (2001).

In Fig. 2.10, there are no clear correlations between T_{max} (the temperature at which the maximum release of hydrocarbons from cracking of organic matter occurs during pyrolysis) and the three parameters (DMDI-1, DMDI-2, and EAI) reported in Table 2.4. Observed lack of significant correlation indicates that these parameters are practically unaffected by thermal maturity of source rock which is a good thing. This is because T_{max} is an indication of the stage of maturation of the preserved organic matter, in rock sources and sediments. A lack of correlation suggests little or no influence of source rock thermal maturity on the measured values of these parameters. Thus, DMDI-1, DMDI-2, and EAI are able to show oil-source rock correlation in a broad range of source rock thermal maturity levels.

The data reported in Fig. 2.11 (Schulz *et al.*, 2001) is another example for the use of correlations. In this figure, DMDI-1, DMDI-2; and EAI were used to distinguish oils from marine and terrigenous organic facies. Facies are a set of rocks with distinct features, regarding any chemical, mineralogical, morphological, or structural features, that make it possible to determine where and when the rock was formed.

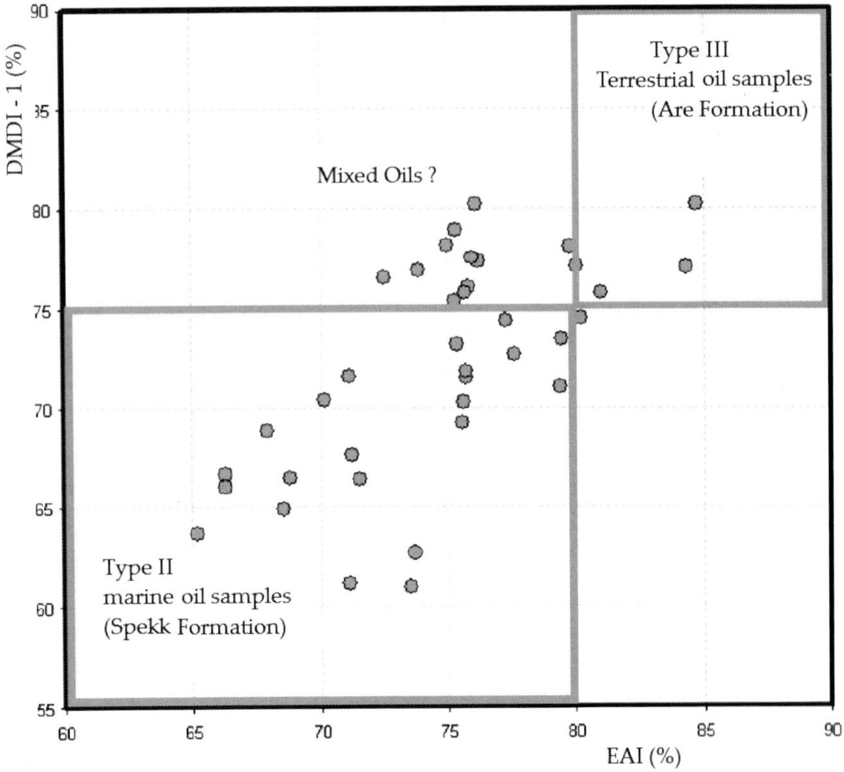

Fig. 2.11. The EAI and DMDI–1 for the Norwegian Continental shelf oil samples. Suggested boundary lines for Type II (Spekk Formation) and Type III (Are Formation) organic facies were taken from Fig. 2.9 (Schulz *et al.*, 2001).

Figure 2.11 exhibits a correlation of percentage of DMDI–1 *versus* percentage of EAI values for Norwegian Continental shelf oil samples.

Most of the samples showed DMDI–1 values in the range of 60–75%. These values are in the range of the Type II marine source rock from the Spekk formation (compare with DMDI–1 values in Fig. 2.10). Furthermore, these samples had EAI values (65 to ~80%) in the range of the same source rock, as well (as seen in Fig. 2.11). A few oil samples showed DMDI-1 and EAI in the range of Type III terrestrial "Are formation" (see upper right rectangle in Fig. 2.10). Several other oils had values consistent with mixed oils from both formations. These findings are in good agreement with the current knowledge about the studied petroleum system. No differences between oil and condensate samples from the same field were observed, indicating that DMDI-1 and EAI parameters are not affected by oil natural fractioning (Schulz *et al.*, 2001).

Indices DMDI-1, DMDI-2, and EAI are determined by diamondoid concentration in petroleum. These concentrations, far from being random, keep close relation with the very important petroleum characteristics. These indices are useful to determine the source where the oil is from. In doing so, they are even act as the "aging proof" of the oil. In addition, DMDI-1 and EAI are able to tell in which environment the source rock was formed — marine or terrestrial, for oils coming from the same shelf! Not even fractioning got in their way, and mixed oils were "denounced", as well.

Schulz and Co-workers (2001) also used DMDI-1 and DMDI-2 to assess correlations of biodegraded and nonbiodegraded oils. Figure 2.12 shows a comparison between two well-known correlation methods:

- Gas Chromatography with Flame Ionization Detector (GC/FID) signals for saturated hydrocarbons,
- Gas Chromatography–Mass Spectrometry with selected ion monitoring (GC/MS/SIM) for $m/z = 191$ of hopane biomarkers, and the values for DMDI-1 and DMDI-2 (GC/MS/SIM for $m/z = 201$ DMDs peak) obtained for oil sample A (nonbiodegraded) and oil sample B (heavily biodegraded), both from the same field.

While chromatographic patterns of saturated hydrocarbons and hopanes were severely altered by biodegradation in oil sample B, DMDI-1 values remained unchanged (65%) and DMDI-2 values slightly increased (from 70 to 72%). Besides, overall DMDs peak quantities and relative intensities are virtually the same in both oils A and B, regardless of differences in their biodegradations. These results, once more, demonstrate the utility of diamondoid-based geological correlations for petroleum characterization in investigating the source of degraded oils. While no correlation could be verified from saturated hydrocarbons or biomarkers data, DMDI-1 and DMDI-2 enabled correlation between oils A and B. Such correlation is also supported by models of oil migration established for the same field of the Norwegian Continental shelf (Schulz *et al.*, 2001).

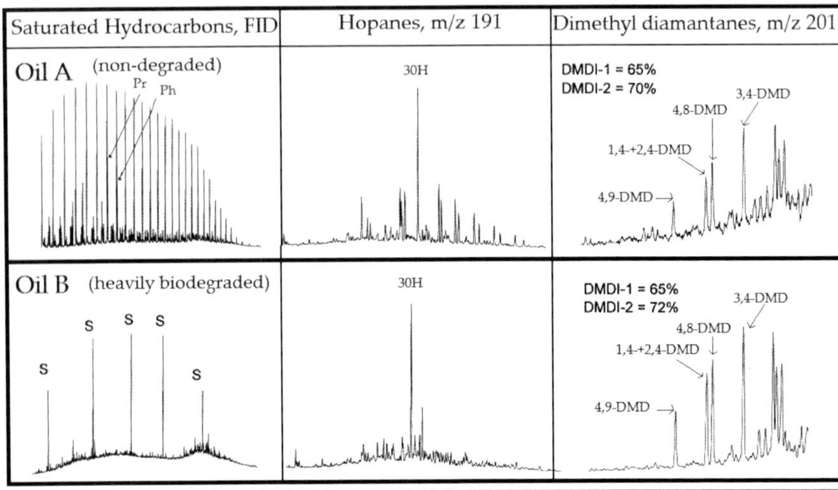

Notations: **S**=*saturated hydrocarbons (internal standard–GC tracer);* **Pr**=*Pristane;* **Ph**=*Phytane (both Pr and Ph are branched acyclic isoprenoid biomarkers);* **30H**=*(C_{30}–hopane); 4,9–DMD=(4,9–Dimethyldiamantane);1,4-+2,4–DMD=(1,4–Dimethyldiamantane)+(2,4–Dimethyldiamantane); 4,8–DMD=(4,8–Dimethyldiamantane); 3,4–DMD=(3,4–Dimethyldiamantane)*

Fig. 2.12. GC/FID signal (saturated hydrocarbons) and mass chromatograms m/z = 191 from hopane biomarkers and *m/z* = 201 from DMDs of two oils from the same field on the Norwegian Continental shelf. Heavily biodegraded Oil B is depleted in hopane biomarkers while DMD peaks appear unaffected by biodegradation, allowing (oil a)–(oil b) correlation (Schulz *et al.*, 2001).

2.4.2. *Susceptibility of diamondoids biodegradation by microorganisms*

In spite of their resistance to biodegradation, diamondoids may be attacked by microorganisms. Wei and co-workers (2007c) investigated diamondoid molecules present in nonbiodegraded and biodegraded oil of similar thermal maturities from San Joaquin Valley, California. Their work showed that these cage compounds are susceptible to biodegradation in the reservoir, but the process is selective and appears to evolve stepwise. The biodegradation ranking reported in Table 2.6 was developed by Peters and Moldowan (1993) and it is based on the extent of destruction of compound classes. According to Table 2.6, biodegredation ranking is closely related with the total diamondoids contents of the crude oils.

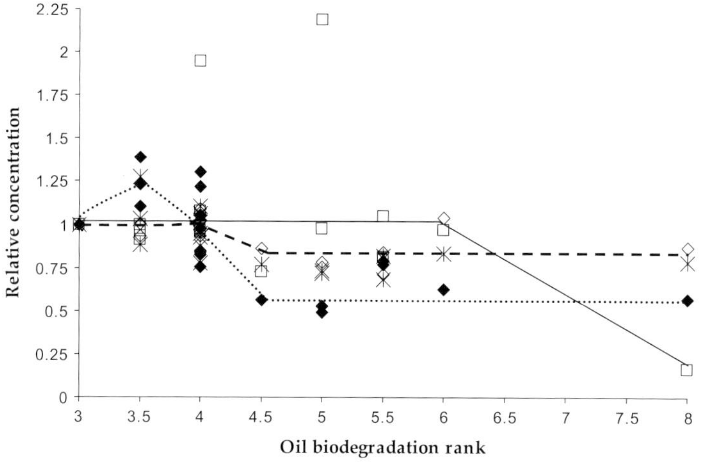

Fig. 2.13. Change in concentration of adamantanes shown by (•) [*which include adamantane,* 1-*methyl,* 2-*methyl*-1-*ethyl-,* 2-*ethyl-,* 1,3-*dimethyl-,* 1,4-*dimethyl-, and* 1,2-*dimethyl-adamantanes*]; diamantanes shown by (◊) [*which include diamantane,* 1-*methyl-, 3-methyl-, 4-methyl-, 1-ethyl-, 2-ethyl-, 4,9-dimethyl-,* 4,8-*dimethyl-,* 1,4-+2,4-*dimethyl and 3,4-dimethyl-diamantanes*]; 3-+4-methyldiamantanes shown by (✳), and biomarkers shown by (□) [C_{30} 17α, 21 β *(H)* — *hopane* + C_{29} $\alpha\alpha\alpha$ *stigmastane 20R, a sterane*]; relative to concentration in sample presenting biodegradation rank equal to 3, from South Belridge oil samples with increasing biodegradation (Wei *et al.*, 2007c). Lines show general trends of data.

Figure 2.13 shows variations of adamantanes, diamantanes and biomarkers concentrations relative to concentration of the sample having the lower rank 3, with biodegradation rank increase, in samples from South Belridge oils. (3,+4-MD) are also plotted in separate and exhibit close resemblance with total diamantanes behavior. (Dahl *et al.*, 1999).

According to Fig. 2.13 at rank 3.5, adamantanes relative concentrations appear to slightly increase, probably due to the loss of other components of higher biodegradability, such as *n*-alkanes and isoprenoids. Diamantanes and biomarker values are around 1, showing no indication of significant variation with increase in rank. At rank 4, all compounds present larger variation in relative concentration, but without a clear indication of significant alteration. Relatively stable behavior of diamondoids regarding biodegradation starts to exhibit alteration at the 4.5–8.0 rank intervals, with a drop in relative concentration of adamantanes and diamantanes to about 0.60 and 0.85, respectively. These results show that adamantanes

Table 2.6. Concentrations of diamondoids in crude oils of different maturity levels. (From: Peters and Moldowan, 1993).

Crude oil field	Bio- degradation rank	Diamondoids [μg/g]			
		Adamantane	Diamantane	(3-,+4-MD)	Total diam's
Race Track	0	2285.2	53.6	14.2	2346
Buena Vista	0–1	212.6	5.1	1.2	221
Fruitvale	2	2091.7	38.6	9.6	2137
McKittrick	2–3	323.4	6.3	1.7	334
South Belridge	3	1188.6	18.5	4.6	1253
South Belridge	3.5[c]	1310.8 to 1648.5	17.3 to 23.0	4.1 to 5.8	1361 to 1699
South Belridge	4[d]	899.0 to 1544.8	15.7 to 22.6	3.6 to 5.1	1017 to 1594
South Belridge	4.5	655.4	15.9	3.5	688
South Belridge	5[c]	580.3 to 689.9	14.0 to 36.7	3.3 to 8.54	647 to 1032
South Belridge	5.5[e]	917.4; 935.7	13.1; 15.5	3.1; 3.74	959; 981
South Belridge	6	747.1	19.2	3.8	868
South Belridge	8	680.9	16.0	3.6	712
Kern River	7	453.2	11.0	2.96	465
Cymric	8	713.0	14.4	4.3	728

Note: The eight South Belridge samples reported in this table are from various parts of the South Belridge Field at different levels of maturation.

[μg/g]: Micrograms of diamondoid per gram of crude oil.

"Biodegradation Rank" is defined based on the extent of destruction of compound class (Peters and Moldovan, 1993): (0–1) = nonbiodegraded. (1–2) = slightly or moderately biodegraded. (2–4) = heavily biodegraded. (four or more) = severely biodegraded.

[c]Range of three samples. [d]Range of seven samples. [e]Values for two samples.

are more affected by biodegradation than diamantanes, but relatively high amounts of both diamondoids can still be detected in severely biodegraded samples. Biomarkers concentration present some fluctuation after rank 4.0, nevertheless, their overall relative concentration seems to remain unchanged, as it would be expected for hopanes and steranes, which do not undergo major biodegradation in ranks lower than six (Peters and Moldowan, 1993). In higher biodegradation rank, a substantial depletion of biomarkers is observed, with concentration at the rank 8 sample been equivalent to *c.a.* 16% of the concentration in the rank 3 sample. Hence, some decrease in the relative concentration of diamondoids could be noticed in less biodegraded oils, while biomarkers degradation occurs in higher rank levels. As rank increases, adamantanes and diamantanes are preserved while biomarkers are degraded.

In order to evaluate and compare the influence of biodegradation rank on diamondoid-based diagnostic ratios among samples from the same oil field, some diamondoid ratios normalized by values found to the less biodegraded sample (rank 3) for South Belridge oils are reported in Fig. 2.14.

General trends for diamantane/adamantane (Dia/Ad, solid line), adamantanes-based (dashed line), and diamantanes-based (dotted line) ratios

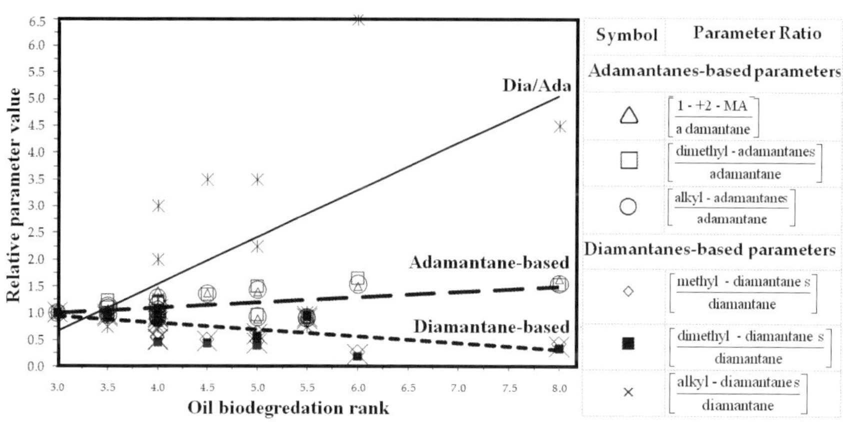

Fig. 2.14. Effect of biodegradation on diamondoid-based parameters (Wei *et al.*, 2007c). Relative parameter values were calculated by dividing the value of the biodegradation parameter for the oil sample by the same value observed for oil samples having biodegradation rank equal to 3, from South Belridge field.

indicate that these ratios are influenced by biodegradation level in different ways. All samples show no significant alteration in biodegradation ranks 3.5. After this point, Dia/Ada ratio presented the highest variation in relative parameter values, with a dramatic increase of 4.5 times in rank 8 sample. A particularly high concentration of diamantane (9.02 ng/g) in the rank 6 sample is partially responsible for an extreme relative value of 6.5. Such increase in Dia/Ada values is a reflection of the difference in the biodegradability of these two caged hydrocarbons, as adamantane is attacked in a greater extent than diamantane. The three calculated adamantanes-based ratios,

[1-+2-*methyladamantanes / adamantane*];
[1,2-+1,3-+1,4-*dimethyladamantanes / adamantane*];
and
[*alkyladamantanes / adamantane*],

show similarities in relative values and exhibit minor elevation at higher ranks, while three diamantanes-based ratios,

[1-+3-+4-*methyldiamantanes / diamantane*];
[3,4-+4,8-+4,9-*dimethyldiamantanes / diamantane*];
and
[*alkyldiamantanes / diamantane*],

Show slight decrease with increase in biodegradation rank. Such results may indicate that alkyl-substitutions have opposite effects on biodegradability of diamondoid cages, making alkyladamantanes less susceptible than adamantane while they make alkyldiamantanes more readily biodegradable than diamantane.

An earlier report on (1-+2-MA)/Ada and (1-+3-+4-MD)/Dia ratios calculated for oils from two Australian sedimentary basins (the Gippsland and the Carnarvon Basins) suggested their use as indicators of oil biodegradation and oil mixing (Grice *et al.*, 2000). As can be seen in Fig. 2.15, for the Gippsland Basin, very little alteration in (1-+2-MA)/Ad ratio could be noticed in nonbiodegraded to moderate biodegraded oils (levels 0 to 3), with ratio values around 6.0. The first important change in this ratio occurs at biodegradation level range 3–4, and the increase in ratio value proceeds to reach 14.5 in the level range 4–5, and 10.2 in level 7.

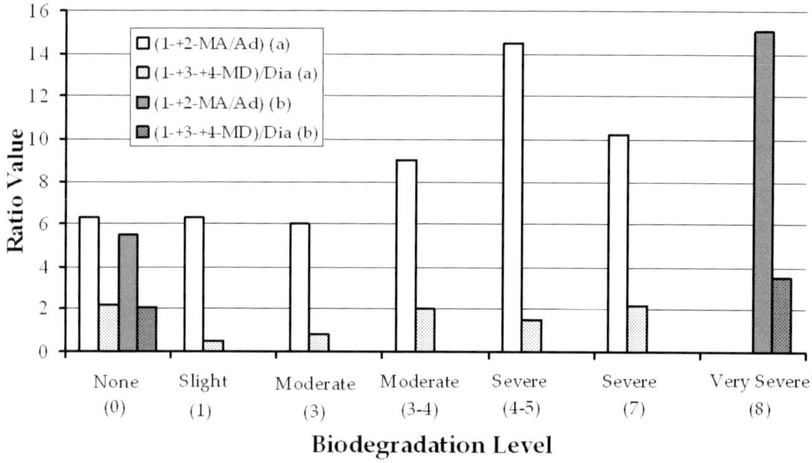

Fig. 2.15. Effect of biodegradation level on 1-+2-*methyladamantanes*/*adamantane* (1-+2-MA/Ad) and 1-+3-+4-*methyldiamantanes*/*diamantane* (1-+3-+4-MD/Dia) ratios for Australian oils from: (a) The Gippsland basin and (b) The Carnarvon Basin (Grice *et al.*, 2000).

These results exhibit similarities with variations found in South Belridge oils, in which (1-+2-MA)/Ada increases *c.a.* 55% in the level range of 3–8 as shown in Fig. 2.13.

In samples from the Carnarvon Basin, the (1-+2-MA)/Ada ratio (grey bars in Fig. 2.15) is almost three times lower in nonbiodegraded samples (biodegradation level 0) than in the severely biodegraded ones (biodegradation level 8). Such fact points out to the importance of differences in microbial attack on alkyl-substituted adamantanes when compared with adamantane.

(1-+3-+4-MD)/Dia ratio showed initial decrease until biodegradation level 3, appearing to reach equilibrium value around 2.0 in severely biodegraded oils from the Gippsland basin (Fig. 2.15, dotted white bars). Very severely biodegraded Carnarvon oils (level 8) showed an elevation of *c.a.* 66% in (1-+3-+4-MD)/Dia ratio in comparison to nonbiodegraded samples from the same basin (dotted grey bars). These results indicate that significant changes in (1-+3-+4-MD)/Dia ratio only occur in extreme biodegradation levels, possibly caused by concomitant degradation of less resistant hydrocarbons and increased concentration of the more resistant diamantanes. Based on the previous discussion, (1-+2-MA)/Ada appears to have a more predictable behavior than (1-+3-+4-MD)/Dia and can be more useful as an indicator of biodegradation extension, particularly in extreme biodegradation levels, in which little information can be retrieved from biomarkers data due to their biological destruction in such oil samples. These findings reinforce the idea of using diamondoid-based ratios for petroleum characterization. (Grice *et al.*, 2000).

Wei and collaborators (2007c) suggested a mechanism for biological degradation of adamantane in petroleum, based on the data published for *in vitro* enzymatic aerobic hydroxylation of adamantane (White *et al.*, 1984) and microbial oxidation of adamantanone (Selifonov, 1992). Their suggested aerobic biological pathway for adamantane degradation is reported in Fig. 2.16.

According to Fig. 2.16 the main biosynthetic intermediates are 4-oxahomo-adamantane-5–1 and 5-hydroxyadamantan-2–1, obtained through the action of camphor 1,2-monooxygenase and camphor 5-monooxygenase enzymes from a gram-negative saprotrophic bacterium named *Pseudomonas putida*, respectively. Nevertheless, geological conditions in petroleum sites are primarily anaerobic, making any aerobic biochemical pathway less probable to occur.

2.4.3. *Diamondoid-based diagnostic ratios in environment science*

Some techniques for correlating relationships between "source rock" and "crude oil" have been applied for the study of environmental pollution. Environmental fingerprinting of biomarkers is used in the assessment of fuel pollutants to characterize the type of the fuel contaminants, to determine their source, and to understand their fate in the environment (Alimi *et al.*, 2003; Christensen *et al.*, 2004; Wang *et al.*, 2006b). Selected diamondoid ratios, distribution or relative abundance, by their turn, may be useful for oil spill source identification (Wang *et al.*, 2006a) and distinguishing sources of refined products as well (Stout and Douglas, 2004; Yang *et al.*, 2006).

All crude oils contain a wide range of hydrocarbons such as light gaseous hydrocarbons, liquid hydrocarbons and heavy ends which may

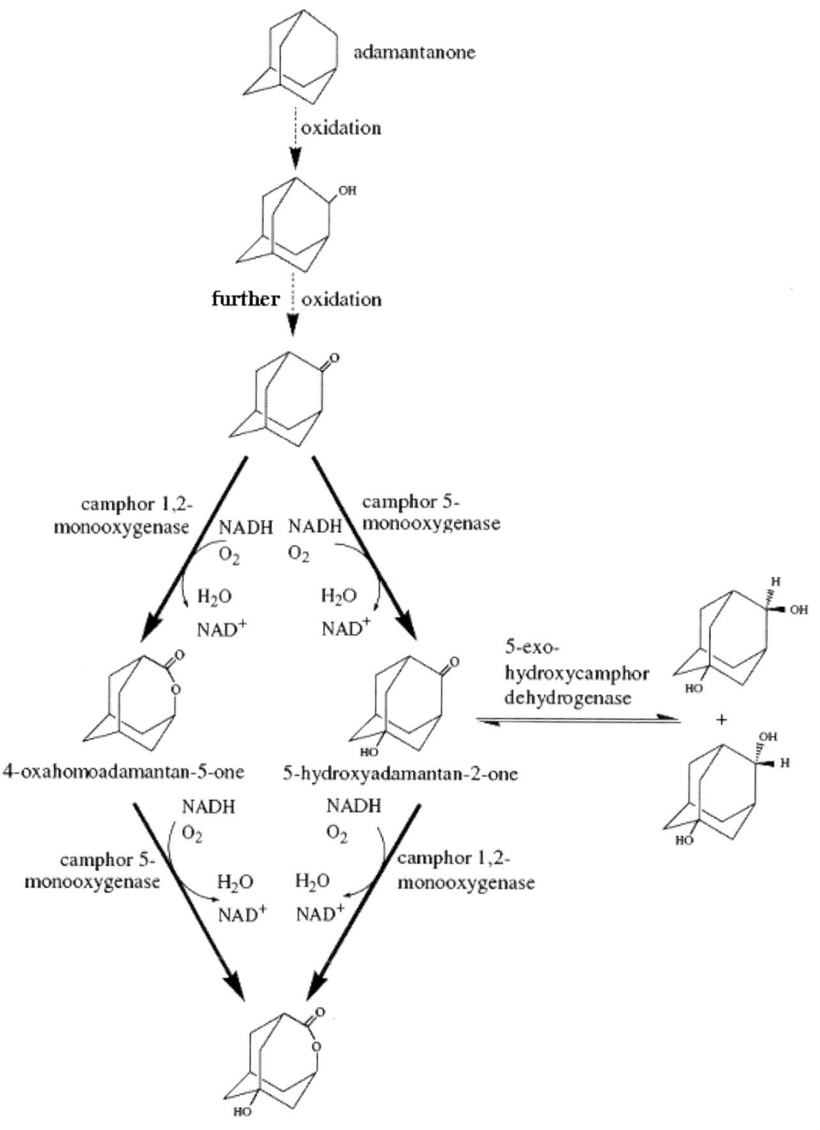

Fig. 2.16. Some possible aerobic biological pathways for adamantane degradation (From Wei *et al.*, 2007c with corrections).

include heavy hydrocarbons, resins and asphaltenes. Crude oils also may contain smaller amounts of sulfur, oxygen, nitrogen, and metals (nickel, vanadium and iron). On the other hand, refined petroleum products are fractions derived by distillation and upgrading of crude oil. Thus, they have differences in their chemical compositions due to the dissimilarities in characteristics of crude oil feed stocks as well as variations in refinery industrial processes.

Variations in chemical compositions of crude oils may result in unique chemical "fingerprints" for each crude oil, providing a basis for identifying the source(s) of spilled oil. GC/MS is the standard analytical method for oil chemical fingerprinting, since it could identify a broad range of polycyclic aromatic hydrocarbons (PAHs) and petroleum markers. In fact, the identification of individual diamondoids is based on the full mass spectral data for structural elucidation and comparison with the literature and MS library data, as well as comparison of GC retention data with authentic standards, determination of refractive index (RI) (Kovats, 1965), and comparison with the literature RI values. GC/MS chromatograms analyses have shown distinct distribution patterns of diamondoids in crude oils and refined products of different origins. Nevertheless, oils of the same origin or from the same area exhibit similar diamondoid signatures (Stout and Douglas, 2004; Yang *et al.*, 2006).

2.4.4. *Diamondoid-based diagnostic ratios for petroleum maturity parameters*

The thermal maturation of sedimentary organic matter is one of the important aspects of petroleum formation and occurrence. The determination of maturity is the keystone for determining the boundary conditions of petroleum generation, coke stability prediction in coals, identification of many geological phenomena (faulting, thrusting, intrusion, unconformity, etc.) and temperature history in the basin evolution (Karlsen *et al.*, 1993; Mango, 1997; Rodriguez and Littke, 2001). There exist a large number of optical, pyrolytic and chemical methods which can be used to assess thermal maturity (Farrimond *et al.*, 1998; Durand, 1980; Tissot and Welte, 1984; Mukhopadhyay, 1994; Peters *et al.*, 2005). These methods differ in approach and in effectiveness to assess various stages of maturation.

The presence of diamondoids is considered to be closely related to the geological maturity of an oil field (Chen *et al.*, 1996; Gao *et al.*, 2001; Shimoyama, 2002). Some investigators have used relative abundances of diamondoids to estimate the degree of geological thermal maturity of source rocks and crude oil, especially in mature and postmature samples (Fang *et al.*, 1996; Chen *et al.*, 1996; Dahl *et al.*, 1999).

The first reported diamondoid-based maturity parameters were Methyladamantane Index (MAI) and Methyldiamantane Index (MDI) as defined by the following expressions (Chen *et al.*, 1996):

$$\mathrm{MAI}(\%) = 1 - \frac{100\mathrm{MA}}{1\mathrm{MA} + 2\mathrm{MA}}, \quad \mathrm{MDI}(\%) = 4 - \frac{100\mathrm{MD}}{1\mathrm{MD} + 3\mathrm{MA} + 4\mathrm{MA}},$$

where,

1MA ≡ 1-methyl-adamantane,
2MA ≡ 2-methyl-adamantane,
1MD ≡ 1-methyldiamantane,
3MD ≡ 3-methyl-diamantane and
4MD ≡ 4-methyl-diamantane.

These two indices, MAI and MDI, were proposed and used to evaluate the maturation and evolution of petroleums from Chinese Basins, through GC/MS investigation of diamondoid and biomarkers content of crude oils, condensates and source rocks from more than five different basins (Chen *et al.*, 1996). At least 30 diamondoids could be identified from these samples: 17 adamantanes, 10 diamantanes and three triamantanes. 1MA was the most abundant diamondoid and 4MD was the most abundant within the diamantane series. Both molecules are substituted at the bridgehead position (See Fig. 2.1). Bridgehead-substituted diamondoids experience less steric interactions than diamondoids substituted at secondary carbons such as 2-methyl-adamantane and 3-methyl-diamantane (Engler *et al.*, 1972); thus, they are expected to possess higher thermal stability and higher concentration in more mature oils.

MAI and MDI were found to increase in source rocks with the increase in vitrinite reflectance (R_0) of the rocks. This was observed in several samples of different lithology that included mudstones in the Tarim Basin

Table 2.7. The relation between "stage of maturation", "level of maturity" and vitrinite reflectance (R_0).

Stage of Maturation[a]		Level of Maturity[b]	R0 [%]
Diagenesis		Immature	
Catagenesis	Mature	Early	
	Mature	Peak	0.65–0.90
	Mature	Late	0.90–1.35
		Postmature	1.35–2.0
Metagenesis		Postmature	2.0–4.0
		Postmature	4.0–5.0

[a]Mukhopadhyay, 1994; [b]Peters and Cassa, 1994.

and dolomites and coals in the Erdos Basin, bearing quite different types of organic matter (Chen *et al.*, 1996).

Vitrinite reflectance (R_0), the major maturity parameter, accurately establishes the effective maximum paleotemperature (the ancient temperature of a geologic process) and its duration at any stage in geological timescale (Robert, 1980; Mukhopadhyay, 1994). Practical applications of R_0 are frequent in various geological activities including fossil fuel exploration, mining, beneficiation, utilization, energy prospecting, etc. Simplicity, precision, and affordability make this method of particular importance for commercial use especially in petroleum, coal, and other industries (Gray, 1991; DeVanney and Stanton, 1994). The vitrinite reflectance (R_0) measurement method is based on irreversible chemical reactions (as a response to increasing temperature and pressure through geological eras) that take place in vitrinite macerals present as dispersed organic particles of sedimentary rocks (Rimmer, 1991; Mukhopadhyay, 1994). The level of maturation or coalification attained is a function of temperature and the duration of heating.

In Table 2.7, we report the relation between "stage of maturation", "level of maturity" and vitrinite reflectance (R_0). Figure 2.17 shows the relations between MAI and MDI with R_0 found for source rocks from Erdos and Tarim Basins.

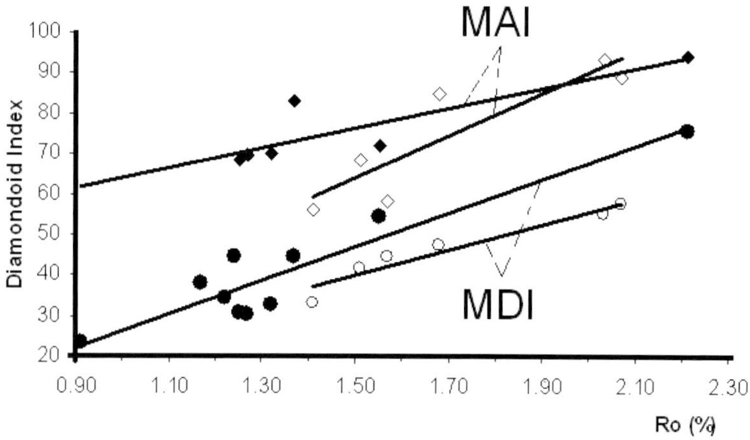

Fig. 2.17. Variations of Methyladamantane Index (MAI) and Methyldiamantane Index (MDI) maturity parameters with Vitrinite reflectance, R_0(%), in oils from: Erdos Basin (○ and ◇) and Tarim Basin (● and ◆) (Chen *et al.*, 1996).

As it is shown in Fig. 2.17, MDI presents good correlation with $R_{0,}$ with its value slightly higher in the Tarim Basin as compared with Erdos Basin values, probably due to the presence of acid catalyst clay mineral in mudstones in higher amounts than in the dolomites and coals from the Erdos Basin. MAI possess higher values than MDI in both basins, with some fluctuations in the 1.35–1.70 range of $R_{0.}$ Part of the inconsistencies in MAI values may be attributed to losses of adamantanes by sublimation during transportation and processing of samples. It is important to note that both MAI and MDI parameters could be measured at various vitrinite reflectance (R_0) values corresponding to late mature (0.9–1.35) and postmature (>1.35) samples (Tables 2.7 and 2.8), in which either biomarkers parameters have already reached equilibrium values or could not be calculated due to total depletion of biomarker molecules. Hence, diamondoid-based parameters, especially MDI, may be useful instruments for evaluation of thermally mature and postmature petroleum's.

The applicability of MDI as a maturity parameter was tested in mature and postmature carbonate source rocks from Majiagou Formations, Shanganning Basin, China (Jinggui *et al.*, 2000) and in crude oils (within the oil window and late mature) of several basins from the continental margin of Brazil (Jesuino, 2005). As shown in Fig. 2.18, no clear

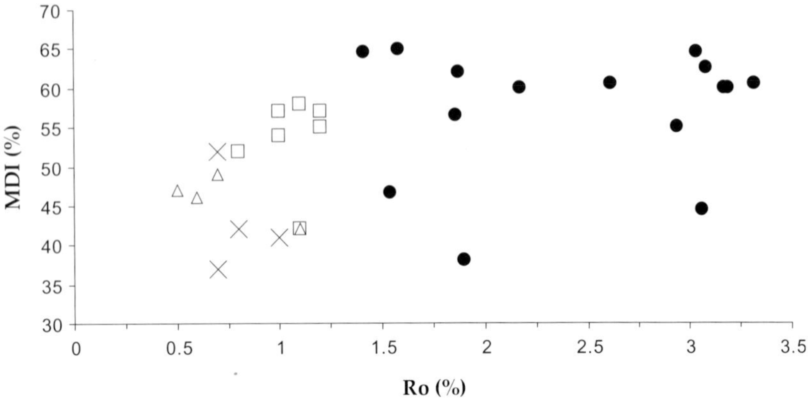

Fig. 2.18. Comparison between Methyladamantane Index (MDI (%)) *versus* R_0(Vitrinite reflectance, %) values calculated for: bitumens from overmature gas-prone carbonate source rocks, Shanganning Basin, China (•) (Jinggui *et al.*, 2000); and crude oils from three sedimentary basins in Brazilian continental margins: LSC (×), LSS (□) and LSS (Δ) (Jesuino, 2005). R_0 was determined as the reflectance of solid bitumen in the whole rock thin section of Shanganning samples and based on α, α, α-C_{29} steranes from Brazilian crude oils, according to Soffer *et al.* (1993).

correlation between MDI and R_0 (%) was observed in these basins. These results indicate that diamondoid parameters are affected by a number of factors, as are diamondoid concentrations, making the extended use of such parameters in different petroleum sites less reliable.

An example of the influence of mineral content on MAI and MDI can be seen in experiments of artificial maturation through hydrous pyrolysis of modern sediments (Table 2.3, Figs. 2.19a and 2.19b). MDI appears to be affected by mineral content and it is suppressed by the presence of $CaCO_3$ In contrast, the presence of montmorillonite (K10 or MS-25) tends to elevate MDI value. Similar behavior is observed to MAI values. In another example, $CaCO_3$ had little influence on yields of (3,+4-MD) and total diamondoids after hydrous pyrolysis in Celestun Lagoon (see Columns b and c of Table 2.3 for Celestun Lagoon and Fig. 2.6 for $CaCO_3$), but severely reduced MDI values of this sample from 26% to 11%. This may be an indication of the distinct action of minerals on generation and release of diamondoids, depending on the type of isomer formed.

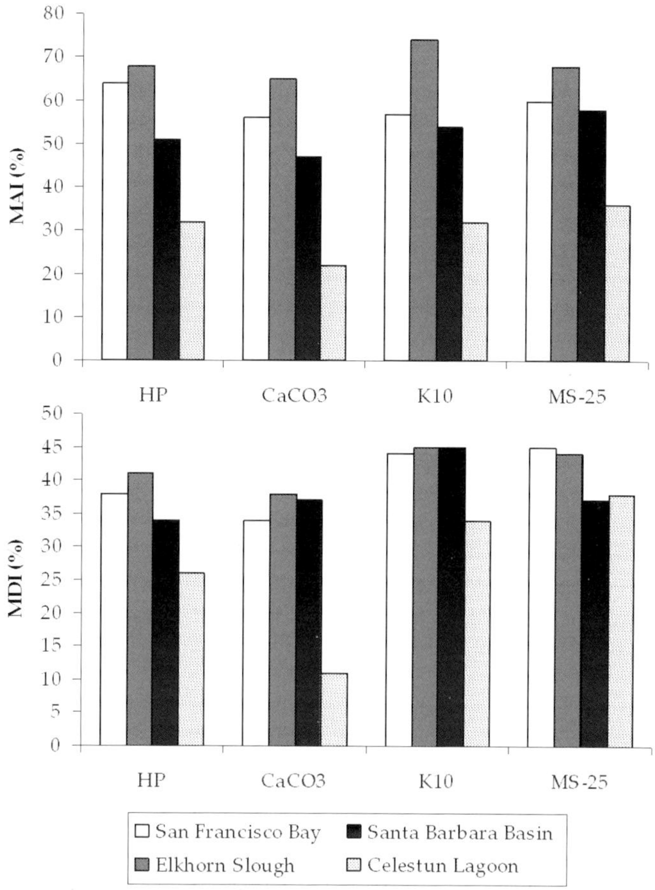

Fig. 2.19. Changes in MAI and MDI ratios as a function of mineralogy in hydrous pyrolysis of modern sediments. HP = hydrous pyrolysis of modern sediments; $CaCO_3$ = hydrous pyrolysis of sediments mixed with $CaCO_3$ (6:1 w/w); K10 = hydrous pyrolysis of sediments mixed with montmorillonite K10 (6:1 w/w); MS-25 = hydrous pyrolysis of sediments mixed with MS-25 (6:1 w/w) (Wei *et al.*, 2006a).

2.5. The Role of Diamondoids in Petroleum and Natural Gas Production Fouling

The relatively high melting points of diamondoids may cause their precipitation in oil wells, transport pipelines and processing equipment during production, transportation and refining of diamondoids-containing petroleum crude oil and natural gas (Escobedo, 1999; Vazquez *et al.*, 1998; Vazquez and Mansoori, 2000; Mansoori, 1997, 2009, 2010; Mansoori *et al.*, 2007). This may then cause fouling of pipelines and other oil processing facilities. Diamondoids deposition and possible fouling problems are usually associated with deep natural underground petroleum reservoirs that are rather hot and at high pressure. Other hydrocarbons with molecular weights in the same range as diamondoids generally have much lower melting points and do not precipitate at high temperatures (Vazquez *et al.*, 1998).

The practice of petroleum production may lead to an environment that favors the reduction of diamondoid solubility in petroleum, their separation from petroleum fluid phase, and their precipitation. Deposition of diamondoids from petroleum streams is associated with phase transitions resulting from changes in temperature, pressure, and/or composition of reservoir fluid. Generally, these phase transitions are solid-gas or solid-liquid transitions. Deposition problems are particularly cumbersome when the fluid stream is, so called, "dry" (i.e., low LPG content in the stream which includes propane C_3H_8 and butane C_4H_{10}). Phase segregation of solids takes place when the fluid is cooled and/or depressurized. In a, so called, "wet reservoir fluid" (i.e., high LPG content in the stream) the diamondoids partition into the LPG-rich phase and the gas phase. Deposition of diamondoids from a wet reservoir fluid is not as problematic as in the case of dry streams (Vazquez *et al.*, 1998; Vazquez and Mansoori, 2000; Mansoori, 1997). As a result, diamondoids may nucleate out of solution upon drastic changes of pressure and/or temperature. Instabilities of this sort in petroleum fluids may potentially initiate deposition of other heavy organic compounds (including asphaltenes, paraffins, and resins) on such nucleation sites (Vazquez *et al.*, 1998; Vazquez and Mansoori, 2000; Mansoori, 1997). Therefore, knowledge about the solubility behavior of diamondoids in organic solvents and dense gases, as reported in Chapter 3, Sec. 3.4.2, becomes important.

Adamantane and diamantane are usually the dominant diamondoids found in petroleum and natural gas pipeline deposits (Vazquez *et al.*, 1998; Vazquez and Mansoori, 2000). This is because diamondoids are soluble in light hydrocarbons at high pressures and temperatures. Upon expansion of the petroleum fluid coming out of the underground reservoir and a drop in its temperature and pressure, diamondoids could deposit.

Deposition of diamondoids can be particularly problematic during production and transportation of natural gas, gas condensates, and light crude oils. However, their presence in petroleum fluids is usually ignored due to their low concentration in these fluids. Nevertheless, in certain cases existence of these molecules in natural gas and gas condensate sources could lead to severe complex system problems of deposition and eventual plugging of flow paths. These nucleation sites may promote interactions among other heavy organic species and serve as their flocculation sites. Thus, it is important to determine whether or not diamondoids are present in a reservoir fluid at a harmful level. For this purpose, the separation, detection and measurements discussed in the next section are used.

2.6. Detection, Measurement and Separation of Diamondoids in Petroleum

Petroleum crude oil, gas condensate and natural gas are generally complex mixtures of various hydrocarbons and nonhydrocarbons with diverse molecular weights. In order to analyze the contents of a petroleum fluid it is a general practice to separate it first to six separate components (known as fractions) which include VOLATILES, INORGANIC MATERIALS, SATURATES, AROMATICS, RESINS, and ASPHALTENES as shown in Fig. 2.20 (Vazquez *et al.*, 1998; Vazquez and Mansoori, 2000; Mansoori, 1997; Mansoori *et al.*, 2008).

Volatiles consist of the low-boiling fraction of crude oil separated at room temperature and under vacuum (~ 27°C and 10 mmHg) from crude oil. The contents of the volatiles fraction can be further analyzed using gas chromatography (GC). Inorganic materials are separated by adding toluene to the vacuum residuum. All the organic materials in the oil will dissolve in toluene and the inorganic materials are filtered out. The remaining five families of components (fractions) are separated from the vacuum residue with the use of a column liquid chromatography [i.e., SARA separation (Jewell *et al.*, 1974)]. Initially the asphaltenes-fraction of the sample is removed by the ASTM D3279–90 separation method (ASTM 2005). Then the saturates-fraction is extracted with *n*-hexane solvent by passing the sample through a liquid chromatography column

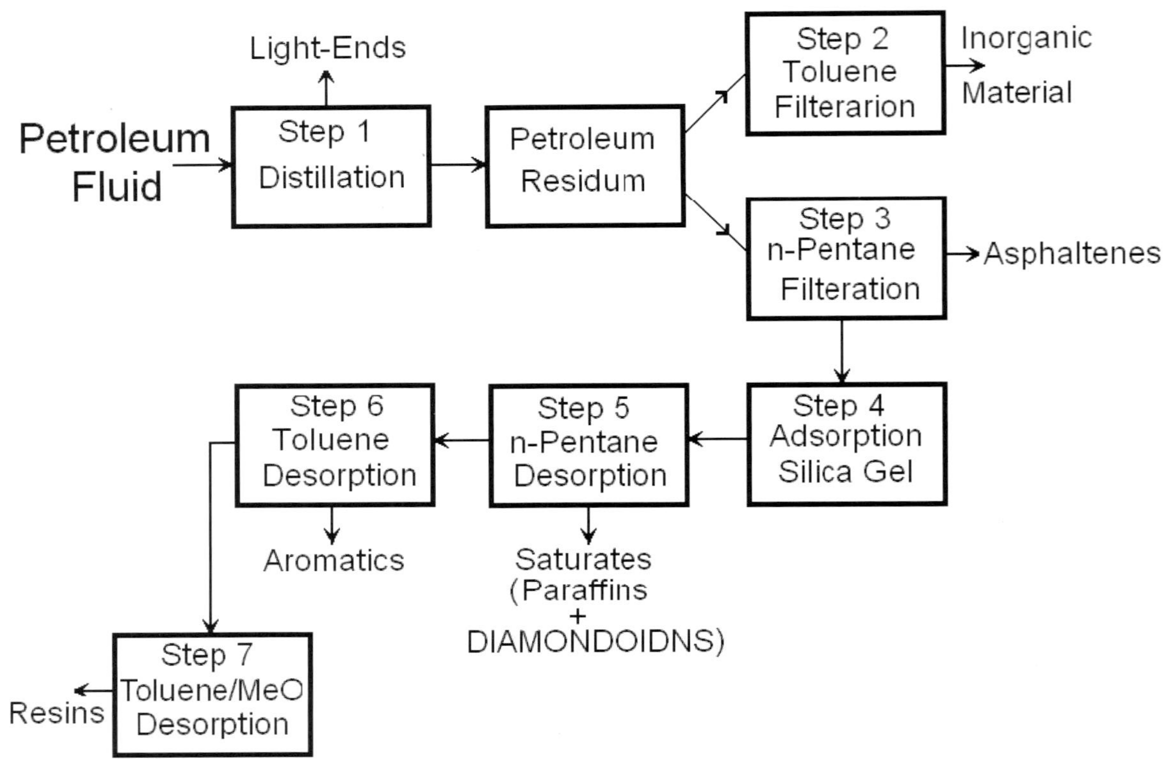

Fig. 2.20. Separation of petroleum fluids into six categories of components. Diamondoids are separated into the SATURATES fraction (Escobedo, 1999).

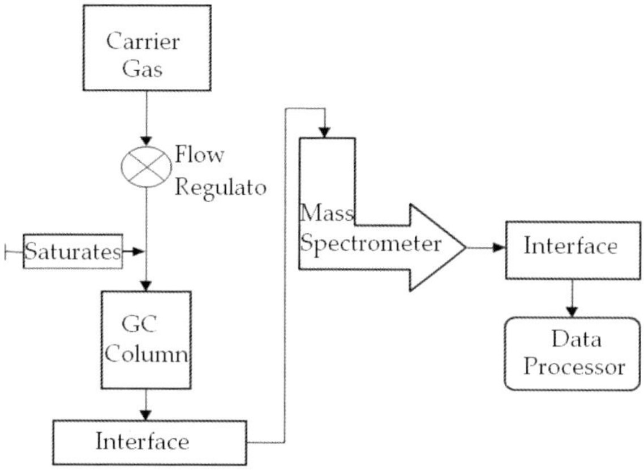

Fig. 2.21. Gas chromatography plus mass spectrometry for identification and measurement of diamondoids content of saturates fraction of petroleum fluids (Escobedo, 1999).

that is packed with silica gel and alumina powder. Since diamondoids are saturated hydrocarbons the analysis to determine their presence and composition in a crude oil must be performed on the saturates-fraction. Mass spectrometry analyses must then be performed on the low-boiling part of the saturates-fraction to determine whether diamondoids are present in it. To achieve this, the saturates-fraction is analyzed through GC/MS (Wingert, 1992; Lin and Wilk, 1995; Escobedo, 1999; Vazquez *et al.*, 1998; Vazquez and Mansoori, 2000; Mansoori *et al.*, 2008) according to Fig. 2.21.

Some representative chromatograms of petroleum fluids containing diamondoids obtained through GC/MS by various investigators are reported in Figs. 2.22–2.26. Adamantane, if present in the sample, will elute between nC_{10} and nC_{11}; diamantane will elute between nC_{15} and nC_{16}; triamantane will elute between nC_{19} and nC_{20}, etc. (Stout and Douglas, 2004). There are a number of other methods for the separation of diamondoids from petroleum fluids or natural gas streams which include the following:

- A gradient thermal diffusion process (Alexander *et al.*, 1990a) is proposed for separation of diamondoids.

- A number of extraction and absorption methods (Alexander *et al.*, 1991; Henderson and Sitzman, 1993) have been recommended for removing diamondoid compounds from natural gas streams.
- Separation of certain diamondoids from petroleum fluids has been achieved using zeolites (Alexander *et al.*, 1990b; Rollmann *et al.*, 1996) and a number of other solid adsorbents.

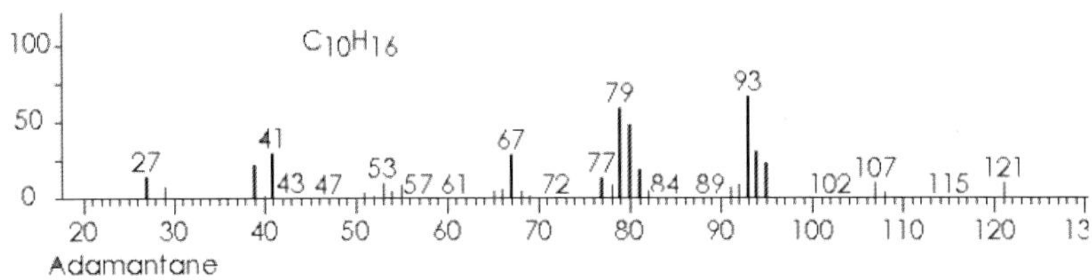

Fig. 2.22. Standard molecular fragmentation spectrum of adamantane (136 = m/z) (Mansoori, 2007).

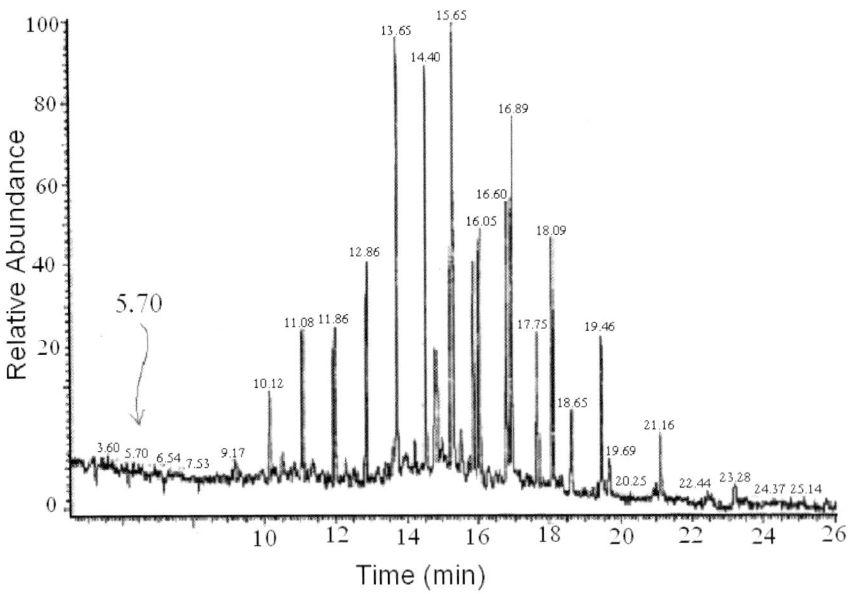

Fig. 2.23. Gas chromatogram of a gas condensate (NGL = natural-gas liquid) sample (Vazquez *et al.*, 1998). The peak with retention time of 5.70 eluted between nC_{15} and nC_{16} is indicative of the probable existence of diamantane in the sample.

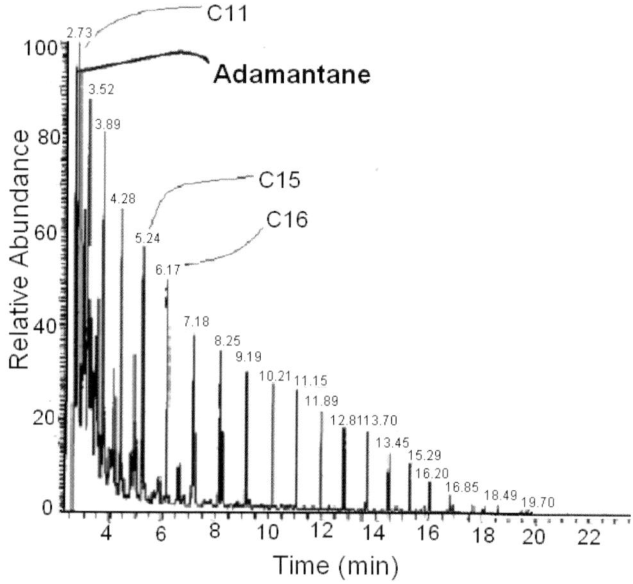

Fig. 2.24. Gas chromatogram of a crude oil sample showing the possible existence of adamantane and diamantane in the sample (Mansoori, 2007).

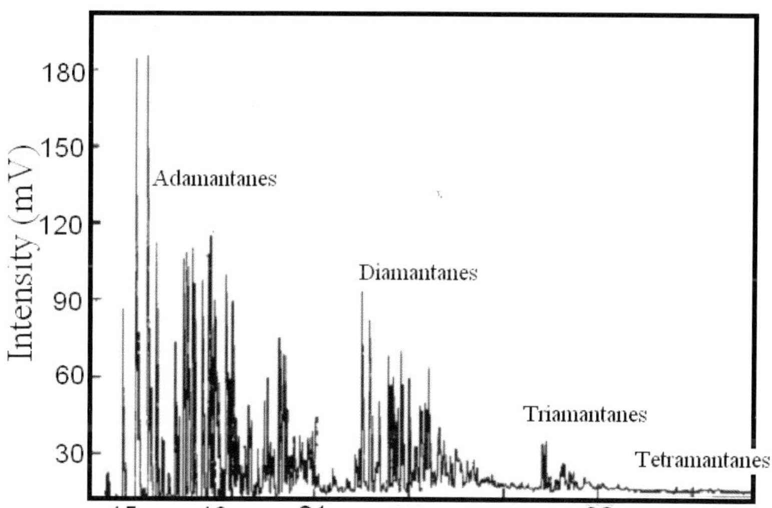

Fig. 2.25. Gas chromatogram of a diamondoid-rich gas-condensate (NGL) sample showing clusters of peaks representing adamantanes, diamantanes, triamantanes, and tetramantanes (Wingert, 1992).

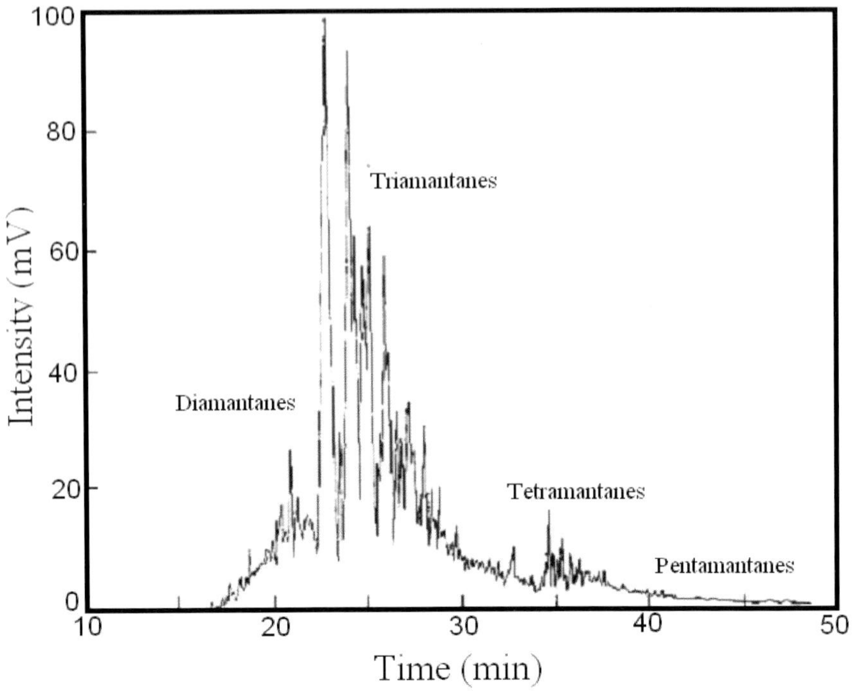

Fig. 2.26. Gas chromatogram from the full-scan CC/MS analysis of a high temperature distillation fraction (343°C) containing diamondoids (Wingert, 1992).

2.7. Concluding Remarks

Diamondoids were originally discovered in petroleum and due to their unique and peculiar nature they are now one of the most important characterizing agents for petroleum and other fossil fuels. In this chapter, we have presented some distinct features and the origin of diamondoids in fossil fuels. Also, we have introduced the methodology of using diamondoids content of petroleum fluids as a geochemical tool in petroleum reservoir fluids characterization. In addition, the effect of diamondoids in petroleum and natural gas production fouling is discussed.

Considering that most of the literature data on naturally occurring diamondoids are about petroleum diamondoids, we have discussed here the origin of petroleum and genesis of diamondoids in order to understand how diamondoids could be present in fossil fuels.

To understand the origin of petroleum, we have discussed about the three known geological stages of petroleum formation that are: "diagenesis" which is the earliest stage of transformation of sedimentary organic matter; "catagenesis" which is the process within which thermal cracking of kerogen produces petroleum hydrocarbons; and finally "metagenesis" which is the transformation stage when an organic material undergoes extensive thermal cracking.

Discussion about diamondoid genesis in petroleum and other fossil fuels starts with the hypothesis that in nature diamondoids are formed from abiogenic carbon in the source rock bed through the Lewis acid catalysis transformation. Concentrations of 3-methyldiamantane + 4-methyldiamantane, [3-,+4-MD], is chosen to represent all diamondoid classes because they are less volatile and relatively abundant in petroleum and other fossil fuels. In addition, similar features between (3-,+4-MD) and other diamondoid yields in fossil fuels is observed. However, experimental evidence reported so far indicate the complexity of diamondoid generation in petroleum. While the available data are useful for understanding of diamondoid formation, the large number of variables involved in the process makes it difficult to model the geological generation of diamondoids based on controlled laboratory experiments.

In this chapter, we also report on diamondoid-based geological correlations for petroleum characterization which are based on geochemical comparisons among crude oils, refined products, and/or extracts from prospective source rocks. From petroleum production industry's point of view such correlations are valuable tools in locating new petroleum sites and extending existing petroleum productions activities.

Since diamondoids, in spite of their resistance to biodegradation, may be attacked by microorganisms we also reported on the susceptibility and biodegradablity by microorganisms and advances made to quantify and predict the mechanisms of their biological degradation. It should be pointed out that, geological conditions in petroleum sites are primarily anaerobic, making any aerobic biochemical pathway less probable to occur.

Considering that pollutions due to fossil fuels (specially oil spills in open waters and land pollution at fuel production and processing sites) is a major problem worldwide, we also presented the diamondoid-based diagnostic ratios developed for applications in environment science.

These techniques have applications in the study of environmental pollution by correlating relationships between a fossil fuel source and the polluted site.

The presence of diamondoids is considered to be closely related to the geological maturity of an underground oil field. To develop techniques for prediction of petroleum formation and occurrence of a number of diagnostic ratios which are diamondoid-based for petroleum maturity parameters are reported in this chapter.

Overall diamondoid-based parameters may be used to assess oil source, maturity and biodegradability, providing supporting information, especially when traditional evaluation data are not available or are difficult to interpret.

When concentration of diamondoids is relatively high in a natural gas or petroleum reservoir they may reach the supersaturation limit and eventual deposition and fouling in transport conduits. This is because of the relatively high melting points of diamondoids. Due to the importance of such fouling incidents, we also report the role of diamondoids in petroleum and natural gas production fouling and the related technologies to deal with such problems.

Considering the fact that the major source of diamondoids are petroleum and other fossil fuels, understanding about the methods of their detection, measurement and separation from petroleum fluids is needed which are also presented in this chapter.

Throughout this chapter, it is possible to realize that the presence of diamondoids in petroleum has become much more than a chemical curiosity since Landa's initial findings in Czechoslovakian crude oils (Landa, 1933) and has advanced to be a resourceful instrument in petroleum science and engineering, and an important family of molecular building blocks in biomedicine, materials science, and nanotechnology which are discussed in the other chapters.

Bibliography

Abdullah, F. H. and Connan, J. (2002). *Org. Geochem.*, **33**: 125–148.

Aczel, T., Gorbaty, M. L., Maa, P. S. and Schlosberg, R. H. (1975). *Fuel.*, **54**: 295–301.

Aczel, T., Gorbaty, M. L., Maa, P. S. and Schlosberg, R. H. (1979). *Fuel.*, **58**: 228–230.

Ahsan, A., Karlsen, D. A. and Patience, R. L. (1997). *Mar. Petrol. Geol.*, **14**: 55–64.

Alimi, H., Ertel, T. and Schug, B. (2003). *Environ. Forensics.*, **4**: 25–38.

Andresen, B., Throndsen, T., Barth, T. and Bolstad, J. (1994). *Org. Geochem.*, **21**: 1229–1242.

ASTM (2005). Book of ASTM Standards, ASTM International, West Conshohocken, PA.

Atlas, R. M. (1975). *Appl. Microbiol.*, **30**: 396–403.

Azevedo, D. A., Tamanqueira, J. B., Dias, J. C. M., Carmo, A. P. B., Landau, L. and Gonçalves, F. T. T. (2008). *Fuel.*, **87**: 2122–2130.

Behar, F., Vandenbroucke, M., Tang, Y., Marquis, F. and Espitalie, J. (1997) *Org. Geochem.*, **26**: 321–339.

Bennett, B. and Olsen, S. D. (2007). *Org. Geochem.*, **38**: 935–956.

Brooks, B. T. (1948). *Am. Assoc. Pet. Geol. Bull*, **32**: 2269–2286

Brosse, E., Riva, A., Santucci, S., Bernon, M., Loreau, J. P., Frixa, A. and Laggoun-Defarge, F. (1990). *Org. Geochem.*, **16**: 715–734.

Cai, C., Worden, R. H., Bottrell, S. H., Wang, L., Yang, C. (2003). China. *Chem. Geol.*, **202**: 39–57.

Carpentier, B., Wilhelms, A. and Mansoori, G.A. (2007). *J. Petrol. Sci. & Eng.*, **58**: 341–343.

Carr, A. D. (2000). *J. Petrol. Geol.*, **23**: 313–343.

Chen, J. H., Fu, J. M., Sheng, G. Y., Liu, D. H., Zheng, J. J. (1996). *Org. Geochem.*, **25**:170–190.

Christensen, J. H., Hansen, A. B., Tomasi, G., Mortensen, J. and Andersen, O. (2004). *Environ. Sci. Technol.*, **38**: 2912–2918.

Clayton, J. L. (1992). p. 16–19. Magoon, L. B. (Ed.), U. S. Geological Survey Bulletin 2007, USA.

Dahl, J. E., Liu, S. G. and Carlson, R. M. (2003b). *Science*, **299**: 96–99.

Dahl, J. E. P., Moldowan, J. M., Peakman, T. M., Cardy, J. C., Lobkovsky, E., Olmstead, M. M., May, P. W., Davis, T. J., Steeds, J. W., Peters, K. E., Pepper, A., Ekuan, A. and Carlson, R. M. K. (2003a). *Angew. Chem., Int. Ed. Eng. Ed.,* **42**: 2040–2044.

Dahl, J. E., Moldowan, J. M., Peters, K. E., Claypool, G. E., Rooney, M. A., Michael, G. E., Mello, M. R., Kohnen, M. L. (1999). *Nature*, **399**: 54–57.

DeVanney, K. F. and Stanton, R. W. (1994). In: *Vitrinite Reflectance as a Maturity Parameter — Applications and Limitations*. P. K. Mukhopadhyay and

W. G. Dow, (Editors), ACS Symposium Series 570, American Chemical Society: Washington, DC, p. 26–38.

Durand, B. (1980). *Kerogen, insoluble organic matter from sedimentary rocks.* (Ed. B. Durand), Editions Technip, Paris.

Ekweozor, C. M. and Telnaes, N. (1990). *Org. Geochem.*, **16**: 401–413.

Engler, E. M., Blanchard, K. R. and Schleyer, P. von, R. (1972). Enhanced axial–equatorial enthalpy differences in the methyl adamantanes. *J. Chem. Soc. Chem. Commun.*, **22**: 1210–1212.

Escobedo, J. (1999), PhD Thesis, University of Illinois at Chicago.

Evans, R. J. and Felbeck, G. T. F. (1983). *Org. Geochem.*, **4**: 145–152.

Fang, H., Sitian, L., Yongchuan, S. and Qiming, Z. (1996). *Org. Geochem.*, **24**: 363–375.

Farrimond, P., Taylor, A. and Telnaes, N. (1998). *Org. Geochem.*, **29** : 1181–1197.

Fort, R. C., Schleyer, P. V. R. (1964). *Chem Rev* **64**: 277–300.

Gao, Z. N., Chen, Y. Y. and Niu, F. (2001). *Geochemical J.*, **35**: 155–168.

Garcia-Valles, M., Vendrell-Saz, M. and Pradell-Cara, T. (2000). *Fuel*, **79**: 505–513.

George, S. C., Volk, H., Dutkiewicz, A., Ridley, J. and Buick, R. (2008). *Geochim Cosmochim Acta* **72**: 844–870.

Gray, R. J. (1991). *Org Geochem* **17**: 535–555.

Grice, K., Alexander, R. and Kagi, R. I. (2000). *Org. Geochem.*, **31**: 67–73.

Hanin, S., Adam, P., Kowalewski, I., Huc, A.-Y., Carpentier, B. and Albrecht, P. (2002). *J. Chem. Soc., Chem. Comm.*, 1750–1751.

Henderson, J. K. and Sitzman, J. R. (1993). US Patent No. 000001185H, May.

Heydari, E. and Moore, C. H. (1989). *Geology*, **17**:1080–1084.

Hold, I. M., Brussee, N. J., Schouten, S. and Damste, J. S. S. (1998). *Org. Geochem.*, **29**: 1403–1417.

Huc, A. Y. (2003). *Oil Gas Sci. Tech.*, **58**: 233–241.

Huizinga, B. J., Tannenbaum, E. and Kaplan, I. R. (1987a). *Org. Geochem.*, **11**: 591–604.

Hunt, J. M. (1979). *Petroleum geochemistry and geology*. W. H. Freeman and Co. — San Francisco, USA.

ICCP. (1998). The new vitrinite classification (ICCP System, 1994). *Fuel.*, **77**: 349–358.

Imuta, K. and Ouchi, K. (1973). *Fuel* **52:** 301–302.

Pradier, B., Bertrand, P., Martinez, L. and Laggoun-Defarge, F. (1991) *Org. Geochem.*, **17**: 511–524.

Jesuino, L. S. (2005). MSc. Dissertation, COPPE, Federal University of Rio de Janeiro, Brazil. Available at www.coc.ufrj.br/teses/mestrado/inter/2005/Teses/JESUINO_LS_05_t_M_int.pdf.

Jewell, D. M., Albaugh, E. W., Davis, B. E. and Ruberto, R. G. (1974). *Ind. Eng. Chem. Fundam.*, **13 :** 278–282.

Jinggui, L., Philp, P., Mingzhong, C. (2000). *Org. Geochem.*, **31**: 267–272.

Johns, W. D. (1979). *Ann. Rev. Earth Planet Sci.*, **7**:183–98.

Karlsen, D. A., Nedkvitne, T., Larter, S. R. and Bjorlykke, K. (1993). *Geochim. Cosmochim. Acta.*, **57**: 3641–3659.

Katz, B. J., Mancini, E. A. and Kitchka, A. A. (2008). *AAPG Bulletin*, **92**(5): 549–556.

Khaddor, M., Ziyad, M., Joffre, J., Ambles, A. (2002). *Chem. Geol.*, **186**: 17–30.

Konyukhov, A. I. (2008). *Lithol Mineral Resources*, **43**: 203–219.

Kovats, E. (1965). In: *Advances in Chromatography* Vol. 1, Giddings, J. C. and Keller, R. A. (Eds.), Marcel Dekker: New York, p. 229–247.

Landa, S., Machacek, V., Mzourek, M. and Landa, M. (1933). *Chim. Ind.* (Publ. No. 506).

Landes, K. K. (1967). *AAPG Bull*, **51**: 828–841.

Li, S., Guo, S. and Tan, X. (1998). *Org. Geochem.*, **29**: 1431–1439.

Lin, R. and Wilk, Z. A. (1995). *Fuel*, **74**: 1512–1521.

Lu, X., Yang, H., Yang, N., Zhao, F. and Ma, Y. (2007). *Petrol Sci.*, **4**: 21–25.

Luterbacher, H.-P. and Bertrand Ligouis, B. (2008). *Organic Petrology*. Eberhard Karls Universität Tübingen website. Available in color at www.uni-tuebingen.de/geo/gpi/ag-luterbacher/mitarbeiter/Liguis/seiten/research.html. Accessed on 07/05/2011.

Machel, H. G. (2001). *Sediment. Geol.*, **140**: 143–175.

Mackenzie, A. S., Patience, R. L. and Maxwell, J. R. (1980). *Geochim. Cosmochim. Acta.*, **44**: 1709–1721.

Magoon, L. B. (1992). U.S. Geological Survey Bulletin 2007, USA.

Mango, F. D. (1997). *Org. Geochem.*, **26**: 417–440.

Mansoori, G. A. (1997). *J. Petroleum Sci. Eng.*, **17**: 101–111.

Mansoori, G. A. (2007). *In* Advances in Chemical Physics, v. 136, chapter 4, Stuart A. Rice, editor. John Wiley & Sons.

Mansoori, G. A. (2009). *Int. J. Oil, Gas and Coal Technology*, **2**(2), 141–167.

Mansoori, G. A. (2010). *ELMI a Sarlar, J. Reservoir and Petroleum Engineering*, SOCAR, pp. 12–23.

Mansoori, G. A., Vazquez, D. and Shariaty-Niassar, M. (2007). *J. Petrol. Sci. and Eng'g.*, **58**(3–4): 375–390.

McKervey, M. A. (1980). *Tet.*, **36**: 971–992.

Mukhopadhyay, P. H. (1994). In: *Vitrinite Reflectance as a Maturity Parameter — Applications and Limitations*. Mukhopadhyay, P. K. and Dow, W. G. (Editors), ACS Symposium Series 570, American Chemical Society: Washington, DC, p. 1–24.

Obermajer, M., Osadetz, K. G., Fowler, M. G. and Snowdon, L. R. (2000). *Org. Geochem.*, **31**: 959–976.

Olah, G. A. (1990). In: *Cage Hydrocarbons*, John Willey & Sons, Hoboken, NJ.

Pepper, A. S. and Corvi, P. J. (1995) *Marine Petrol. Geol.*, **12**: 291–319.

Peters, K. E. and Moldowan, J. M. (1993). The Biomarker Guide: Interpreting Molecular Fossils in Petroleum and Ancient Sediments. Prentice Hall, Englewood Cliffs, NJ.

Peters, K. E. and Cassa, M. R. (1994). In: *The petroleum system — from source rock to trap*, L. B. Magoon and W. G. Dow (Editors), American Association of Petroleum Geologists: Tulsa, p. 93–117.

Peters, K. E., Walters, C. C. and Moldovan, J. M. (2005). *The Biomarker Guide,* Vol. 1, 2nd. Ed., Cambridge University Press, Cambridge, UK.

Price, L. C. (1993). *Geochim. Cosmochim. Acta.*, **57**: 3261–3680.

Radke, M. (1988). *Mar. Petrol. Geol.*, **5**: 224–236.

Radke, M., Welte, D. H. and Willsch, H. (1982). *Geochim. Cosmochim. Acta.*, **46**: 1–10.

Requejo G. (1994). *Org. Geochem.*, **21**: 91–105.

Richnow, H. H., Jenisch, A. and Michaelis, W. (1992). *Org. Geochem.*, **19**: 351–370.

Rimmer, S. M. (1991). *Org. Geochem.*, **17**: 265–265.

Robert, P. (1980). In *Kerogen, insoluble organic matter from sedimentary rocks.* (Ed. B. Durand), Editions Technip, Paris, p. 386–414.

Rodriguez, J. F. R. and Littke, R. (2001) *Mar. Pet. Geol.*, **18**: 995–1028.

Rollmann, L. D., Green, L. A., Bradway, R. A., and Timken, H. K. C. (1996). *Catalysis Today*. **31**: 163–169.

Sassen, R. and Post, P. (2008). *Org. Geochem.*, **39**: 147–151.

Saxby, J. D., Chatfield, P., Taylor, G. H., Fitzgerald, J. D., Kaplan, I. R. and Lu, S.-T. (1992). *Org. Geochem.*, **18**: 373–383.

Schaeffer, P., Harrison, W. N., Keely, B. J. and Maxwell, J. R. (1995). *Org. Geochem.*, **23**: 541–554.

Schleyer, P. V. R. (1957). *J. Am. Chem. Soc.*, **79**: 3292–3292.

Schulz, L. K., Wilhelms, A., Rein, E. and Steen, A. S. (2001). *Org. Geochem.*, **32**: 365–375.

Scott, A. C. (2002). *Int. J. Coal Geol.*, **50**: 119–134.

Seewald, J. S. (2003). *Nature*, **426**: 327–333.

Seifert, W. K. and Moldowan, J. M. (1986). *In: Methods in Geochemistry and Geophysics Series* **24** — *Biological Markers in the Sedimentary Record*, R. B. Johns, ed., Elsevier, Amsterdam, p. 261–290.

Seiwerth, R. (1996). Prelog's Zagreb school of organic chemistry (1935–1945). *Croat. Chem. Acta.*, **69**: 379–397.

Selifonov, S. A. (1992). *Biochem. Biophys. Res. Comm.*, **186**: 1429–1436.

Shimoyama, A., Yabuta, H. (2002). *Geochemical.*, *J* **36**: 173–189.

Sofer, Z., Regan, D. R., Muller, D. S. (1993). In: XII Congreso de Geológico Argentino y II Congreso de Exploración de Hidrocarburos, Actas I, 407–411.

Springer, M. V., Garcia, D. F., Gonçalves, F. T. T., *et al.* (2010). *Organic Geochemistry*, **41**: 1013–1018.

Stout, S. A. and Douglas, G. S. (2004). *Environ. Forensics*, **5**: 225–235.

Tennenbaum, E. and Kaplan, I. R. (1985). *Geochim. Cosmochim. Acta.*, **49**: 2589–2604.

Tennenbaum, E., Ruth, E. and Kaplan, I. R. (1986). *Geochim. Cosmochim. Acta.*, **50**: 805–812.

Tissot, B. P. and Welte, D. H. (1984). *Petroleum Formation and Occurrence*, Springer-Verlag, New York, USA.

Trolio, R., Grice, K., Fisher, S. J., Alexander, R. and Kagi, R. I. (1999). *Org. Geochem.*, **30**: 1241–1253.

Vakili-Nezhaad, G. R., Mansoori, G. A. and Kiani-Nassab, K. (2005). *Int'l J. Nanoscience & Nanotechnology*, **1**: 65–74.

vanGraas, G. W. (1990). *Org. Geochem.*, **16**: 1025–1032.

Vazquez, D., Escobedo, J. and Mansoori, G. A. (1998). In *Proceedings of the EXITEP 98*, International Petroleum Techniques. Exhibition, Placio de Los Deportes; 15–18 November 1998, Mexico City, Mexico.

Vazquez, D. and Mansoori, G. A. (2000). *J. Petroleum Sci. & Eng.*, **26**: 49–55.

Vijayakumar, C. T., Vinayagamoorthi, S., Fink, J. K., Sivasamy, P. (2006). *J. Anal. Appl. Pyrolysis.*, **76**: 191–197.

Wang, Z., Stout, S. A., Fingas, M. (2006b). *Environ. Forensics*, **7**: 105–146.

Wang, Z., Yang, C., Hollebone, B. and Fingas, M. (2006a). *Environ. Sci. Technol.*, **40**: 5636–5646.

Waples, D. W. and Curiale, J. A. (1999). In: Beaumont, E.A., Foster, N.H. (Eds.), *Exploring for Oil and Gas Traps. American Association of Petroleum Geologists*, Tulsa, pp. 8–1–8–71. Available at http://64.233.167.104/search?q=cache:UF0dp5uY2k0J:search.datapages.com/data/specpubs/beaumont/ch08/images/ch08.pdf+Waples+and+Curiale+and+petroleum+and+correlations&hl=en&ct=clnk&cd=10&gl=us. Accessed in 05/04/2011.

Wei, Z., Moldowan, J. M., Dahl, J., Goldstein, T. P. and Jarvie, D. M. (2006b). *Org. Geochem.*, **37**: 1421–1436.

Wei, Z., Moldovan, J. M., Fago, F., Dahl, J. E., Cai, C. and Peters, K. E. (2007a). *Energy & Fuels*, **21**: 3431–3436.

Wei, Z., Moldowan, J. M., Jarvie, D. M. and Hill, R. (2006c). *Geology*, **34**: 1013–1016.

Wei, Z., Moldowan, J. M. and Paytan, A. (2006a). *Org. Geochem.*, **37**: 891–911.

Wei, Z., Moldowan, J. M., Peters, K. E., Wang, Y. and Xiang, W. (2007c). *Org. Geochem.*, **38**: 1910–1926.

Wei, Z., Moldowan, J. M., Zhang, S., Hill, R., Jarvie, D. M., Wang, H., Song, F., Fago, F. (2007b). *Org. Geochem.*, **38**: 227–249.

White, R. E., McCarthy, M., Egeberg, K. D. and Sligar, S. G. (1984). *Arch. Biochem. Biophys.*, **228**: 493–502.

William, J. A., Bjoroy, M., Dolcater, D. L. and Winters, J. C. (1986). *Org. Geochem.*, **10**: 451–461.

Wingert, S. W. (1992). *Fuel*, **71**: 37–43.

Worden, R. H., Smalley, P. C. and Oxtoby, N. H. (1996). *Geochim. Cosmochim. Acta.*, **60**: 3925–3931.

Xianming, X., Wilkins, R. W. T., Zufa, L. and Jiamo, F. (1998). *Org. Geochem.*, **28**: 669–676.

Yang, C., Wang, Z. D., Hollebone, B. P., Fingas, M. and Landriault. (2006). *Environ. Forensics*, **7**: 377–390.

Yang, C., Wang, Z. D., Hollebone, B. P., Peng, X., Fingas, M. and Landriault, M. (2006). *Environm. Forensics*, **7**: 377–390.

Zhang, S. C., Huang, H. P., Xiao, Z. Y., Liang, D. G. (2005). *Org. Geochem.*, **36**: 1215–1225.

3

Physical Properties of Diamondoids

3.1. Introduction

In this chapter, we report the physical properties of diamondoids including their molecular, electronic, optical, and thermodynamic properties in microscopic and macroscopic scales.

Diamondoid molecules, which possess unique features and essentially have a hydrogen-terminated diamond structure, were originally obtained through extraction from residual petroleum in small quantity (Landa, 1933). Currently, some of these molecules could be synthesized in relatively large amounts. They have diverse applications such as the feed for nanotechnology as molecular building blocks, in nanomedicine, in polymer synthesis and many other applications (Mansoori, 2007). Indeed, different properties of diamondoids have been investigated including their electronic and atomic structures (Richardson *et al.*, 2005; McIntosh *et al.*, 2004; Drummond *et al.*, 2005; Lu *et al.*, 2005; Willey *et al.*, 2005; Fokin *et al.*, 2005; Lenzke *et al.*, 2007; Xue and Mansoori, 2008), spectroscopic properties (Wang *et al.*, 2008; Willey *et al.*, 2008; Filik *et al.*, 2006; Ramachandran and Manogaran, 2007; Oomens *et al.*, 2006; Jensen, 2004; Marsusi *et al.*, 2009) and thermodynamic properties (Kabo *et al.*, 2000; Kong *et al.*, 2003; van Ekeren *et al.*, 2006; Yashkin *et al.*, 2008; Bazyleva *et al.*, 2008; Bazyleva *et al.*, 2005; Konstantinova and Kurbatova, 2006; Reiser *et al.*, 1996; Sandrock, 1982; Karásek *et al.*, 2008a; Karásek *et al.*, 2008b; Nunes *et al.*, 2007; Smith and Teja, 1996). On the other hand, optical properties such as refractive index and dispersion are indispensable for optoelectronic device applications.

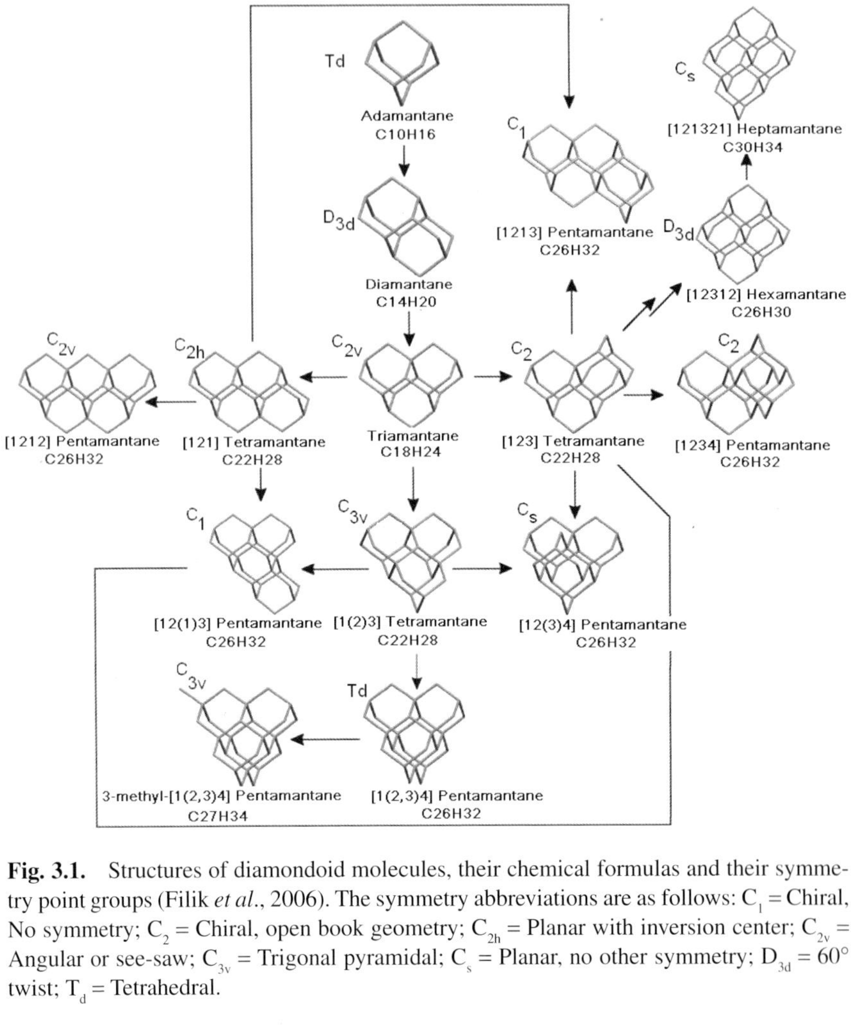

Fig. 3.1. Structures of diamondoid molecules, their chemical formulas and their symmetry point groups (Filik *et al.*, 2006). The symmetry abbreviations are as follows: C_1 = Chiral, No symmetry; C_2 = Chiral, open book geometry; C_{2h} = Planar with inversion center; C_{2v} = Angular or see-saw; C_{3v} = Trigonal pyramidal; C_s = Planar, no other symmetry; D_{3d} = 60° twist; T_d = Tetrahedral.

Figure 3.1 shows the relationship between molecular structures of lower diamondoid molecules.

Observe that adamantane (a), diamantane (b), and triamantane (c), are the only isomers possible to be constructed of one, two, or three adamantane units, respectively. On the other hand, the addition of further units (eight distinct ways) can produce [121]tetramantane (d), [1(2)3]tetramantane (e), and both enantiomers (optical isomers) of the chiral [123]tetramantane (f).

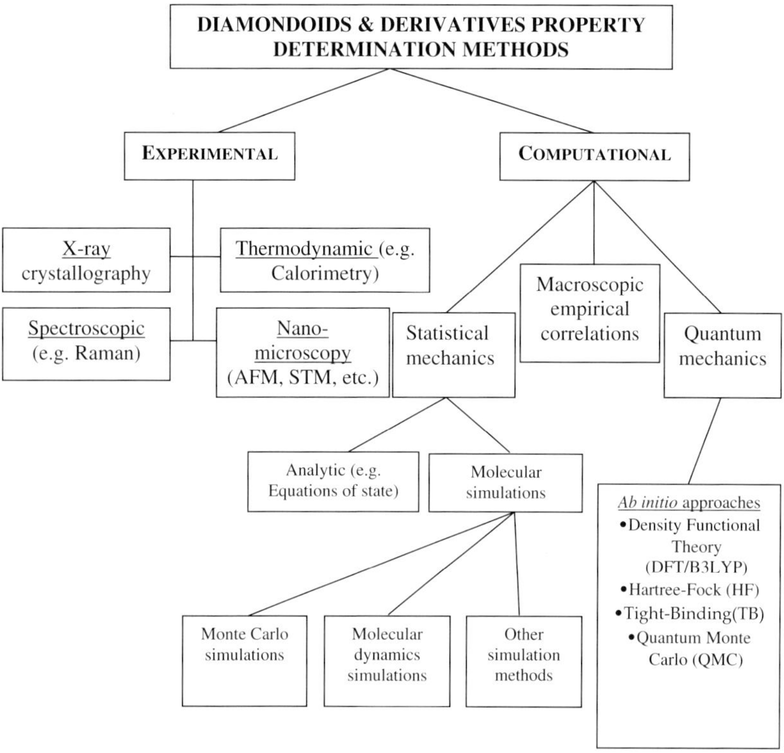

Fig. 3.2. Schemes of determining diamondoid properties. Although experimental methods, especially X-ray crystallography, are standard methods, computational approaches tend to be faster and provide high quality information. Predictive techniques, such as *ab initio* quantum mechanics calculations, provide accurate geometries and electron distribution properties for diamondoid molecules.

The physical properties of a molecular system may be investigated by using experimental and/or computational methods. Figure 3.2 shows a scheme of experimental and computational property determination methods for diamondoid molecules. In this chapter, we will present experimental methods and data only. However, Chapter 5 is dedicated to computational methods which are employed in prediction and correlation of diamondoid physical properties.

Physical properties data of diamondoids have been reported in the literature in limited amounts and mostly for the lower molecular weight diamondoids (Kabo *et al.*, 2000; Fort and Schleyer, 1964; Clark *et al.*, 1979; Reiser *et al.*, 1996; Hedberg, 1948; Fort, 1976; Boyd *et al.*, 1971; Cullick and Magouirk, 1994; Poot *et al.*, 2003, Poot and Loos, 2004; Buttler *et al.*, 1971; Mansson *et al.*, 1970; Westrum, 1961; Westrum *et al.*, 1978; Spinella *et al.*, 1978; Jochems *et al.*, 1982; Bratton *et al.*, 1967; Clark *et al.*, 1975; Steele and Watt, 1977). The three lowest diamondoids (adamantane, diamantane, and triamantane) are chemically and thermally stable compounds and strain-free. They possess higher melting points (MPs) in comparison to other hydrocarbons and organic compounds. For example, the MP of adamantane is estimated to be ~269°C, yet it sublimes easily, even at atmospheric pressure and room temperature. Diamantane and triamantane MPs are ~236.5°C and ~221.5°C, respectively. Table 1.1, in Chapter 1, includes some melting point and other physical properties data for diamondoids.

Table 3.1. Carbon binding environments, evaporation temperatures (t_{evap}), experimental ionization potentials (IP) and HUMO-LUMO band gap (in eV) for seven lower diamondoid molecules and bulk diamond.

		Bulk	Surface				Band gaps[2]
Diamondoid	**Chemical formula**	**C**	**CH**	**CH_2**	**t^1_{evap} (°C)**	**IP[1] (eV)**	**SXE–XAS (eV)**
Adamantane	$C_{10}H_{16}$	0	4	6	24	9.23 ± 0.12	6.03 ± 0.3
Diamantane	$C_{14}H_{20}$	0	8	6	62	8.80 ± 0.06	5.82 ± 0.1
Triamantane	$C_{18}H_{24}$	1	10	7	85	8.57 ± 0.08	5.68 ± 0.1
[121]tetramantane	$C_{22}H_{28}$	2	12	8	—	—	5.60 ± 0.1
[123]tetramantane	$C_{22}H_{28}$	2	12	8	175	8.18 ± 0.13	5.65 ± 0.1
[1(23)4] Pentamantane	$C_{26}H_{32}$	6	8	12	153	8.07 ± 0.12	5.51 ± 0.1
Hexamantane	$C_{26}H_{30}$	2	18	6	—	—	5.54 ± 0.1
Bulk diamond	$C_{(diamond)}$	—	—	—	—	—	5.47 ± 0.1

[1]Lenzke *et al.* (2007).
[2]Willey *et al.* (2006).

3.2. Spectrometric Properties

3.2.1. *Soft-X-ray emission (SXE) and X-ray absorption (XAS) spectroscopy*

Soft-X-ray emission (SXE) and X-ray absorption spectroscopy (XAS) techniques allow describing electronic structures of diamondoids. Diamondoid nanocrystals have different electronic properties when compared to group IV semiconductors such as silicon (Si) and germanium (Ge). These semiconductor nanocrystals exhibit quantum confinement effect (see Sec. 5.3.2), with decreasing sizes leading to increasing HOMO-LUMO band gap through evolution in both the conduction and valence bands (van Buuren *et al.*, 1998; Bostedt *et al.*, 2004). XAS, also called "near edge X-ray absorption fine structure spectroscopy" (NEXAFS), from the carbon 1*s* level probes the unoccupied states and indicates the relative energy position of the lowest unoccupied orbitals (LUMO). X-ray absorption data shows that diamondoid molecules exhibit negligible shifting in the LUMO, and these states are dominated by the hydrogen surface termination that must be considered as an integral part of the electronic structure, as shown in Fig. 3.3 (Willey *et al.*, 2006). On the other hand, Soft X-ray-emission (SXE) spectroscopy data, also shown in Fig. 3.3, allows probing the highest occupied molecular orbital (HOMO) sp^3 hybridized states in diamondoid molecules, by fluorescent photons observation emitted as valence states decay into the carbon 1*s* core-hole (Endo *et al.*, 2003; Willey *et al.*, 2006).

Figure 3.4 shows the carbon *K*-edge absorption data for the diamond clusters series from adamantane to hexamantane and a bulk diamond crystal as the reference. These measurements were performed on diamondoids in the gas phase which allows investigation of ideal interaction-free systems. This data is useful as the basis for testing the theoretical models. During the experimental process, diamondoids are brought into the gas phase by sublimation at room temperature or by gentle heating up to 200°C with a temperature-stabilized-in-vacuum furnace. The diamondoid beam is guided to the X-ray interaction region with a heated nozzle to avoid recondensation, and also no carrier gas is used. So, a

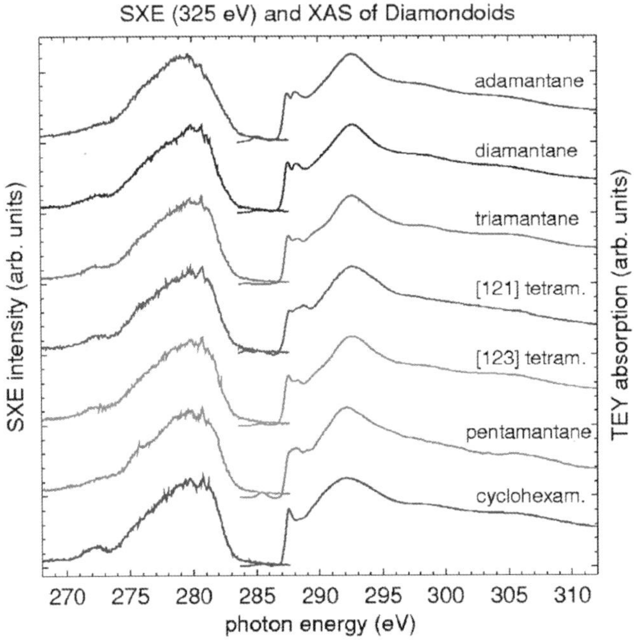

Fig. 3.3. Soft-X-ray emission spectra acquired with excitation energy 325 eV on the left; X-ray absorption of the diamondoids (condensed-phase) series on the right (Willey *et al.*, 2006).

monochromatized photon beam is crossed with the diamondoid beam under the aperture of a time-of-flight (TOF) mass spectrometer. XAS spectra are measured by photon energies scanning through the C 1*s* absorption edge. This way, the diamondoids are charged and all ionic fragments are produced from direct ionization, autoionization or secondary processes of core-hole annihilation, and detected with the TOF mass spectrometer. Thus, the recorded total ion yield approximates the absorption cross section and from the XAS spectra, atomic environment and electronic structure of the excited atom can be assessed (Stöhr, 1996).

Diamondoids possess only three basic configurations: carbon bound to only other carbon atom (C), carbon bound to three other carbon atoms and one hydrogen (CH), and carbon bound to two carbon and two hydrogen atoms (CH_2) (see Table 2.1, Sec. 2.1). In a bulk diamond, carbon atoms (C) can be interpreted as bulk-coordinated whereas the atoms bound to hydrogen (CH and CH_2) can be considered as surface atoms.

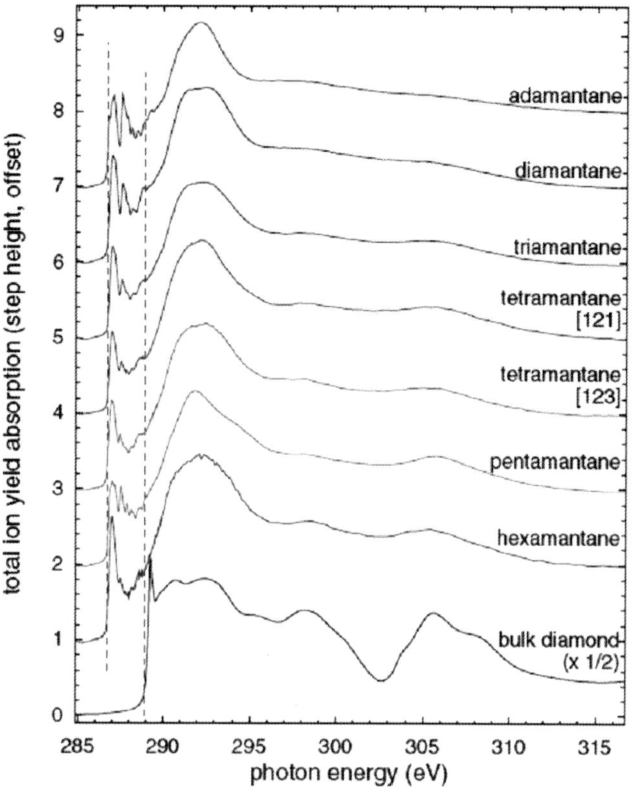

Fig. 3.4. The carbon *K*-edge absorption of the adamantane to hexamantane, all in gas phase, and a bulk diamond as the reference. (Willey *et al.*, 2005).

As it is observed in Fig. 3.4 diamondoids exhibit sharp spectral features at the 287–289 eV region. Such signals do not appear in bulk diamond spectrum. This spectral intensity is in Rydberg's energy range, C–H valence states, and mixed Rydberg-valence states of gaseous hydrocarbon molecules (Stöhr, 1996). In addition, these features are comparable in energy to the intramolecular excitation within the C–H bond of hydrogen terminated diamondoid (111) and (100) surfaces (Graupner *et al.*, 1999; Hoffman *et al.*, 1998).

Thus, the measured pre-edge features can be attributed to the C–H valence states (π^*) that show sharp vibrational structure, which can be considered surface states of the diamandoids. These π^* states are pulled below Fermi level (E_F) by the electron-hole Coulomb interaction (Stöhr, 1996).

On the other hand, the unoccupied densities of states (UDOS) attributed to the σ* states of the molecular C–C bonds are observed as a broad resonance centered at 292 eV. These σ* states, well-featured in diamondoids, evolve through nanodiamond to the bulk crystal (Fig. 3.4) where they constitute the first band including the conduction band minimum. However, contrary to other group IV semiconductors (i.e., Si and Ge), the (C–C bonds) σ* resonance absorption onset does not exhibit a size dependent blueshift, as expected within the quantum confinement model (see the next section). All theoretical approaches have pointed out the existence of this model in diamondoid molecules. It is interesting to note the C 1*s* core exciton resonance (π* states) at ~289,5 eV and the large dip at ~303 eV, assigned to a second absolute gap in the diamond band structure, as observed only in the bulk diamond spectrum (Morar *et al.*, 1985).

SXE spectroscopy probes the highest occupied molecular orbitals (HOMO) in diamondoids (Endo *et al.*, 2003; Willey *et al.*, 2006), whereas XAS also provides measures of the lowest unoccupied molecular orbitals (LUMO or UDOS) (Willey *et al.*, 2006). Figure 3.3 shows SXE (on the left) and XAS (on the right) spectra of the condensed-phase diamondoid series. For these measurements, solid diamondoid powders were pressed into clean indium foil. The lower diamondoids (adamantane–triamantane) sublime readily in vacuum at room temperature; therefore all diamondoids were cooled in vacuum to the temperature of liquid nitrogen before spectroscopy experiments.

The XAS measurements used total electron yield (TEY), with upstream clean gold grid for normalization (Stöhr, 1996). In this case, the photon flux was adjusted to minimize photochemical changes in XAS spectra, and required measurements over multiple experimental runs. Moreover, highly oriented pyrolytic graphite C *K*-edge absorption was measured to calibrate the photon energy output of the beam-line monochromator. The XAS spectra are similar in shape to gas-phase diamondoid absorption spectra (Fig. 3.4); however they present a weak vibronic fine structure in the lowest energy states at 287–289 eV. These features are below the bulk absorption (σ* states centered at ~292 eV) onset, and are similar in energy to hydrogen-terminated diamond surface states (Grauper *et al.*, 1999). It can be noted that through the diamondoid series, the absorption onset, within experimental error (~0.05 eV over multiple experimental runs) does not change significantly. Thus, these data clearly show that LUMO are dominated by the hydrogen surface termination and quantum confinement effects do not influence these states in diamondoid molecules. On the other hand, SXE spectra (shown on the left side of Fig. 3.3) exhibit a similar shape for all the diamondoid series reported here. The weak peak observed at ~272 eV is attributed to states with 2*s* character (Endo *et al.*, 2003) whereas the more intense fluorescent features closer to the highest occupied orbitals arise from states with 2*p* character. Thus, fluorescent photons, emitted as valence states decay into the C 1*s* core hole, map the energy of the highest occupied states (Willey *et al.*, 2006).

The intersection of the linear extrapolation of the edge to the baseline has generally been the method used in indication of the HOMO, as in some reports on quantum confinement effects in semiconductor nanocrystals (van Buuren *et al.*, 1998; Bostedt *et al.*, 2004). However, hydrocarbons can be unstable under intense X-rays required to produce sufficient carbon fluorescence (Zharnikov and Grunze, 2002). Thus, the method of inflection position of the SXE is the most suitable to investigate possible shift in highest occupied states in series of diamondoids, since X-ray beam damage effects can obscure the intersection of the SXE with the background (gray shaded area in Fig. 3.5). Note, in Fig. 3.5, that the intensity gradually arises at ~282 eV (indicated by the arrow), and it can be attributed to sample damage under soft X-ray beam.

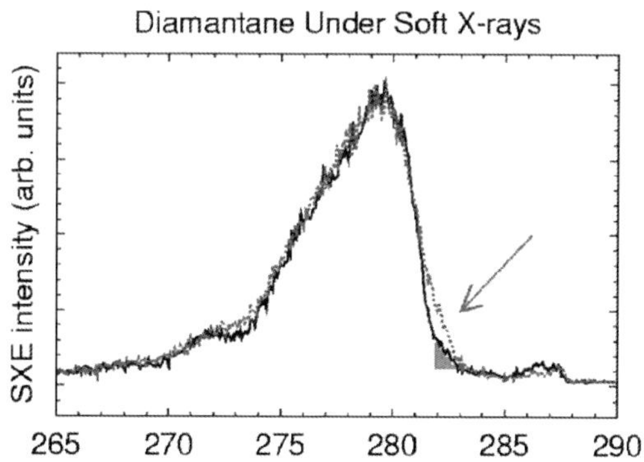

Fig. 3.5. Soft X-rays spectra of diamantane with minimal photochemical changes (solid line) and under the beam continuously for 300 s (dotted line) showing spurious intensity (arrow) appearing near the HOMO states (Willey *et al.*, 2006).

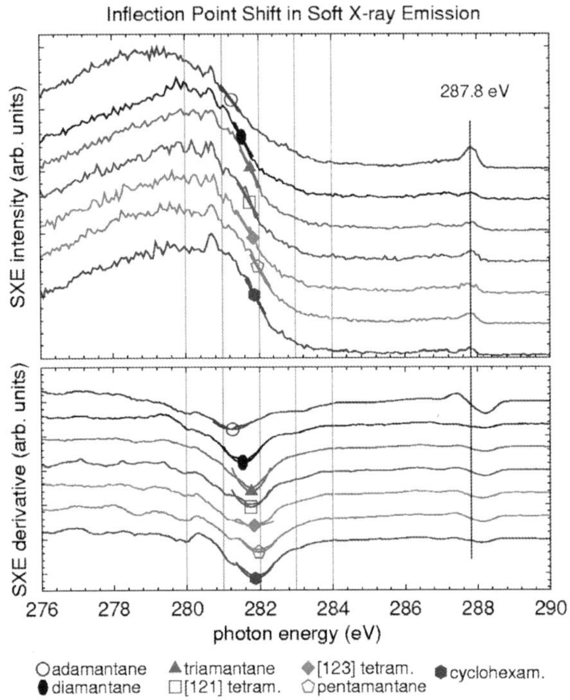

Fig. 3.6. Top figure: SXE spectra, showing the inflection point and a line fit through the data near the inflection point. Bottom figure: The smoothed derivatives of the SXE spectra, with the minima (the inflection point) indicated. The excitation elastic line is visible at 287.8 eV (Willey *et al.*, 2006).

Figure 3.6 shows the inflection point shift in SXE spectra in the diamondoid series. For better visualization and analysis the inflection positions were determined by minima values of the first-derivatives of the SXE spectral functions (lower panel in Fig. 3.6). The inflection positions and respective lines are superimposed on the Soft X-ray spectral data in the upper panel of Fig. 3.6. Note that the inflection points sharply evolve and shift toward lower energy with decreasing size throughout the diamondoid series, relative to the constant excitation energy, as seen on the elastic line at ~287.8 eV. This excitation energy corresponds to LUMO features associated with the hydrogen surface termination, which can be used to obtain the relative energy shift of HOMO states in the diamondoid series, from adamantane to hexamantane. In addition, the measured LUMO energy is constant throughout the diamondoid series, with a difference smaller than 0.05 eV between diamondoids, not affecting therefore the measured precision of the occupied states. Thus, size-dependent HOMO states shift are due to quantum confinement (QC) effects observed in diamondoid molecules. Theoretical predictions also confirm QC effects in diamondoids, see Sec. 5.3.2. Table 3.1 shows HOMO-LUMO gaps, in eV, measurements from SXE–XAS spectroscopy.

As mentioned before, diamondoid molecules have unique properties with potential applications in nanotechnology. The availability and ability to selectively functionalize this special class of nanomaterials opens new possibilities for surface modification, mainly for high-efficiency field emitters in molecular electronics, as seed crystals for diamond growth, or as robust mechanical coatings. In addition, higher diamondoids (molecules with four or more adamantane cages) are promising sources of applications on energy field as fuel additives (Hermann, 2004). The properties of self-assembled monolayers (SAMs) of diamondoids are thus of fundamental interest for a variety of emerging nanotechnology applications. Willey *et al.* (2008) have investigated the effects of thiol substitution position and polymantanes order (see Fig. 3.7) on diamondoid SAMs on gold using near-edge X-ray absorption fine structure spectroscopy (NEXAFS) and X-ray photoelectron spectroscopy (XPS). Diamondoid electronic properties are an interesting blend between macroscopic diamond and small sp^3-bonded hydrocarbon molecules (Willey *et al.*, 2005; Wang *et al.*, 2008; Willey *et al.*, 2006). As mentioned before, the diamondoids LUMOs stem from the hydrogen surface termination and do not shift in energy as a function of size (Willey *et al.*, 2005), in contrast to Si and Ge (van Buuren *et al.*, 1998;

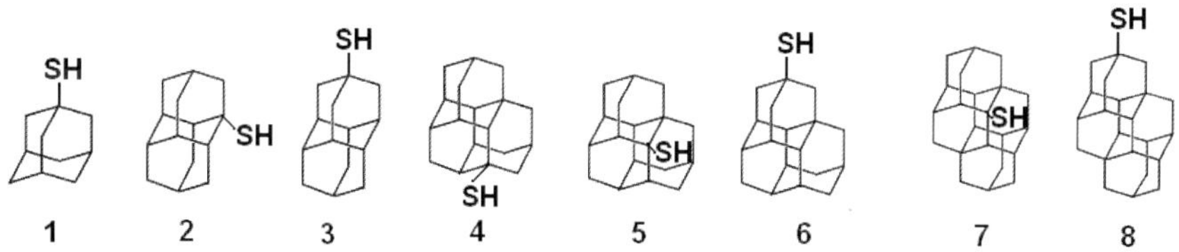

Fig. 3.7. Thiolated diamondoids: (1) Adamantane-thiol; (2) diamantane-1-thiol; (3) diamantane-4-thiol; (4) triamantane-3-thiol; (5) triamantane-2-thiol; (6) triamantane-9-thiol; (7) [121]tetramantane-2-thiol; and (8) [121]tetramantane-6-thiol (Yang *et al.*, 2007).

Bostedt *et al.*, 2004). On the other hand, diamondoids HOMOs, exhibit sharp size-dependent shifts (Willey *et al.*, 2006) similar to other group IV elements (van Buuren *et al.*, 1998; Williamson *et al.*, 2004). Thus, this demonstrates the ability to tune HOMO-LUMO gaps and other electronic properties based on diamondoid size (Lenzke *et al.*, 2007). Quantum calculations (see Secs. 5.3.2 and 5.3.3) also support the observed quantum confinement effects (McIntosh *et al.*, 2004; Drummond *et al.*, 2005; Lu *et al.*, 2005) and have predicted negative electron affinities (NEAs) in some of the higher diamondoids (Drummond *et al.*, 2005). NEAs and monochromatic electron emission have been demonstrated for [121]tetramantane-6-thiol (#8 in Fig. 3.7) when on Au and Ag surfaces by Yang *et al.* (2007). In this case, the intimate connection between the metal and the diamondoid appears to play a key role in the emission process. Section 3.3.1 presents further details concerning this investigation.

Figure 3.8 shows the NEXAFS data for the series of thiolated diamondoids of Fig. 3.7 on gold surfaces. NEXAFS spectra were acquired at different incidence angles 20°, 30° and 90° as pointed out on #8 (corresponding to [121]tetramantane-6-thiol). The difference between the acquired spectra and the spectrum at 20° is plotted just below the acquired spectra in order to accentuate the angular-dependent resonances. In all diamondoids, the C–H σ*/R* resonances are present at about 287–289 eV, whereas the broad C–C σ* resonance is centered about 297 eV (Stöhr, 1996). Note that these two manifolds of resonances exhibit angular dependence and are used to determine the molecular orientation. Although in some of the diamondoids, at least two C–H σ* resonances are clearly resolvable (e.g., #**6** in Fig. 3.8), the manifolds of these two surface terminations overlap and are thus considered together here for all thiolated diamondoids. It is worth to mention that all the diamondoids also exhibit the emergence of the bulk diamond second gap band structure, characteristic of the diamondoids, at about 303 eV (Willey *et al.*, 2005).

Both [121]tetramantane-2-thiolthiols (#**7** in Fig. 3.8) and [121]tetramantane-6-thiol (#**8** in Fig. 3.8) exhibit the strongest angular dependence. Tetramantane have stronger intensity variation due to its prevalence of C–C bonds generally aligned with the long axis and a prevalence of bonds

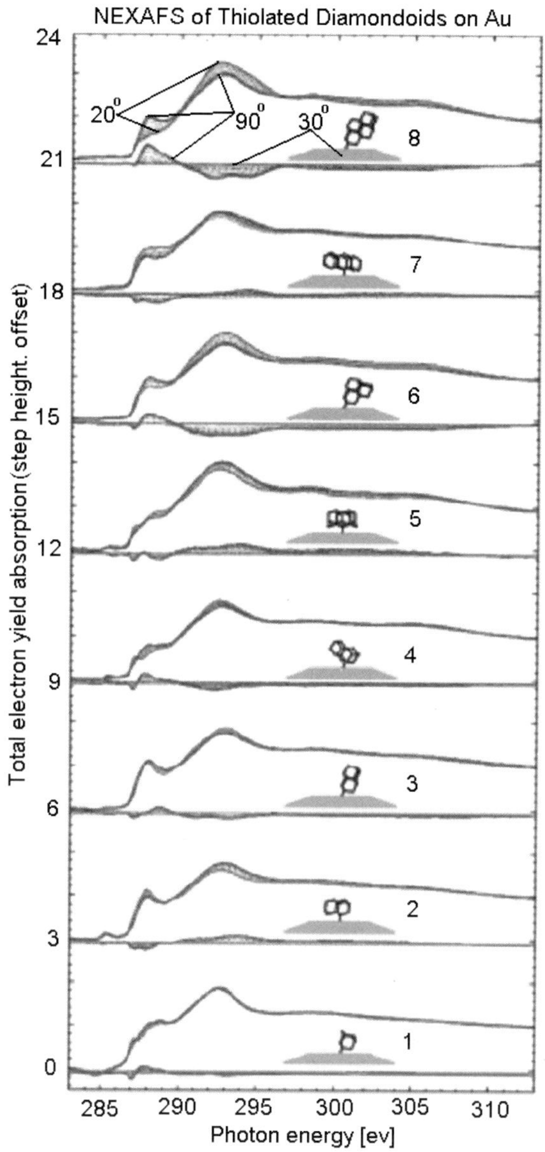

Fig. 3.8. Near-edge X-ray absorption fine structure spectra (NEXAFS) for the series of diamondoid thiolates 1–8 shown in Fig. 3.7. For each molecule, the NEXAFS traces at 20°, 30°, 40°, 55°, 70°, and 90° incidence angles were measured. Due to the small scale of this figure the 20°, 30°, and 90° incidence angles are distinguishable as indicated in #8. The lower traces have the 20° trace subtracted to emphasize the angular dependence (Redrawn from Yang *et al.*, 2007).

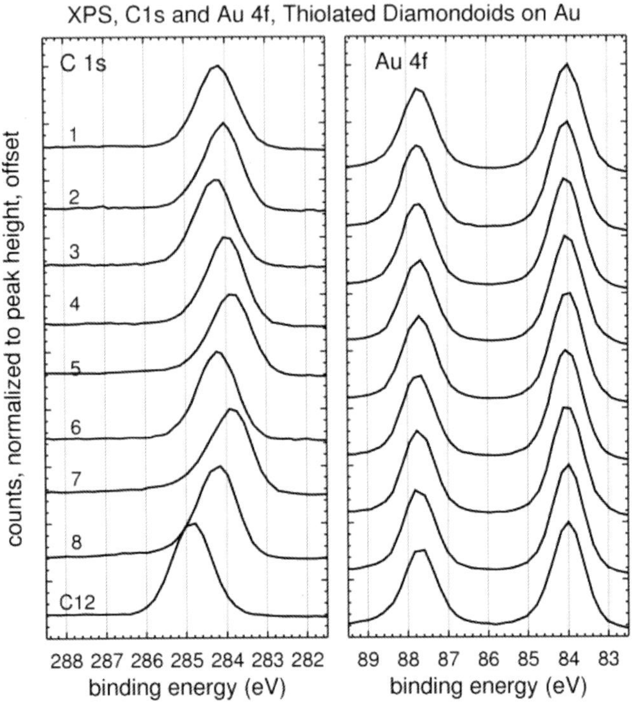

Fig. 3.9. C 1s and Au 4f XPS for the series of diamondoid thiolates no's 1–8 of Fig. 3.7 and the corresponding photoelectron peaks acquired from dodecane thiolate/Au as labeled "C12" (Yang *et al.*, 2007).

in a plane orthogonal to the long axis (C–H bonds) compared to smaller thiolated diamondoids. The XPS of C 1s and S 2p core-level provides further information about the surface attachment and structure of the monolayers.

Figure 3.9 shows normalized C 1s and Au 4f XPS spectra for the series of thiolated diamondoids on Au as well as the corresponding photoelectron peaks acquired from dodecane thiolate/Au (reference SAM), labeled C12. Note that the Au photoelectron peaks (right pane) do not exhibit appreciable changes in shape or full width at half-maximum. These electrons are thus inferred to represent the bulk gold, where the Fermi level is fixed with respect to the electron analyzer, and are used to calibrate the binding energy scale of the other photoelectron spectra. Adamantane thiolate (#1 in Fig. 3.9) displays a C 1s binding energy that

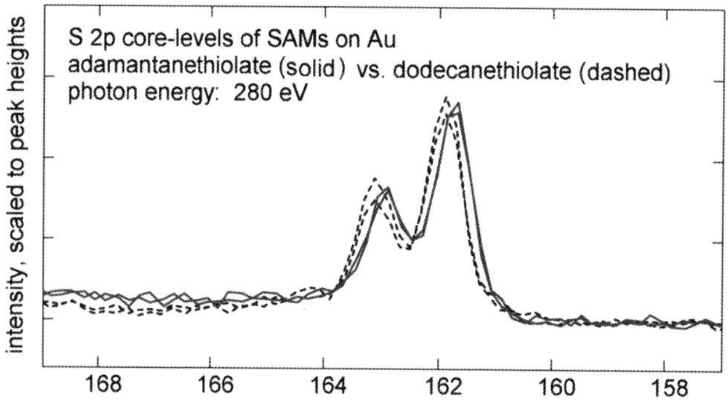

Fig. 3.10. S 2p X-ray photoelectron spectra for adamantane-thiol (#1 in Fig. 3.7) on gold (grey line) and dodecane-thiol on gold (black line). Photon energy of 280 eV. The S 2 binding energy for the adamantane-thiol is about 0.16 ± 0.04 eV lower than the dodecane-thiol on Au. A linear background has been subtracted, and peak heights are normalized to emphasize this small binding energy difference. (Yang *et al.*, 2007).

is 0.067 ± 0.05 eV lower than dodecane thiolate, in agreement with other studies (Dameron *et al.*, 2007). This is much larger than the 0.3 eV difference seen between unfunctionalized adamantane and cyclohexane (Klünder, 2007), indicating more than chemical shift contributions from structure alone.

Moreover, the S 2p binding energy of adamantane thiolate also is lower than dodecane thiolate by 0.16 ± 0.04 eV, as seen in Fig. 3.10. This small binding energy difference is resolvable using highly bright, 280 eV photons. Thus, both carbon and sulfur photoelectrons exhibit binding energies that depend upon diamondoid and thiol attachment position. Variation of these binding energies among the diamondoid thiolates is an indication of their variable electronic interactions with the Au substrate. A possible explanation for the differences in binding energies is because of the following mechanisms: (1) the C 1s shifts to lower binding energy are roughly proportional to the number of hydrogen atoms in close proximity to the gold, causing a misalignment of Fermi levels between the gold and carbon frameworks of diamondoids in the monolayer; (2) the S 2p shifting is consistent with the strained nature of the gold-thiolate bond. The NEXAFS data for prostrate diamondoid thiols shows the S–C bonds to be nearly normal to the surface, exhibit the largest shifts towards lower S 2p binding energies (Wiley *et al.*, 2008).

3.2.2. *Ionization potential and photoion yield spectra of diamondoids*

In this section, we present the experimental data of ionization potentials (IPs) and photoion yield spectra for the diamondoid series from adamantane to pentamantane. IPs (work function for macroscopic structures) are fundamental parameters for all materials. Such data may play important roles for all potential technological applications as well as for benchmarking theoretical predictions (see also Sec. 5.3.3). Table 3.1 shows experimental IPs and evaporation temperatures data for seven lower diamondoid molecules together with the data for bulk diamond.

The ionization potentials are determined from photoion yield measurements of diamondoids in the gas phase using photoion-yield spectroscopy apparatus (Lenzke *et al.*, 2007). Photoion is a cation produced through photoionization.

Figure 3.11 displays the photoion yield spectra for adamantane through pentamantane. Additionally, cyclohexane data was measured as a reference since it is closest related monocyclic hydrocarbon molecule to diamondoids. The spectra are the average of multiple individual scans.

The ionization energy of diamondoids is determined by using the common linearization procedure. Observe in Fig. 3.11 that it is indicated as linear extrapolation of the lowest photon energy features (solid lines) to the base line (dashed lines). In fact, this is a standard method for determining the adiabatic ionization potentials and has been successfully applied to diverse systems such as fullerenes (Hertel *et al.*, 1992), metal oxides (Janssens *et al.*, 2003) and metal clusters (deHeer, 1993).

3.2.3. *Vibrational spectroscopy of diamondoids*

Vibrational spectroscopy can be relatively complicated. In order to study vibrational structure of isolated nanosystems Raman spectroscopy is used. It relies on inelastic (or Raman) scattering of monochromatic light, usually from a laser in the visible, near infrared, or near ultraviolet range. The Raman effect is observable as shifts in energy of photons which are related to the vibrational states of a sample. To observe the shifts, we need to have a light source in which all of the photons possess very narrow band

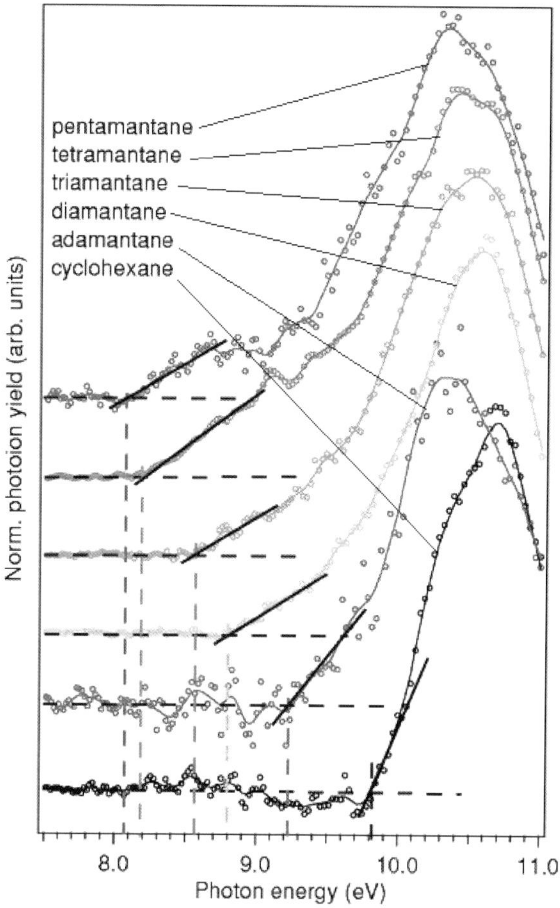

Fig. 3.11. Photoion yield spectra for diamondoid molecules and cyclohexane as reference. The extrapolation of the ionization potentials is indicated with solid and dashed lines (Lenzke *et al.*, 2007).

of energy frequencies. The laser can produce the light in a small concentrated and very intense beam with a very narrow band of frequencies. According to quantum mechanics only certain well-defined frequencies and atomic displacements are allowed known as the *normal modes of vibration* of the molecule (Kittel, 2005; Jorio *et al.*, 2001).

Laser Raman spectroscopy is a quite popular technique used to probe the vibrational, rotational, and other low-frequency modes in materials.

In particular, it is possible to carry out experiments with diamondoids in solid phase and to divide, with a good approximation, their vibrational modes into intramolecular and intermolecular contributions.

The intramolecular investigations focus on the high wavenumber region, i.e., above 300 cm^{-1}, of the spectrum and, in general, use theoretical calculations performed on a single molecule to aid with the vibrational mode assignment (Filik *et al.*, 2006; Bistricic *et al.*, 2002; Jensen, 2004). It is worth to mention that computational methods have been used as a good tool in the study of physical properties of diamondoid molecules (see Chapter 5).

On the other hand, intermolecular studies focus on the low wavenumber region, i.e., below 300 cm^{-1}, and are concerned with the motion of molecular units in the crystal relative to each other (Filik *et al.*, 2006). In general, these investigations are performed over a range of temperatures in order to observe also solid–solid phase transitions (Jenkins and Lewis, 1980). A good example is the plastic transition that takes place in lower diamondoids. All three solids have phase transitions to a plastic state; adamantane above 209 K, diamantane above 447 K and triamantane above 428 K (Jenkins and Lewis, 1980) (see Sec. 3.4.1 for further detail). Plastic crystals exhibit a high-temperature pre-melting plastic phase in which the molecules show considerable orientational disorder but positional order (Timmermans, 1961). The molecules are generally globular in shape and the crystal lattice is, in its high temperature form, highly symmetric (cubic or hexagonal) (Jenkins and Lewis, 1980). At low temperature, crystalline adamantane has the space group D_{2d}^{4} with two molecules per primitive unit cell, whereas the isolated molecule has T_d (tetrahedral) symmetry and the site symmetry is S_4. On the other hand, at temperatures above 209 K a phase transition to the plastic phase takes place.

3.2.3.1. *Raman spectroscopy of diamondoids*

A general study of the Raman spectra of a large family of diamondoids has been reported by Filik *et al.* (2006). In this investigation, the effects of variations in structure and symmetry on the intramolecular vibrations of a selection of diamondoid molecules, from adamantane to [121321] heptamantane, are analyzed by Raman spectroscopy. In addition, quantum calculations of vibrational frequencies and Raman intensity, using density functional theory (DFT) at the B3LYP level (see Sec. 5.2.1), were performed to assignments of the vibrational modes. In the following sections, we present some of the results reported by Filik *et al.* (2006).

3.2.3.2. *Raman spectra of adamantane, diamantane, and triamantane*

Adamantane, with 24 atoms, has $3n-6 = 66$ degrees of vibrational freedom. However, only 22 vibrational modes are Raman actives. Figure 3.12 depicts experimental and calculated (DFT/B3LYP) Raman spectra of adamantane, diamantane, and triamantane. A comparative analysis between experimental and calculated spectra shows that only half of these vibrational modes are intense enough to be experimentally observable.

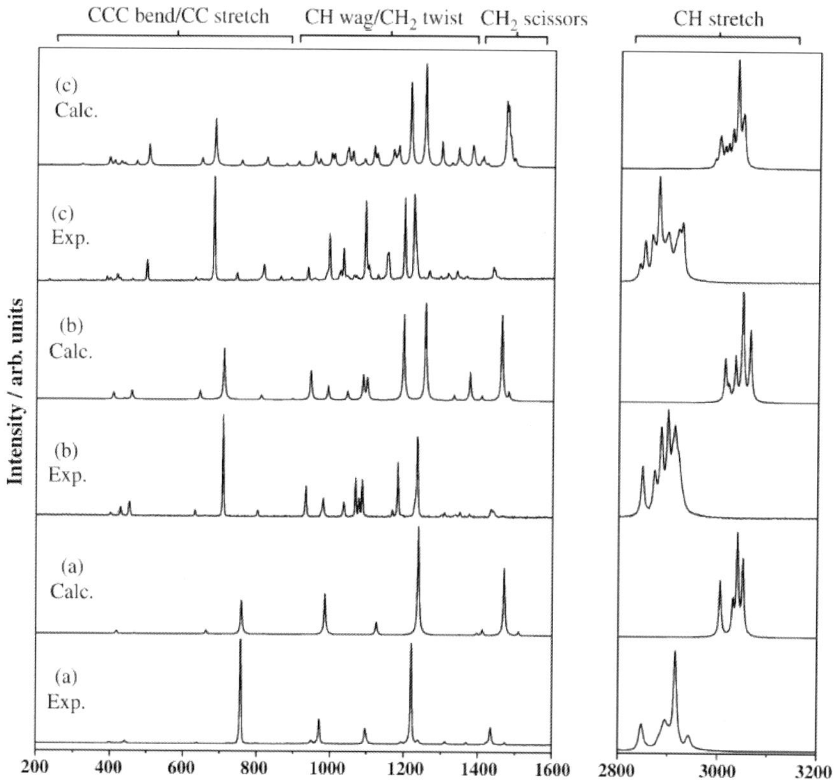

Fig. 3.12. Experimental and calculated (DFT/B3LYP) Raman spectra for lower diamondoids: (a) adamantane, (b) diamantane, (c) triamantane (Filik *et al.*, 2006.

The CH stretch region consists of six broad vibrations, three of which are so close in energy that they are unresolvable (experimental wavenumber 2828 cm^{-1}, calculated wavenumber 3005 cm^{-1}). The vibration with highest Raman intensity is a fully symmetric mode (experimental wavenumber 2916 cm^{-1}, calculated wavenumber 3039 cm^{-1} and intensity 654 $Å^4$/amu). On the other hand, the lower wavenumber region contains five strong Raman bands, the most intense of which being the doubly-degenerate CH_2 twist mode (experimental wavenumber 1220 cm^{-1}, calculated wavenumber 1238 cm^{-1} and intensity 2×46 $Å^4$/amu). The other visible modes are the fully symmetric CC stretch (breathing) mode (experimental wavenumber 757 cm^{-1}, calculated wavenumber 759 cm^{-1} and intensity 27 $Å^4$/amu), a CC stretch/CCC bend mode (experimental wavenumber 971 cm^{-1}, calculated wavenumber 985 cm^{-1} and intensity 3×12 $Å^4$/amu), a CH_2 rock/CH wag mode (experimental wavenumber 1097 cm^{-1}, calculated wavenumber 1126 cm^{-1} and intensity 3×4 $Å^4$/amu) and finally a CH_2 scissor mode (experimental wavenumber 1435 cm^{-1}, calculated wavenumber 1469 cm^{-1} intensity and 2×29 $Å^4$/amu).

Diamantane crystallizes in the space group T_h^6 at room temperatures with four molecules per primitive cell occupying sites of symmetry S_6. The isolated molecule has symmetry D_{3d}, with 96 vibrational modes, but only 27 Raman active vibrational modes. In a close inspection of the experimental and calculated spectra (Fig. 3.12), it is possible to observe 22 visible Raman signals. Comparison between adamantane and diamantane reveals similar broad high wavenumber CH stretch peaks and the sharper lower wavenumber peaks. In addition, on the CH stretch region it can be observed that both adamantane and diamantane show six Raman signals, produced by similar nuclear displacements, but in diamantane only two modes are close enough to be unresolvable, and hence giving the extra peak in the spectrum. On the other hand, a reasonable correlation between the peak positions in the spectra of adamantane and diamantane seems to take place at the lower wavenumber range, but there are more peaks present in the diamantane spectrum. For example, the single peak at 1220 cm^{-1} in the adamantane spectrum is replaced by two signals in the diamantane spectrum. The nuclear displacements associated with these vibrations consist of CH_2 twisting motions strongly mixed with CH wagging modes.

Triamantane has 120 vibrational modes and space group D_{2h}^{24} with four molecules per primitive cell. The isolated molecules have symmetry C_{2v} and their site symmetry is C_2. It is interesting to note that the addition of another adamantane unit to diamantane reduces the molecular symmetry to C_{2v}, so all modes are now Raman active, producing 120 possible signals in the Raman spectrum (Fig. 3.12). From adamantane to diamantane the addition of one unit had little effect, and in both only some limited symmetries were Raman active. Thus, the increase in triamantane molecule size and decrease in its symmetry means having 20 intense CH stretch vibrations within the 100 cm^{-1} range, which produces a very poorly resolved experimental signal.

On the other hand, in the lower frequency region, the agreement between the experimental and calculated data appears to decrease. It can be observed that for adamantane and diamantane only the major inconsistencies in the calculated intensities were the underestimation of the breathing modes (757 cm^{-1} in adamantane, 708 cm^{-1} in diamantane) and the overestimation of the CH_2 scissor modes (1435 cm^{-1}) relative to the ~1220 cm^{-1} CH_2 twist modes. As the complexity of the Raman spectra increases, the deficiencies in the calculated intensities become more prominent. Thus, the calculated spectrum for triamantane still shows the same discrepancies as the adamantane and diamantane calculations. However, there is also disagreement in the intensity of vibrations in the region around ~1100–1300 cm^{-1}. This type of disagreement is not completely unexpected for calculations performed on molecules of this size due to the limitations in the size of the basis set (Bistricic *et al.*, 2002). The most prominent signals in the low wavenumber region are two close peaks at 1197 and 1222 cm^{-1} similar to those observed in the diamantane spectrum both of which are assigned to CH wag motions, and the 681 cm^{-1} cage deformation (CCC bend, CC stretch) which, again, has an analogous peak in diamantane. In addition to these modes, there is an abundance of weaker modes throughout the 400–1500 cm^{-1} region produced by the larger structure of triamantane and the absence of symmetry-forbidden vibrations (Filik *et al.*, 2006).

The addition of one adamantane unit to triamantane can occur in four distinct ways, producing two distinct molecules and a pair of enantiomers. Further addition of an adamantane unit to tetramantane molecules produces several pentamantane derivatives.

For the sake of simplicity, in the next two sections, we present Raman studies reported by Filik *et al.* (2006) for only [121]tetramantane and [123]tetramantane, [1212] pentamantane and [1234]pentamantane as well as [12312]hexamantane and [121321]heptamantane.

3.2.3.3. *Raman spectra of* [121]*tetramantane and* [1212] *pentamantane*

[121]Tetramantane ($C_{22}H_{28}$, C_{2h} point group), trivially called "*anti-tetramantane*", has 144 vibrational modes, producing a maximum of 72 Raman fundamental signals (Fig. 3.13). The majority of the remaining polymantanes are large (≥ 50 atoms) with low symmetry (≤ C_{2v}). Therefore, their CH stretch and CH_2 twist/CH wag modes are ignored due to the growing complication in extracting information from the increasingly large number of peaks in these regions. Instead, the focus is on the very low wavenumber modes (sub 800 cm^{-1}) as these are still easily resolvable and should be characteristic of the unique structural features of each diamondoid.

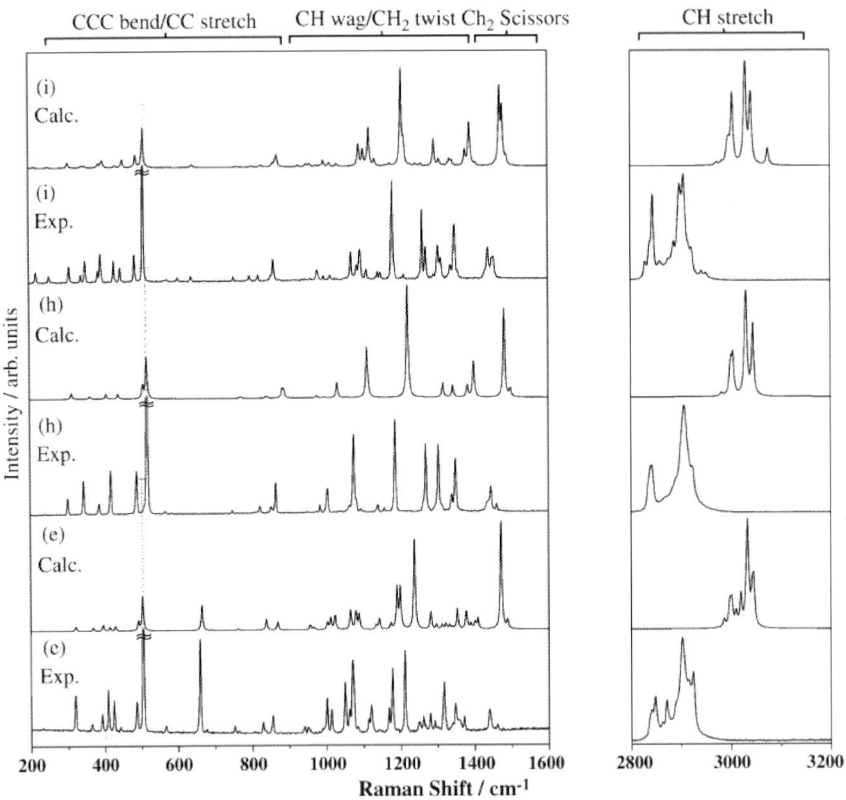

Fig. 3.13. Experimental and calculated Raman spectra for tetramantane (d) and [1212] pentamantane (g). Spectra are normalized to the most intense signal at ~1200 cm^{-1}. Higher intensity peaks are sown as dashed lines (Filik *et al.*, 2006).

Note that diamondoids are, in general, identifiable by using this lower wavenumber region as their molecular fingerprint. Adamantane through triamantane exhibit only one very intense mode below 800 cm^{-1}, so can be distinguished by the specific wavenumber of each mode. All others should be identifiable by a few relative peak intensities and positions.

[121]Tetramantane exhibits a very strong peak at 680 cm^{-1}. By closely looking at the area around this peak and comparing with diamantane and triamantane, it can be observed that there is always a very weak accompanying lower peak ~50 cm^{-1}. The appearance of this pair of peaks is attributed to the rod-shaped structure of diamondoids whose prefix contain only the alternating numbers one and two, as [121]tetramantane and [1212] pentamantane (Filik *et al.*, 2006). Note that for the series of molecules, diamantane, triamantane and [121]tetramantane, (b–d) in Fig. 3.1, an adamantane unit is added to the end of an earlier structure, it has the effect of "growing" the molecules in one direction. On the other hand, the nuclear displacements from the DFT calculations show that again the ~680 cm^{-1} vibration is the same breathing mode across a single adamantane unit, and its accompanying peak is also due to a CCC bend/CC stretch deformation at a different angle, as shown in Fig. 3.14, so both can only take place when the molecule possesses rod-like structure.

[1212]pentamantane (rod-shaped) is obtained by the addition of another adamantane unit onto the end of [121]tetramantane, and it has 168 vibrational modes all being Raman active due to the C_{2v} point group (Fig. 3.13). A close inspection at lower wavenumber regions of the two previous molecules, triamantane and [121]tetramantane, it is apparent that apart from some extra structure around the 426 cm^{-1} peak in [121]tetramantane, the only obvious difference is the relative intensity of the peaks at ~680 cm^{-1} and ~500 cm^{-1}, with the former being far more intense in triamantane. Moreover, the spectrum of [1212]pentamantane also satisfies the rule mentioned above, as it has a peak at ~680 cm^{-1} with a weaker companion lower peak at ~50 cm^{-1}, which is characteristic of a rod-like structure.

It is interesting also to observe the occurrence of a down-shifting in wavenumber of the 352 cm^{-1} peak present in [121]tetramantane to 325 cm^{-1} in [1212]pentamantane. Calculations show that this mode involves stretching of the molecule in the "growth" direction, thus, a reduction in wavenumber is expected as the molecules becomes longer. Nevertheless, this mode is surprisingly absent or weak in triamantane compared to [121]tetramantane

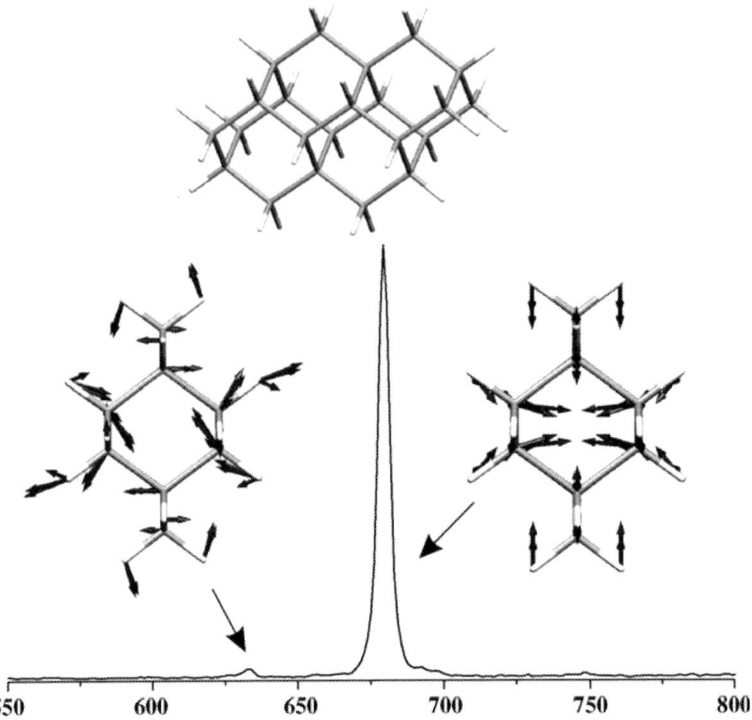

Fig. 3.14. Nuclear displacements of the 680 cm^{-1} vibrational mode and its companion mode in [121]tetramantane (Filik *et al.*, 2006).

and [1212]pentamantane. The wavenumber of this particular mode may be a good indication of length of these linear-structure (rod-shaped) diamondoids. The Raman spectra of [1(2)3]tetramantane, [123]tetramantane, [1234] pentamantane, [1(2,3)4] pentamantane, 3-methyl-[1(2,3)4]pentamantane, [12(3)4]pentamantane, [12(1)3]pentamantane and [1213]pentamantane are also discussed by Fillik *et al.* (2006).

3.2.3.4. *Raman spectra of* [12312]*hexamantane and* [121321]*heptamantane*

[12312]Hexamantane ($C_{26}H_{30}$, D_{3d}), trivially called cyclohexamantane, has 162 vibrational modes of which only 44 modes are Raman active. The highest intensity mode in the lower frequency region is again a mixed CH wag/

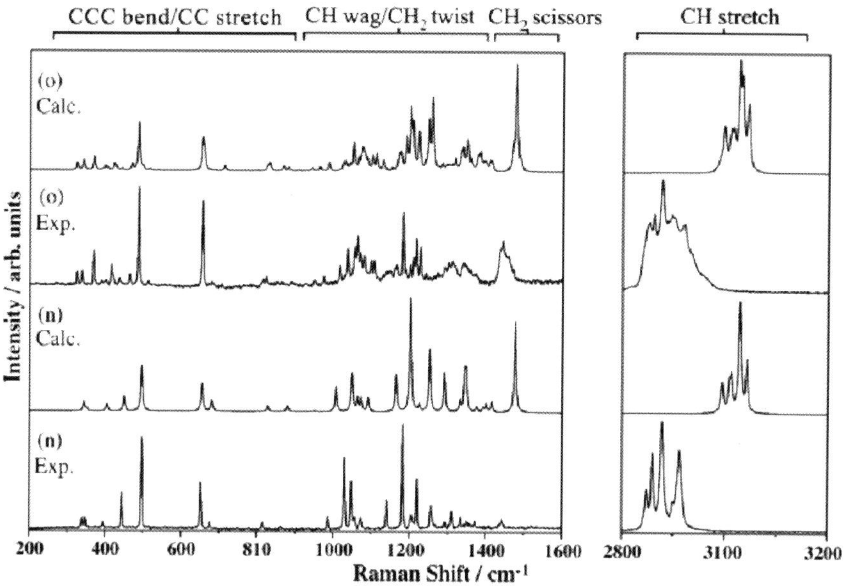

Fig. 3.15. Experimental and calculated Raman spectra for [12312]hexamantane (n), and [121321]heptamantane (o) (Fillik *et al.*, 2006).

CH_2 twist mode at ~1200 cm^{-1}, as seen in Fig. 3.15, which is similar to those observed in the higher symmetry molecules. On the other hand, the fingerprint region again has a peak at 651 cm^{-1} and 498 cm^{-1} due to the same CCC bend/CC stretch deformations parallel to and perpendicular to the three-fold rotational axis as observed before in [1(2)3]tetramantane. [12312] Hexamantane has a two-dimensional isotropic disc-shaped polymantanes structure, and that is clearly reflected in its Raman spectrum (Fig. 3.16).

[121321]Heptamantane ($C_{30}H_{34}$, C_s), in turn, is produced by addition of another adamantane unit to [12312]hexamantane. This molecule has 186 vibrational modes all of them Raman active (Fig. 3.15). It is worth mentioning that the Raman spectra of [12312]hexamantane and [121321] heptamantane are reasonably similar. Note also that there are many more peaks in the [121321]heptamantane spectrum but it does not have an extra 142 Raman active vibrational mode in relation to [12312]hexamantane. So this is not surprising. It might be expected that as the diamondoid molecules become larger the addition of extra adamantane units would have less effect upon their Raman spectrum, until the only signal is the 1332 cm^{-1} phonon of the diamond crystal. But that is much greater than

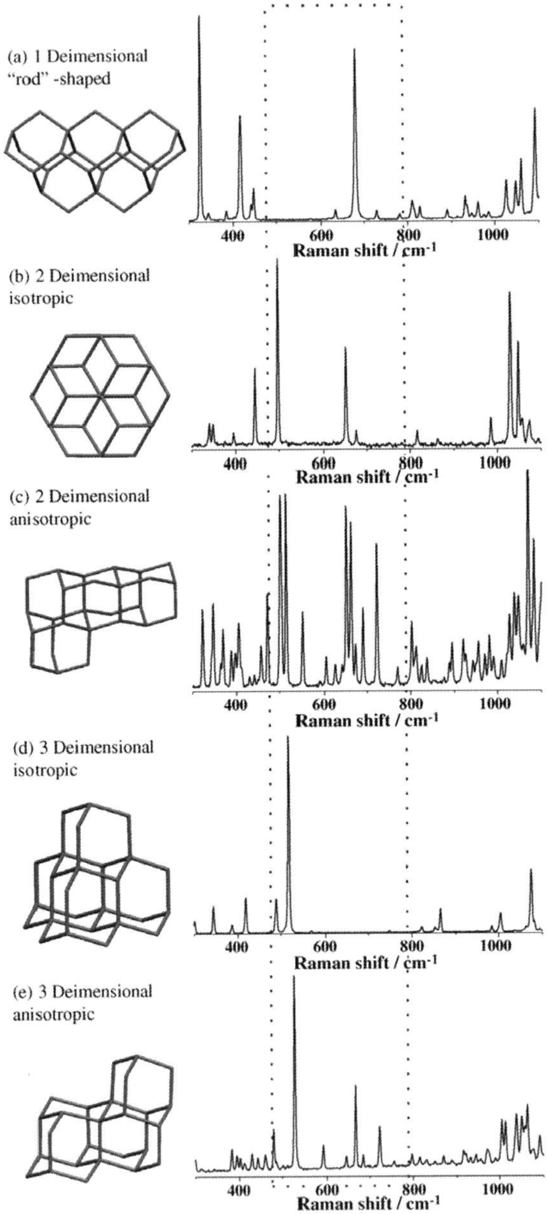

Fig. 3.16. Structural fingerprint region of: (a) one-dimensional rod-shaped [1212]pentamantane; (b) two-dimensional isotropic disc-shaped [12312]hexamantane; (c) two-dimensional anisotropic [1213]pentamantane; (d) three-dimensional isotropic [1(2,3)4] pentamantane; (e) three-dimensional anisotropic [1234]pentamantane (Filik *et al.*, 2006).

the size of the molecules dealt with here. [121321]Heptamantane has low symmetry, but it has the same strong vibrational modes in the fingerprint region as [12312]hexamantane and is also two-dimensional.

3.2.3.5. *Infrared spectroscopy of diamondoids*

In this section, we present the infrared spectra for a number of diamondoid molecules, as measured using the ATR-FTIR (Attenuated Total Reflectance-Fourier Transform Infrared) spectroscopy technique and an analysis based on DFT calculations of their vibrational spectra. ATR-FTIR spectroscopy is a useful technique to reveal details about the molecular structure of diamondoids by producing their vibrational spectra. ATR is the technique which enables samples to be examined directly in the solid or liquid state without special preparation. In ATR-FTIR, the infrared spectrum of a liquid or solid sample is recorded by passing a beam of infrared light through the sample. Fourier transform instrument is used to measure all wavelengths at once. From this, a transmittance or absorbance spectrum can be produced, showing at which IR wavelengths the sample absorbs.

The systems investigated are adamantane, diamantane, triamantane, [121]tetramantane, [123]tetramantane, [1(2,3)4]pentamantane, [12(3)4] pentamantane, [12312]hexamantane, and 3-methyl-[1(2,3)4]pentamantane. Oomens *et al.* (2006) reported this study using ATR-FTIR technique with the exception of adamantane. Adamantane in gas-phase data is taken from the NIST Web Book (webbook.nist.gov) and it is reported in Fig. 3.17. All spectra were obtained from pure diamondoid samples in solid powdered forms, and measurements were performed at room temperature. In addition, the intensities of the ATR-FTIR absorbance spectra are corrected for the penetration depth of the evanescent (exponentially vanishing) wave, which is proportional to the wavelength. Geometry optimization and computation of the harmonic vibrational frequencies were performed using the DFT/B3LYP approach and Dunning´s double zeta basis set with polarization functions 95(d,p). This method has been shown to give reliable results for numerous organic compounds, such as a large variety of polycyclic aromatic hydrocarbons (Bauschlicher *et al.*, 1997).

Figures 3.18–3.25 show infrared spectra of diamantane through 3-methyl-[1(2,3)4]pentamantane, compared "back-to-back" with the

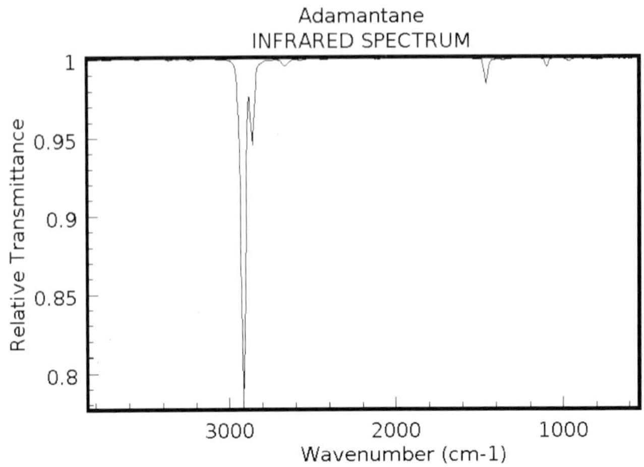

Fig. 3.17. Infrared spectrum of adamantane in the gas phase (http://webbook.nist.gov).

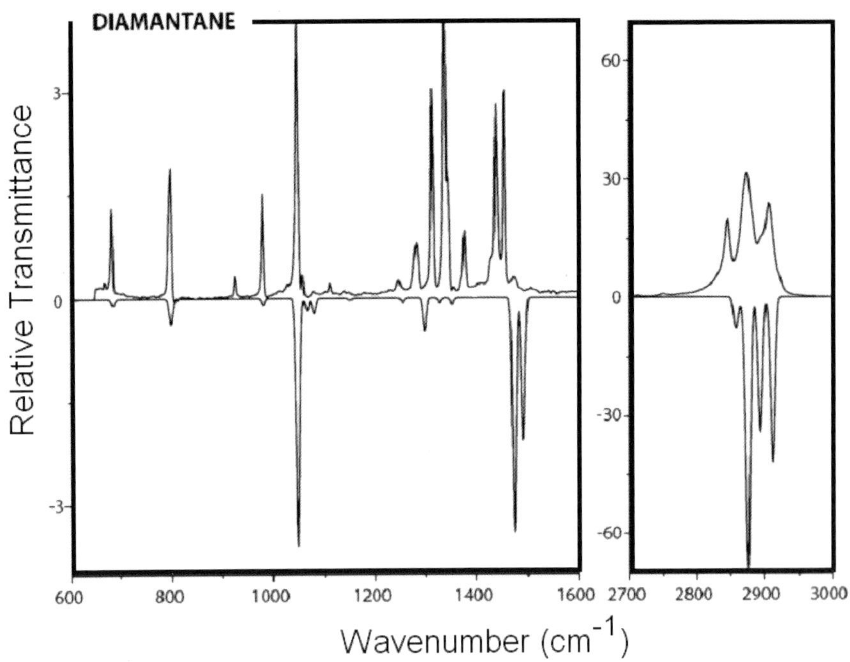

Fig. 3.18. Infrared spectrum of diamantane (Oomens *et al.*, 2006).

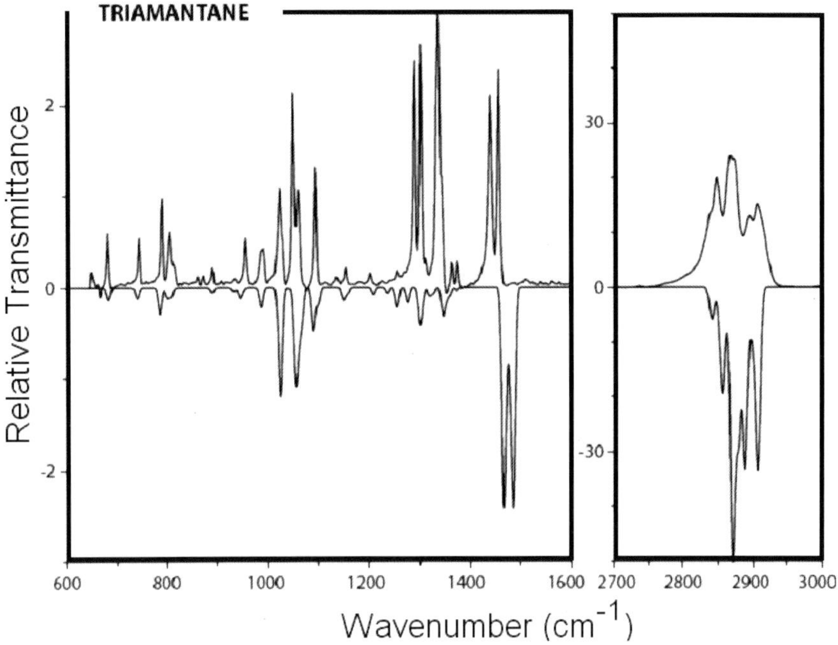

Fig. 3.19. Infrared spectrum of triamantane (Oomens *et al.*, 2006).

spectra calculated using DFT/B3LYP theory. The infrared spectra of the more symmetric species are much simpler relative to those of the asymmetric ones, as expected, even though they are larger in size. This can be observed, for instance, by comparing the spectra of [123]tetramantane (C_2) (Fig. 3.21) with those of [1(2,3)4]pentamantane (T_d) (Fig. 3.22) or [12312]hexamantane (D_{2d}) (Fig. 3.24). On the other hand, it can be noted that the spectra of two different molecules belonging to the same symmetry group can be quite different. Compare, for example, the spectra of adamantane (Fig. 3.17) and [1(2,3)4]pentamantane (Fig. 3.22) (both T_d symmetry) or diamantane (Fig. 3.18) and [12312]hexamantane (Fig. 3.24) (both D_{3d} symmetry). Nevertheless, despite the dense spectrum for the lower symmetric diamondoids, the experimental *versus* theoretical match in the mid-infrared frequency range is usually acceptable. Note, for instance, the spectra of [123]tetramantane (Fig. 3.21) and [12(3)4]pentamantane (Fig. 3.23), recorded at higher resolution because of the highly

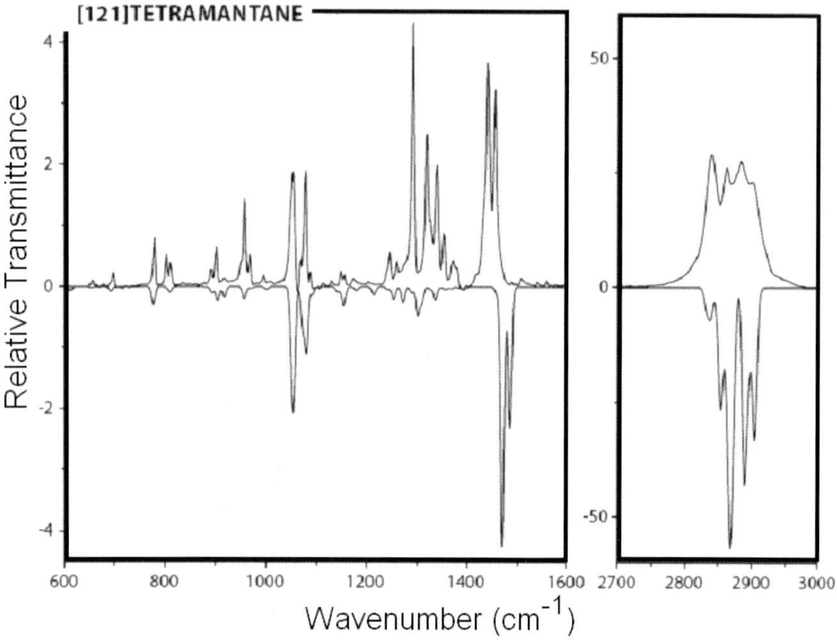

Fig. 3.20. Infrared spectrum of [121]tetramantane (Oomens *et al.*, 2006).

congested mid-infrared spectrum, present good agreement with the DFT/B3LYP calculation.

A closer inspection of all the spectra reveals clearly large similarities between them. The general appearance of the spectra could be considered as the spectral fingerprint of the diamondoid class of molecules. In this analysis, four distinct groups of bands can be recognized in all diamondoids spectra. Visualization of the calculated vibrational displacements allows one to roughly assign a general mode description pertinent to the four groups, as seen in Table 3.2.

A group of modes in all spectra, appearing as a doublet around 1450 cm^{-1}, is attributed to two or more CH_2 scissoring modes. In [123]tetramantane, the low symmetry causes further splitting into about four bands, whereas in adamantane and [12312]hexamantane only a single band was observed. The latter, however, the computational calculation predicts three bands with nonvanishing intensity within 1 cm^{-1}. Note that in all cases the

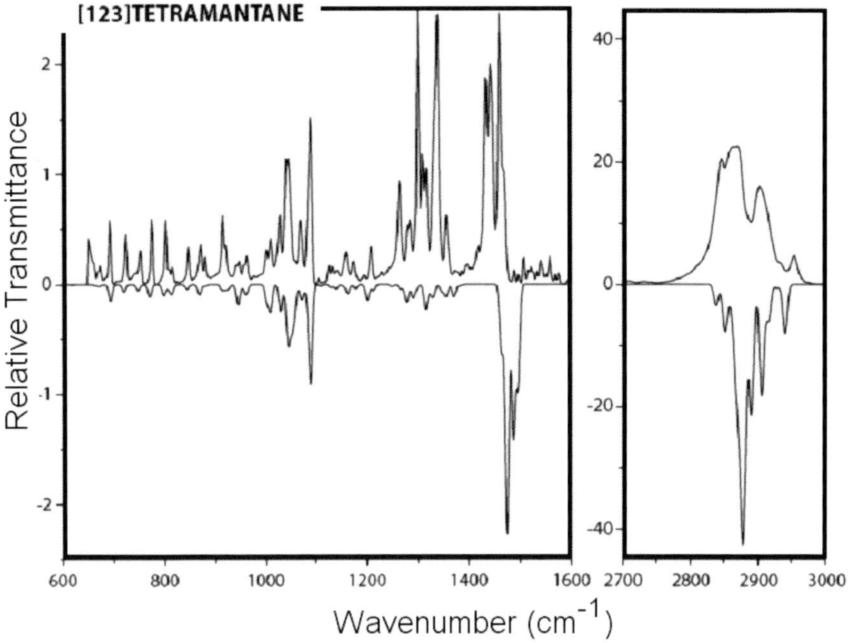

Fig. 3.21. Infrared spectrum of [123]tetramantane (Oomens *et al.*, 2006).

bands are well reproduced by theory, except for a consistent blue shift of the calculated frequencies of around 35 cm^{-1}.

The bands ranging between 1000 and 1400 cm^{-1} have mostly CH bending and CH_2 rocking, wagging, and twisting character. Toward the red end of this group, some CC stretching character is found. Finally, in the long wavelength part of the spectra, below 1000 cm^{-1}, mostly skeletal deformation modes are found and these modes have been classified as belonging to Group D in Table 3.2. Comparing the bands in Groups C and D to the DFT calculations, a striking observation is the underestimation of their intensities (except some around 1050 cm^{-1}) by the DFT calculations. The reason for this discrepancy, that appears to be a general feature in all diamondoid spectra, is at the moment unclear. A possible influence of the basis set used was investigated for the [121]tetramantane molecule (see Fig. 3.26). However, the underestimation of intensities in Groups C and D is consistently reproduced upon using basis sets ranging in size from 3-21G–cc-pVDZ.

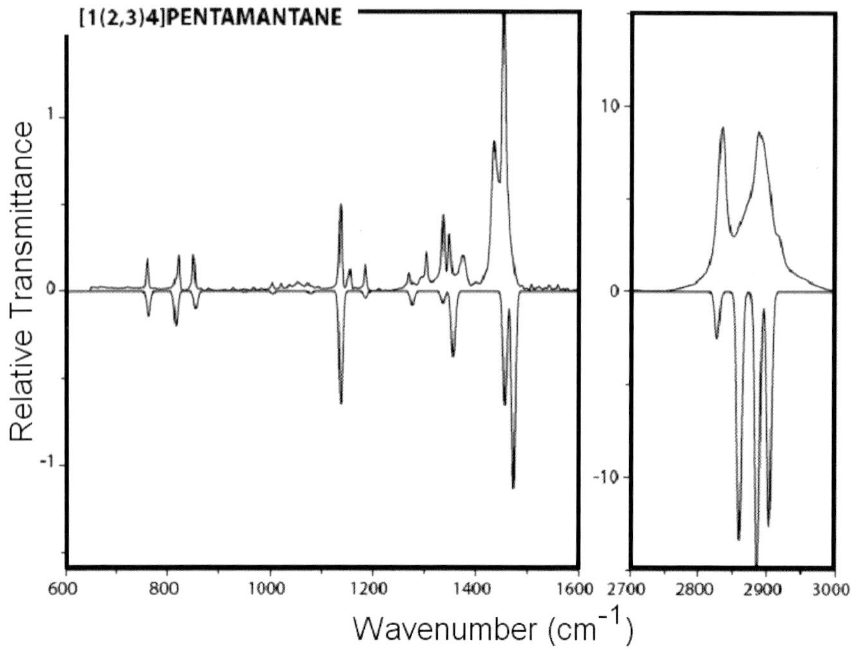

Fig. 3.22. Infrared spectrum of [1(2,3)4]pentamantane (Oomens *et al.*, 2006).

In summary, a global inspection of the spectra shows that the match between theoretical and experimental frequencies is reasonable for most of the bands in the mid-infrared range. Nevertheless, this does not apply to the relative intensities, which are not so well reproduced. Although intensities were corrected for the wavelength dependency of ATR, there remains a significant discrepancy (a factor of ≈ 3) between the experimental and computed intensity ratios for the mid-IR versus the 3 μm spectral ranges. Judging from the increased linewidth and the more Lorentzian-like lineshape, it seems that this discrepancy results at least partly from saturation in the 3 μm range (Oomens *et al.*, 2006). However, it is well known that while DFT calculation generally provides a good prediction of the frequencies, intensities are not so reliably predicted. Moreover, interaction of the molecules with their slid environment may affect the intensities differently for different vibrational modes. Indeed, for a number of polyaromatic molecules, Joblin *et al.*, (1994) found that

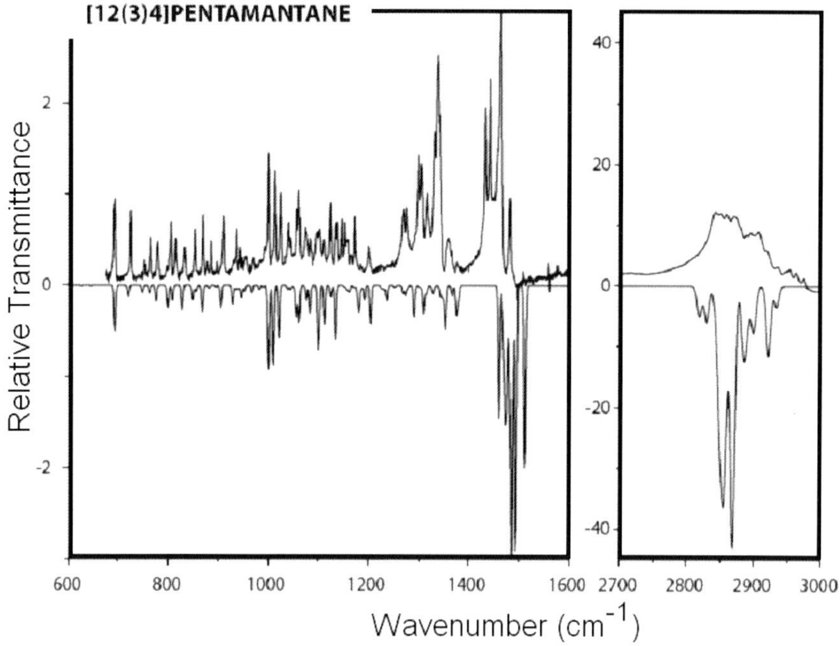

Fig. 3.23. Infrared spectrum of [12(3)4]pentamantane (Oomens *et al.*, 2006).

the CH stretching modes were attenuated by a factor of three in the solid-phase with respect to the gas-phase. Although such an effect would indeed explain here the observed discrepancies, only when gas-phase diamondoid spectra become available can such a conclusion could be truly substantiated.

Figure 3.25 displays the infrared spectrum of 3-methyl-[1(2,3)4]pentamantane, which is obtained by substitution of an H atom by a CH_3 group, causing, thus, a reduction in symmetry from T_d to C_{3v} in this modified diamondoid. We can observe some significant differences on general appearance of these diamondoid spectra. In the CH stretching region around 3 μm, the calculation by Oomens *et al.* (2006) produced CH stretching modes localized on the methyl group at 2930.1 cm^{-1} and 2860.5 cm^{-1}, corresponding to the CH_3 stretching modes parallel and perpendicular to the threefold symmetry axis, respectively. These modes can be well observed in the experimental spectrum, particularly when it is

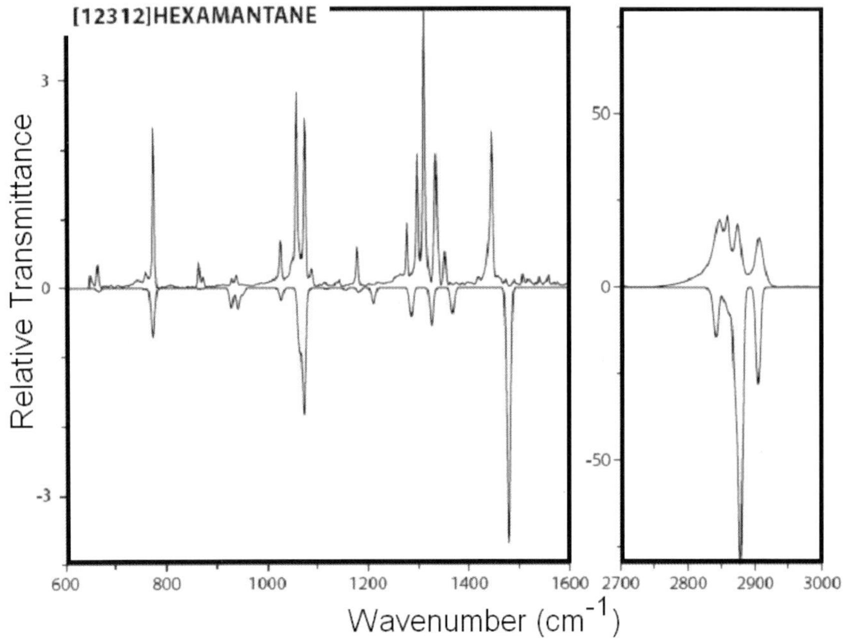

Fig. 3.24. Infrared spectrum of [12312]hexamantane (Oomens *et al.*, 2006).

compared with the experimental spectrum of the nonmethylated [1(2,3)4] pentamantane (see inset in Fig. 3.25). Though weak, additional bands are observed for the methylated species at 2857 cm^{-1} and ~2945 cm^{-1}. At the red end of the CH stretching manifold, there appears to be another additional band at ~2820 cm^{-1}, but the calculation indicates that this band is due to the three degenerate CH stretching modes in the (111) bulk lattice plane which are red-shifted by about 10 cm^{-1} in the methylated molecule with respect to the nonmethylated one.

Comparison of the 3-methyl-[1(2,3)4]pentamantane spectrum (Fig. 3.25) with that of [1(2,3)4]pentamantane (Fig. 3.22) clearly shows that additional bands appear (see the 3 μm spectrum in Fig. 3.25 and Fig. 22), but inspection of the corresponding normal modes indicates that those additional bands are due to nonlocalized modes. Therefore, it appears that these differences are mainly induced by the reduction of symmetry from T_d

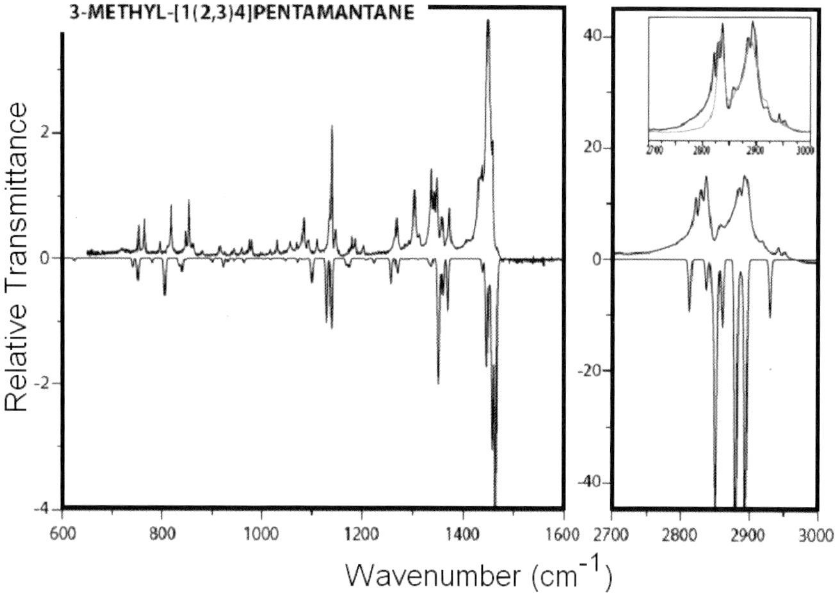

Fig. 3.25. Infrared spectrum of 3-methyl-[1(2,3)4]pentamantane. A comparison with the 3 *μm* spectrum of its nonmethylated counterpart is shown in the inset (Oomens *et al.*, 2006).

Table 3.2. General assignment pertinent to diamondoid molecules studied in four distinct spectral ranges.

Group	Range (cm^{-1})	General description
A	2,900	CH stretch
B	1,450	CH_2 scissor
C	1,000–1,400	CH_2 rock, wag, twist
D	≤ 1,000	Skeletal deformation

([1(2,3)4]pentamantane) to C_{3v} (3-methyl-[1(2,3)4]pentamantane). Some local modes of the methyl group are computed at 1369.7 cm^{-1} (CH_3 umbrella) and 1457.0 cm^{-1} (CH_3 scissor), but they are not easily recognized in the experimental spectrum, probably due to overlapping diamondoid bands (Oomens *et al.*, 2006).

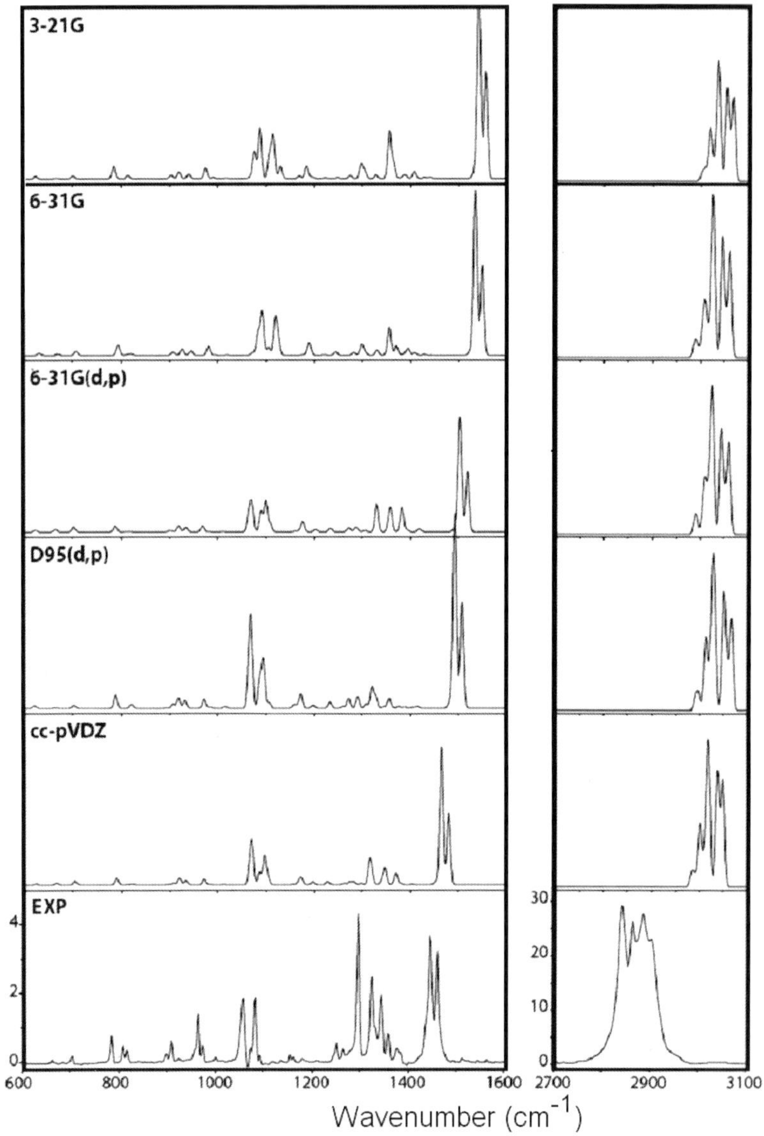

Fig. 3.26. Infrared spectrum of [121]tetramantane compared to (unscaled) spectra using different basis set (Oomens *et al.*, 2006).

3.3. Optical Properties

3.3.1. *Diamondoids as electron photoemitters*

Since the discovery of diamondoids in petroleum, together with selective functionalization techniques (Dahl *et al.*, 2003; Tkachenko *et al.*, 2006; Schrener *et al.*, 2006, Mansoori, 2007), diamondoids have attracted interest because of their potential to combine the properties of diamond and nanomaterials. In addition to the optical, mechanical, and thermal properties of bulk diamond, hydrogen-terminated diamond surfaces have negative electron affinity (NEA). The electron affinity of an atom or molecule is defined as the change in its energy when an electron is added to it producing a negative ion. NEA materials are useful for the development of electron emitters like the design of efficient cathodes that can supply electrons to the vacuum with little energy loss (Paoletti and Tucciarone 1997).

The difficulties in emission uniformity, electron injection, and electron transport have hindered the use of bulk diamond for this purpose. The primary challenge for electron emitters remains to find a material that would realize uniformly large, highly efficient, and highly monochromatic electron emission (Yang *et al.*, 2007). Thus, diamondoids are interesting candidates for electron emission, since they are essentially fully hydrogen-terminated diamond clusters. Quantum Monte Carlo (QMC) and Density functional theory (DFT) calculations (see Chapter 5) predict NEA properties for diamondoids up to 1 nm in size (Drummond *et al.*, 2005; Wang *et al.*, 2008).

In this section, we will present some experimental results reached by Yang *et al.* (2007) about NEA features of [121]tetramantane-6-thiol assembled on Ag or Au substrates, which yields large-area SAMs. Photoemission spectroscopy (PES) and near-edge X-ray absorption fine structure (NEXAFS) measurements were performed to analyze electron-emission properties and the molecular orientation of [121]tetramantane-6-thiol SAMs, respectively. Figure 3.27 shows the total electron yield NEXAFS spectra of [121]tetramantane-6-thiolate on Au. These spectra resemble those from bulk and gas-phase [121]tetramantane, indicating a high-purity film of tetramantanethiol absorbed on the surface. The angular

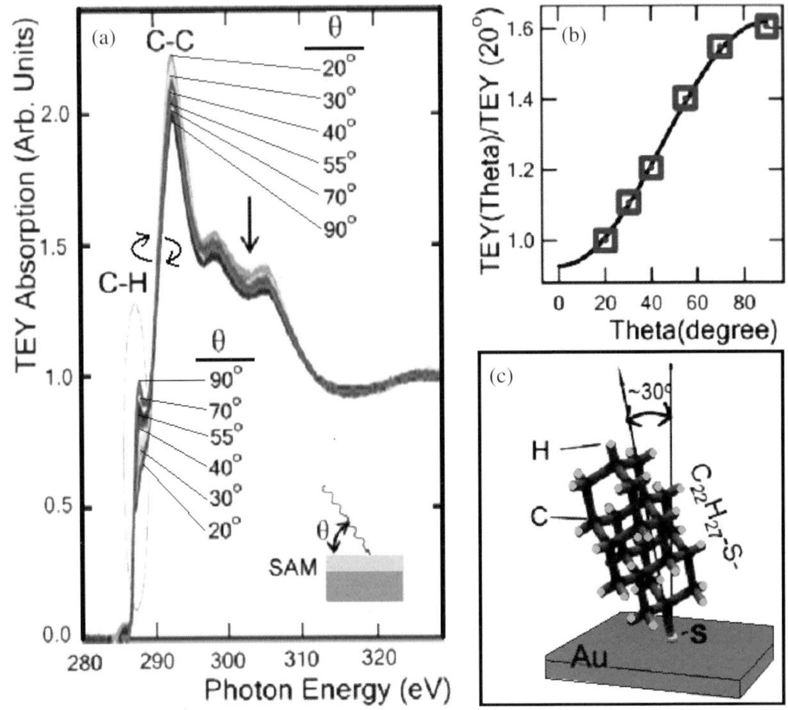

Fig. 3.27. (a) Polarization-dependent total electron yield NEXAFS spectra collected on [121]tetramantane-6-thiol SAMs prepared on Au. Total electron yield (TEY) is plotted against photon energy at different beam incident angles. The leading-edge peak at 287.6 eV (oval) is assigned to transitions from the C 1s core level to the unoccupied (C–H) σ* orbitals, and the peak at about 292.5 eV is assigned to (C–C) σ* orbitals (Stöhr, 1996; Willey *et al.*, 2005). The second gap indicated by the arrow is the characteristic signature of diamondoids (Willey *et al.*, 2005). (b) Comparison between experiments and theoretical simulations. Squares represent the experimental ratio of (C–H) σ* spectral weight between data at different angles and that at 20°. The solid line is the calculated ratio based on the molecular geometry as shown in (c), with a polar angle of 36.5° (Yang *et al.*, 2007).

dependence seen in the NEAXAFS spectra implies that the tetramantanethiol forms well-ordered monolayers with a preferential orientation. Absorption intensity of a NEAXAFS resonance from a C 1s orbital depends directly on alignment between the linearly polarized incident X-rays and the transition dipole moment into a particular unoccupied orbital (Stöhr, 1996; Hähner, 2006). The angular dependence is simulated by summing over each atomic center with transition dipole moments oriented along the bonds (Stöhr, 1996; Hähner, 2006). Thus, these

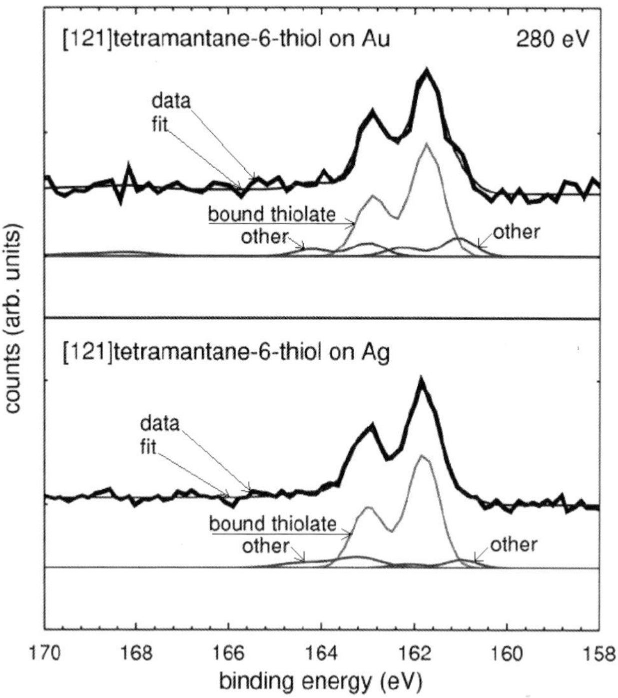

Fig. 3.28. The S 2p core-level X-ray photoelectron spectra (XPS) of [121]tetramantane-6-thiol SAMs on Au (top panel) and on Ag (bottom panel). In this figure, "data" lines are experimental results. "fit" lines are least-square fittings to the data based on Au-thiolate and Ag-thiolate sulfur. "other" lines constitute S species, unbound elemental and oxidized sulfur, which only have insignificant contributions (www.sciencemag.org/cgi/content/full/316/5830/1460/DC1).

simulations are in agreement with the experimental data (Fig. 3.27b) when the tetramantane-6-thiol, with the z axis of the molecule defined by the S–C bond, is tilted b 30 ± 10° from the surface normal (Fig. 3.27c). In addition, the affinity of the thiol for the metal leads to thiolate-bound monolayers (Ulman, 1996; Shaporenko *et al.*, 2005; Love *et al.*, 2005), as confirmed by S 2p core-level XPS (Fig. 3.28) (Castner *et al.*, 1996; www.sciencemag.org/cgi/content/full/316/5830/1460/DC1).

Figures 3.29 and 3.30 depict Photoemission spectra (PES) of [121] tetramantane-6-thiol SAMs grown on Ag and Au substrates, respectively as reported by Yang *et al.* (2007). An emission peak appears for both surfaces at about 1 eV kinetic energy, the onset of the spectra at low kinetic

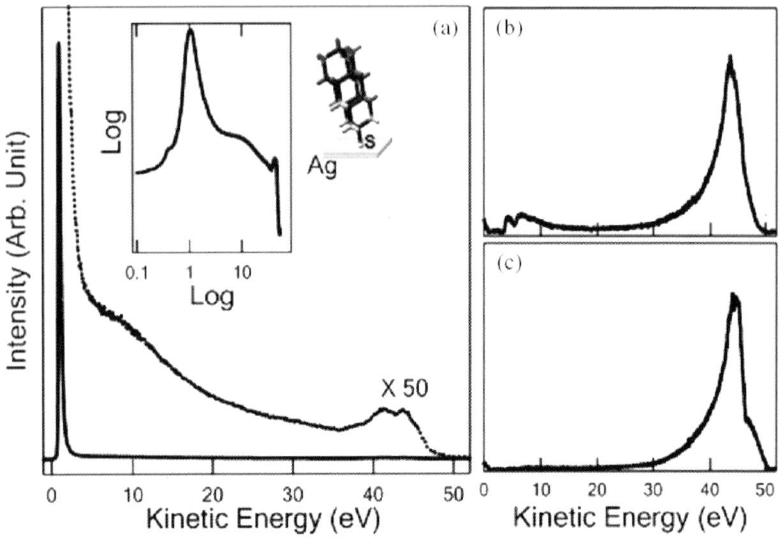

Fig. 3.29. (a) Photoelectron spectra of [121]tetramantane-6-thiol SAM grown on Ag substrates, collected with 55 eV photon energy. The strong peak at 1 eV kinetic energy contains 68% of the total photoelectrons. The dotted line is a 50 times enlargement of valence band features. The inset shows the same spectra in a double logarithm plot. (b) Photoelectron spectrum collected on [121]tetramantane-6-thiol SAM covered by *in situ* sublimed fullerene. The strong electron-emission peak disappears after fullerene coverage. (c) Photoelectron spectrum collected on an annealing to 550°C. The difference between the spectrum in (c) and a pure Ag PES spectrum could be partially due to residual S atoms still bound to the surface after annealing (Yang *et al.*, 2007).

energy. The intensity of the peaks exceeds all the valence band features. For SAMs grown on Ag and Au, the sharp peak comprises about 68% and 17% of the total electron yield, respectively. Note that this peak intensity is several times as strong as that found for hydrogen-terminated diamond surfaces. Even with a logarithm plot (insets), one can still see a sharp feature instead of the typical exponential decay of secondary electrons in this energy range. Two different techniques to cover or remove the monolayer *in situ* were applied, just to make sure that this unusual electron emission originates from the diamondoid monolayers. Coating the diamondoid SAM with one monolayer of fullerene (C_{60}), which was evaporated onto the SAM surface, caused the sharp emission feature to vanish (Fig. 3.29b). The valence band of the fullerene covered surface is neither from tetramantane-6-thiol nor from fullerene (Yang *et al.*, 2003), but the

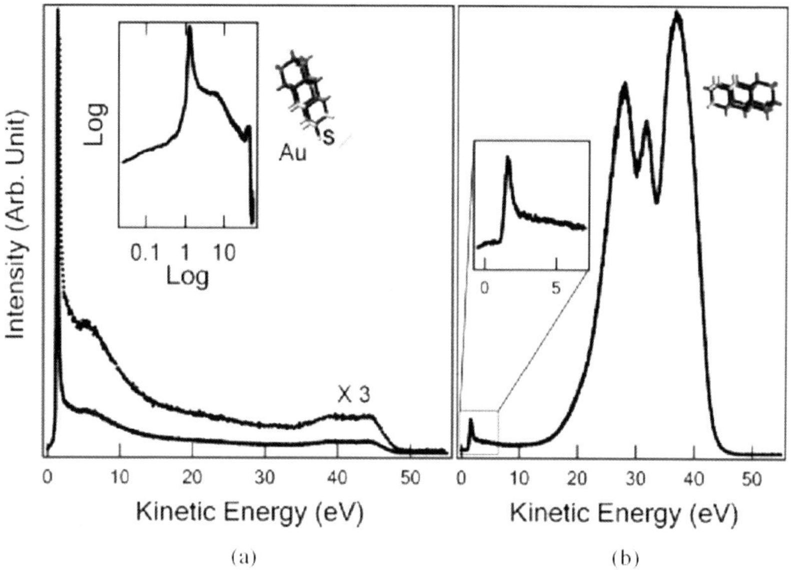

Fig. 3.30. (A) Photoelectron spectra of [121]tetramantane-6-thiol SAM grown on Au substrate. The sharp peak at 1 eV contains about 17% of the total photoelectrons. The inset shows a double logarithm plot. (B) Photoelectron spectra of unsubstituted [121]tetramantane films prepared *in situ* on Au substrate. The inset is an enlargement of the low-kinetic energy part of the spectrum showing only a small peak (Yang *et al.*, 2007).

origin of these features is not clear at present. Note that the diamondoid SAM was also removed by *in situ* annealing a SAM sample to 550°C. Since the thermal stability of conventional alkane thiol SAMs is ~70°C (Bain *et al.*, 1989), this treatment should remove the diamondoid, and as shown in Fig. 3.29c, the low-kinetic energy peak completely disappears after annealing. The importance of a monolayer of functionalized diamondoid versus a thin film of diamondoid is concluded by comparing [121] tetramantane-6-thiol SAMs with unsubstituted [121]tetramantane films. Figure 3.30b depicts the PES spectrum of the unsubstituted [121]tetramantane film. The spectrum shows a small peak at low kinetic energy, in sharp contrast with the data for [121]tetramantane-6-thiol SAMs (Fig. 3.30a). Thus two factors may contribute to this difference. One is the poor electron conductivity within the thicker films versus that through the monolayers, and the other is the role that the thiol groups play in the SAM samples. Notably, this result indicates that the strong electron emission

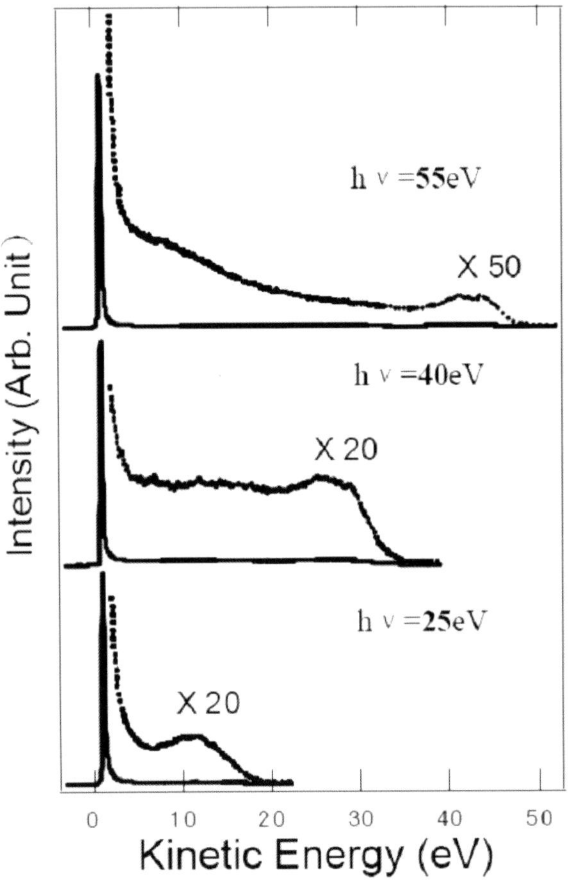

Fig. 3.31. Photoelectron spectra of [121]tetramantane-6-thiol SAM on Ag collected with 25 eV, 40 eV, and 55 eV photon energy. Note that the unusual emission peak at 1 eV persists. A quantitative analysis of the peak intensity upon photon energy is nontrivial because of the unavoidable higher harmonic components of synchrotron photon beam. (www.sciencemag.org/cgi/content/full/316/5830/1460/DC1).

does not occur solely from the diamondoid surface, but that the metal substrate is also intimately involved in the process.

The sharp peak observed on PES spectra of [121]tetramantane-6-thiol SAMs grown on Ag (Fig. 3.29a) or Au (Fig. 3.30a), at 1 eV kinetic energy, remains at the same energy position even varying photon excitation energy (Fig. 3.31). These rules out the possibility of core-level excitations and suggests that sharp feature is not from electrons directly excited by photons, but from electrons accumulated at an intrinsic energy level of the molecules.

Many studies in NEA materials use PES techniques, and a sharp feature at a low-kinetic energy threshold is often evident in the spectra of NEA (Paoletti and Tucciarone, 1997; van der Weide and Nemanich, 1993; James and Moll, 1969; Himpsel *et al.*, 1979; Pate, 1986; Eden *et al.*, 1967). Thus, the emission peak presented here provides direct evidence that certain functionalized diamondoids are NEA materials, consistent with the Quantum Monte Carlo (QMC) and Density Functional Theory (DFT) calculations (Drummond *et al.*, 2005; Wang *et al.*, 2008). Further evidences of NEA features on diamondoid SAMs are observed when potassium (K) metal is slightly evaporated onto the SAM sample. In the PES spectrum of a K-covered [121]tetramantane-6-thiol SAM on an Ag substrate, the sharp peak retains its high intensity and occurs at the same energy position as in the PES spectrum without K (Fig. 3.32). This is another indication of NEA because K deposition onto a positive electron affinity semiconductor will lead to a shift of the low-kinetic energy cutoff and strong enhancement of the secondary electron background.

Note that on a typical NEA surface, electrons excited into unoccupied states relax to the bottom of the conduction band as a result of inelastic scattering, a process normally referred to as the secondary cascade. A number of secondary electrons will then accumulate at the bottom of the conduction band. Thus, for a surface with positive electron affinity (as occurs in almost all untreated semiconductor surfaces), these accumulated electrons cannot escape. For an NEA surface, these accumulated electrons can be emitted directly because the vacuum level lies below the bottom of the conduction band. As a result, a peak will be observed at the low-kinetic energy threshold in PES (Paoletti and Tucciarone, 1997; van der Weide and Nemanich, 1993; James and Moll, 1969; Himpsel *et al.*, 1979; Pate, 1986; Eden *et al.*, 1967).

On the other hand, on diamondoid SAM surfaces, there is only a single layer of diamondoid molecules. The detailed mechanism responsible for their highly monochromatic emission is unknown at this stage. However, one may consider that photoexcited electrons lose energy by creating phonons in the molecules, but this would likely lead to the destruction of the molecules.

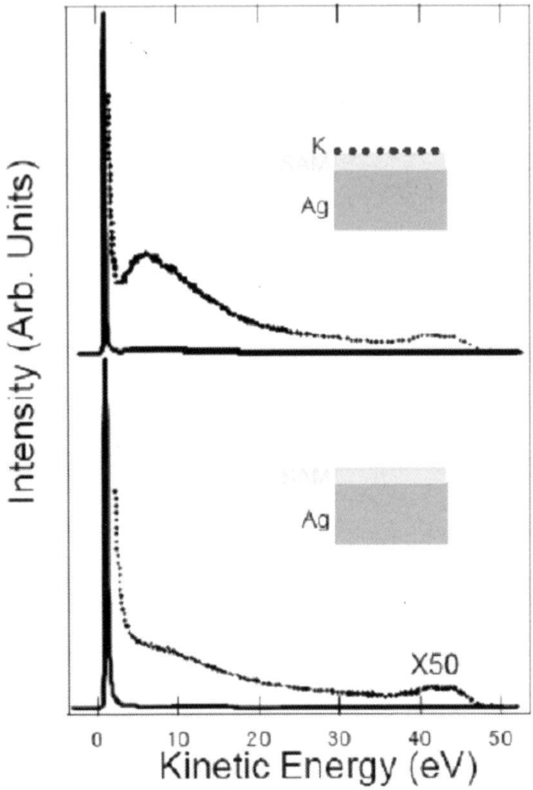

Fig. 3.32. Photoelectron spectra of [121]tetramantane-6-thiol SAM with (upper) and without (lower) potassium deposition. No energy shift was observed for the sharp peak, and only a small increase of the secondary electron background can be seen (www.sciencemag.org/cgi/content/full/316/5830/1460/DC1).

A plausible scenario is that the photoexcited electrons come from the substrate. These electrons first thermalize in the metal, producing many more low-energy electrons. Electrons with energies above the diamondoid conduction-band minimum may get transferred to diamondoid molecules, reach the bottom of the conduction band by creating phonons, and get emitted. This scenario is shown schematically in Fig. 3.33.

Another difference between a typical semiconductor (Paoletti and Tucciarone, 1997; van der Weide and Nemanich, 1993, James and Moll, 1969; Himpsel *et al.*, 1979; Pate, 1986; Eden *et al.*, 1967) and a diamondoid

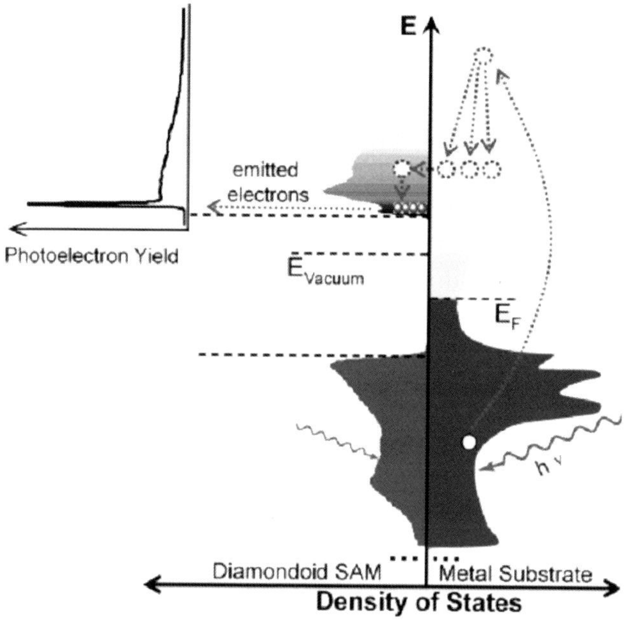

Fig. 3.33. Schematic of the electron-emission process on diamondoid SAM surfaces. E_F is the Fermi level of the metal substrate, sitting in the energy gap of diamondoid. The vacuum level (E_{Vacuum}) is below the conduction-band minimum of diamondoid, a characteristic of NEA. The dotted line depicts the high-probability electron emission. At first, electrons in metal substrate are excited by photons into unoccupied states above E_F. Second, the excited electrons effectively thermalize in the metal, producing more electrons with lower energy. Third, electrons with energy above the conduction band minimum are transferred to diamondoid moieties. These electrons further lower their energies by exciting phonons in the molecules, and they accumulate at the bottom of the conduction band. Finally, because of NEA, electrons accumulated at the bottom of the conduction band emit into vacuum spontaneously and generate a peak at the low-kinetic energy threshold. Electron emission takes place also at high-kinetic energy levels, but with much lower photoelectron yield (Yang *et al.*, 2007).

SAM surface with NEA features is that the latter shows a spike in the spectra rather an exponential rise of the secondary tail toward the threshold, suggesting that a single energy level, resulting from the molecular nature of nanometer-sized diamondoids, and/or a strong resonance process are involved.

In summary, the results reached by Yang *et al.* (2007) suggest that diamondoid monolayers may have promising utility. In addition that functionalized diamondoids can be easily grown into large area SAMs with

NEA properties, they naturally circumvent the long-standing electron-conductivity issues encountered for wide-gap bulk NEA semiconductors (Paoleti and Tucciarone, 1997; Bell, 1973). On a diamondoid SAM surface, electron conduction from the electron reservoir (metal substrates) to the emission surface is through a single molecule, which successfully avoids the low-conductivity problem and enhances the electron emission. Moreover, the possibility of different functionalizations (Schreiner *et al.*, 2006) allows one to optimize the NEA and other properties of diamondoids. Many technical issues, of course, need to be addressed before diamondoid SAMs can be used as electron emitters. Diamondoids provide intrinsic advantages over bulk diamond because of their special molecular characteristics. For instance, narrow energy distribution of the electronic states.

Moreover, investigations made by Wang *et al.* (2008) on diamondoid monolayers on Au substrate, using scanning tunneling microscopy (STM) and spectroscopy, reveal that the diamondoid electronic structure and electron-vibrational coupling exhibit unique and unexpected spatial correlations characterized by pronounced nodal structure across the molecular surfaces. The analysis of individual [121]tetramantane diamondoids was performed using the STM tip to manipulate the diamondoids from the edge of an island onto an empty Au(111) terrace at T = 7K. Figure 3.34(c and d) shows STM topographs of the same individual diamondoid taken with different samples biases. The image taken at sample bias V = +2 V (Fig. 3.34c), which probes the unoccupied local electronic density of states (LDOS), exhibits pronounced line nodes across the molecular surface. In contrast, the image taken at –2 V (Fig. 3.34d) has less pronounced features, revealing a much weaker spatial dependence of the occupied electronic states. On the other hand, [121]tetramantane was found to have a number of different orientations on Au(111).

Figure 3.35a depicts STM images of three tetramantane diamondoids lying on Au(111) with different molecular orientations, as seen by the distinctive spatial patterns of the line nodes. Note that the diamondoids can be switched between molecular orientations by dragging and rotating the tetramantane using the STM tip (Wang *et al.*, 2008). Figure 3.35d shows the calculated (DFT) isosurfaces of the HOMO and LUMO electronic wavefunction square at a value corresponding to 50% of the charge

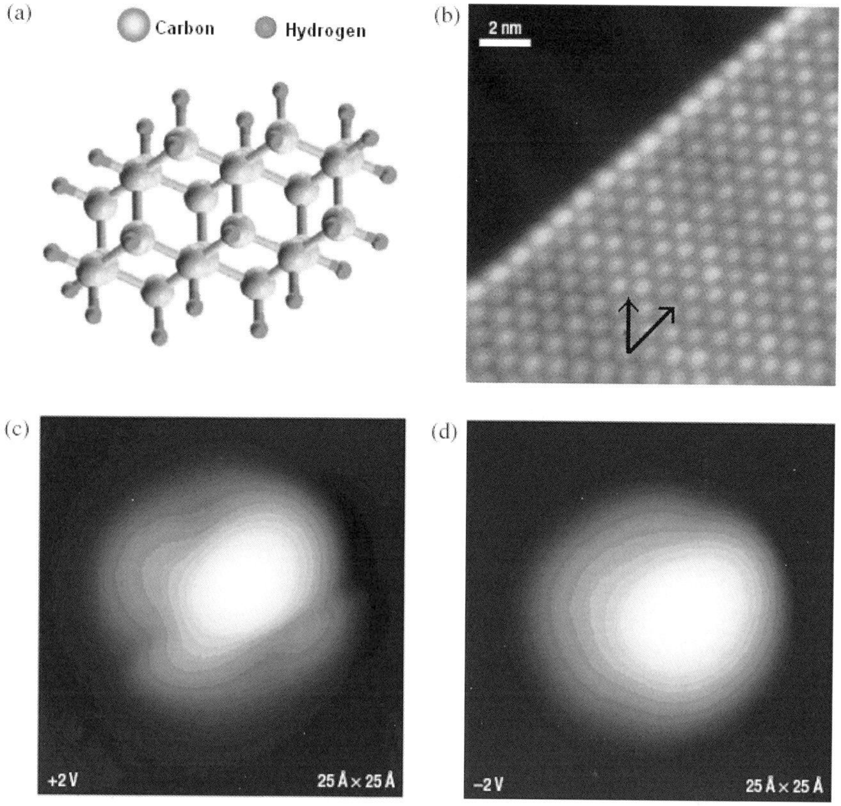

Fig. 3.34. STM topography of [121]tetramantane molecules. (a) The structural model of [121]tetramantane ($C_{22}H_{28}$). (b) Constant-current topograph of a sub-monolayer of tetramantane on Au(111) taken with sample bias V = 2.0 V and tunneling current I = 50 pA. The long axis of the tetramantane molecules is aligned along the atomic step edge of the Au(111) surface. The arrows indicate the basis vectors of the ordered structure. (c) and (d) are STM topographs (25 Å × 25 Å) of an individual tetramantane molecule taken with sample bias of +2.0 V (c) and −2.0 V (d) (Wang *et al.*, 2008).

of each state for an isolated tetramantane molecule with a particular molecular orientation. The occupied electronic states are concentrated on the centers of the C–C bonds, reflecting the spatial localization of the sp^3 bonding orbitals. In fact, these confined states are difficult to detect using scanning tunneling microscope, since the STM tip dimensions are several Angstroms larger than the diamondoid molecule size. This explains the weak spatial features of the negative-bias image (Fig. 3.34d). On the other

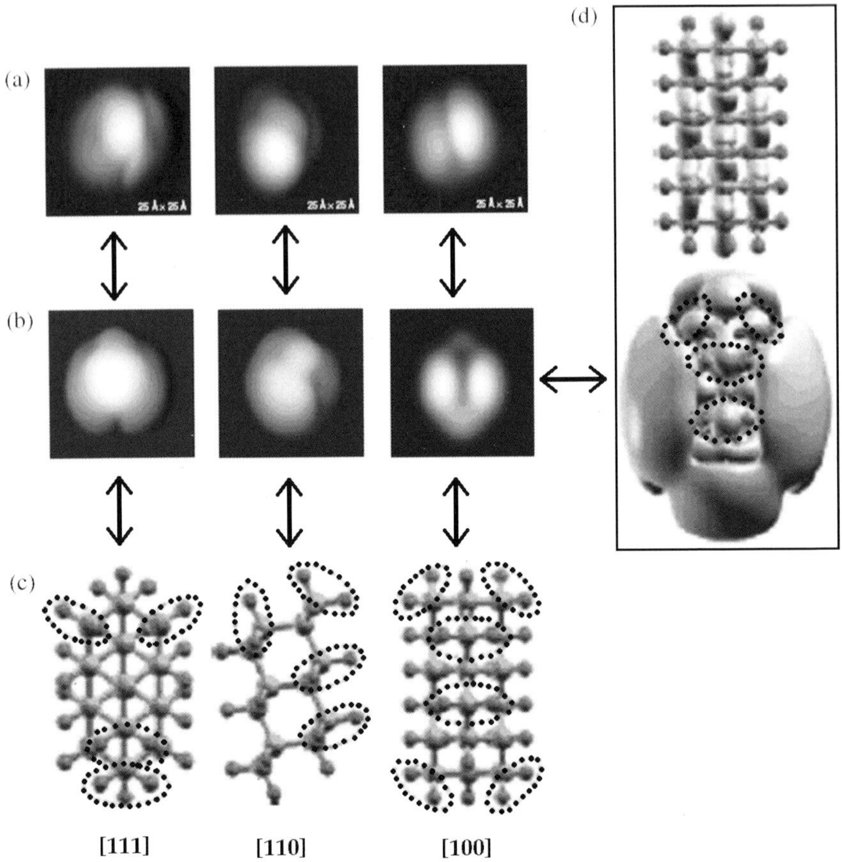

Fig. 3.35. STM images and DFT simulations of individual [121]tetramantane molecules. (a) Positive-bias (V = +2.0 V) STM images of three individual tetramantane molecules with different molecular orientations. (b) DFT-simulated LUMO-level STM topographs of the three isolated [121]tetramantane molecules. (c) Top view of the schematic structure of the three tetramantane molecules lying on the substrate with their *z* axis orientated along the diamond crystallographic [111], [110] and [100] directions, respectively (Wang *et al.*, 2008).

hand, DFT calculation of LUMO orbital is much more delocalized in space and exhibits pronounced spatial variations similar to those seen experimentally (Wang *et al.*, 2008). Note that the extended LUMO state has been previously shown to be responsible for the NEA and anomalous quantum confinement effect (see Sec. 5.3.2) of diamondoids (Drummond *et al.*, 2005). Closer comparison of the tetramantane LUMO isosurfaces and the diamondoid molecular structure uncovers an intriguing trend: The strong spatial variation of the electronic states is closely related to the nature of the surface hydrogen terminations. The large, delocalized LUMO state density exists only at the 12 singly hydrogenated CH sites, and forms smooth patches in these CH-rich areas. The eight doubly hydrogenated CH_2 sites (circled by dashed lines in Fig. 3.35c), in contrast, exhibit little electronic density and form depressions in the LUMO isosurface plot. This is caused by the peculiar formation of the LUMO orbitals on the CH_2 sites.

Figure 3.36 shows the schematic energy diagram of the isolated and on Au substrate [121]tetramantane diamondoid. The DFT calculation of HOMO-LUMO gap of the isolated molecule is found to be 5.2 V, which is an underestimation of the true quasiparticle gap (Hybertsen and Louise, 1986). The electron affinity and ionization potential were obtained by computing the total energy of singly charged molecules, and were calculated to be –0.3 eV and 7.6 V, respectively. Note that the HOMO-LUMO gap and the electron affinity/ionization potential gap are in good agreement with the quantum Monte Carlo calculations as reported by Drummond *et al.* (2005). Thus, diamondoid molecules have an intrinsic negative affinity, which makes them promising candidates for electron-emission devices.

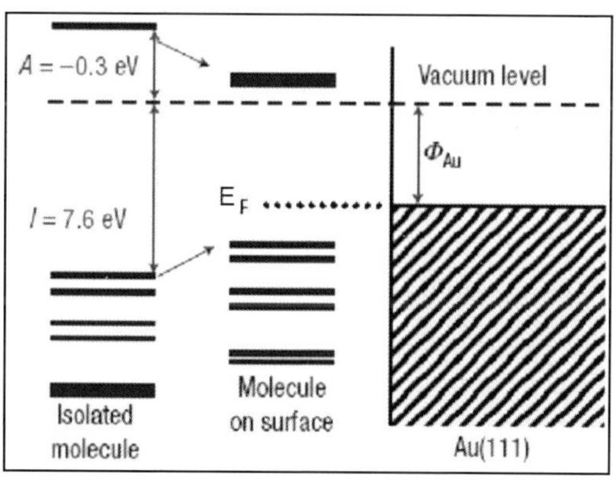

Fig. 3.36. Schematic electronic energy diagram of [121]tetramantane (Wang *et al.*, 2008).

The investigations reported by Wang *et al.* (2008) show that the important electronic and vibrational properties of diamondoid molecules are determined by microscopic differences in the surface hydrogen termination, where the singly hydrogenated CH and doubly hydrogenated CH_2 sites behave very differently. The pronounced nodal (antinodal) features in the STM topography originate exclusively from the CH_2 sites. Thus, this information should help to structurally optimize the effect of C–H stretch vibrations on electron transport in the future molecular devices. The selective substitution of H atoms on the singly or doubly hydrogenated sites may lead to very different functionalities for electronic devices, taking advantage of novel diamondoid properties such as negative electron affinity, tuned electrical conductivity, optical absorption, and heat flow.

3.3.2. *Refractive index measurements of diamondoids*

Most of the investigations on diamondoid molecules have concentrated on their electronic and atomic structures, as reported in the previous sections. On the other hand, it is rather difficult to perform conventional optical characterization of the optical properties of these small systems, which normally exist in powder forms. However, optical properties, such as refractive index and dispersion, are indispensable for optoelectronic device applications (Choi *et al.*, 2008). In this section, we report the available experimental results for refractive index of diamondoid molecules.

The optical properties have been investigated as a function of the polymantane order of micrometer-sized molecular diamondoids powders (Choi *et al.*, 2008). Here, the Becke method (see Cooper, 1983) was used to determine the refractive index. A Becke line is a band or rim of light visible along a crystal boundary in plane-polarized light (Choi *et al.*, 2008). During focusing the line moves into or away from the grain boundary, depending on the mismatch between the grain and the surrounding medium. In these experiments, microscopy (Nikkon, Portola) with interference filters of 486 nm (F-line), 589 nm (D-line) and 656 nm (C-line) were used to determine the refractive index at various wavelengths. Also, oils with various refractive indices with an accuracy of $\pm$ 0.002 were used to match the refractive index of the diamondoid samples. The refractive index measurements were taken at the C-line, D-line, and F-line at 25°C, with absolute error approximately within ± 0.004. In this study, the measurements were carried out on 20 samples for each diamondoid in order to obtain an average refractive index value.

Figures 3.37(1a–c) show the Becke line moving from the inside out, indicating that the oil was of a higher refractive index than the diamondoid grain. Figures 3.37(2a–2c), in turn, show the different relief conditions, necessary to observe the Becke line, for diamantane in the oil of refractive index 1.626 with 486, 589 and 656 nm wave-length optical filters. Observe that the grain boundaries almost disappear completely when the refractive index of the grain and the oil become quite close (Fig. 3.37(2b)). The Becke line is quite pronounced and the direction of the movement through the focus is clear, as seen in Fig. 3.37. This indicates whether a higher or lower refractive index fluid is needed to find the low relief condition. Figures 3.37 and 3.38 illustrate how dramatically the relief

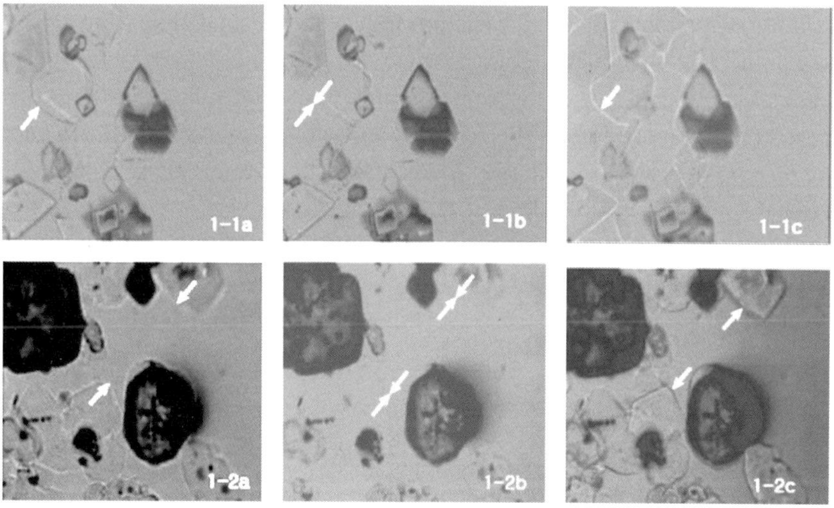

Fig. 3.37. Diamantane-grain microscopy. Becke line images moving outward, indicating accordance with the refractive index of immersion oil. Figures 1a–c show the Becke line moving from the inside out, indicating that the oil was of a higher refractive index than the diamondoid grain. Figures 2a–c, in turn, show the different relief conditions, necessary to observe the Becke line, for diamantane in the oil of refractive index 1.626 with 486, 589, and 656 nm wavelength optical filters. (Choi *et al.*, 2008).

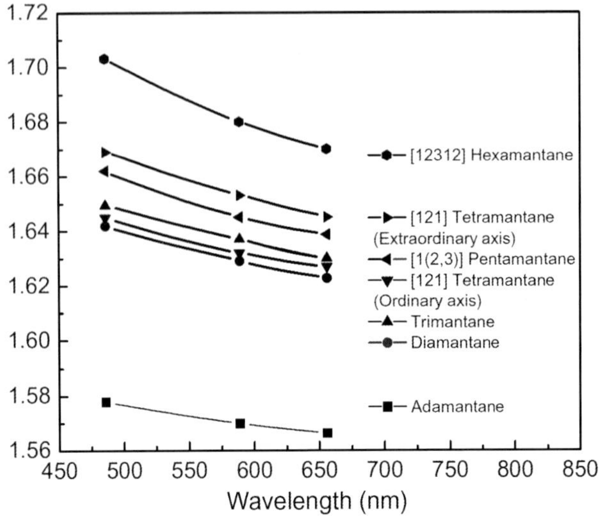

Fig. 3.38. Wavelength dependence of refractive index (n) for molecular diamondoids (Choi *et al.*, 2008).

condition can change with a variation in wavelength and shows the condition indicating that the matching refractive index oil has been found. Thus, by making successive measurements and changing index matching oils based on the Becke line of the previous measurement, it is possible to zero in on each crystal's refractive index at each wavelength (Choi *et al.*, 2008).

Figure 3.38 summarizes the refractive index measurements for six different diamondoids, comprising adamantine to hexamantane at three different wavelengths. The data of each diamondoid is well fitted to Cauchy equation (Born and Wolf, 1999):

$$n = A + \frac{B}{\lambda} + \frac{C}{\lambda^2}, \tag{3.1}$$

where A, B, and C are coefficients obtained by a nonlinear regression analysis, as listed in Table 3.3. Note that refractive indices of higher-molecular weight diamondoids more strongly dependent on wavelength, as can be seen in Fig. 3.38. To further illustrate this observation, Fig. 3.39 presents the variation of Abbe number, $v_d = (n_D-1)/(n_F-n_C)$, as a function of polymantane order. Here n_D, n_F, and n_C are the refractive indices at the

Table 3.3. Cauchy equation (Eq. 3.1) coefficients from adamantine to hexamantane (Choi *et al.*, 2008).

Diamondoid	**Chemical formula**	*A*	*B*	*C*
Adamantane	$C_{10}H_{16}$	1.5521	6.81E–15	–2.09E–28
Diamantane	$C_{14}H_{20}$	1.5914	1.54E–14	–8.20E–28
Triamantane	$C_{18}H_{24}$	1.6004	1.51E–14	–8.40E–28
[121]tetramantane	$C_{22}H_{28}$	1.5746	3.45E–14	–3.60E–27
[1(2,3)4]pentamantane	$C_{26}H_{32}$	1.6159	8.38E–15	5.90E–28
[12312]hexamantane	$C_{26}H_{30}$	1.6259	1.99E–14	–3.82E–28

D-line (589 nm), F-line (486 nm) and C-line (656 nm), respectively. Note that the Abbe number decreases with increasing polymantane order (Fig. 3.39). In fact, the v_d for the adamantine is 49 and is about 20.5 for [12312]hexamantane, i.e., [12312]hexamantane is more dispersive than adamantine.

On the other hand, Fig. 3.38 shows that, in addition to the general trend that refractive index increases with increasing polymantane order, two interesting aspects should be observed. The overall polymantane order dependence may all be understood in terms of the Lorentz–Lorentz equation (Garside *et al.*, 1968):

$$\frac{R_m}{V_m} = \frac{(n^2 - 1)}{(n^2 + 2)}, \tag{3.2}$$

which indicates that the refractive index *n* at 589 nm is dependent on two factors: The atomic packing degree, which is related to molar or molecular volume (V_m), and molecular polarization, which is measured by the molar or molecular refractivity (R_m). Thus an increase in ratio between R_m and V_m leads to an increase in the overall refractive index, *n*. Note that, for each diamondoid, V_m can be obtained from the available density and the average molecular weight. Therefore, the experimental data from the refractive index can be used to elucidate some fundamental insights into the electronic and optical structures of diamondoid molecules. Using the refractive index data from Fig. 3.38 and the known specific volume, one

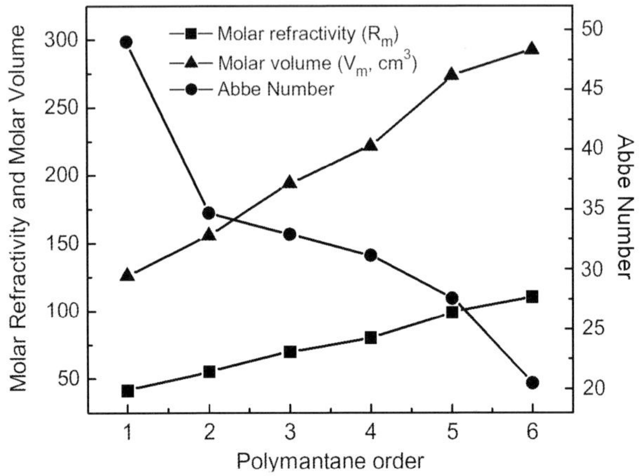

Fig. 3.39. Variation of molar volume (V_m) and molar refractivity (R_m) as a function of polymantane order (Choi *et al.*, 2008).

can determine the molar refractivity from the Lorentz–Lorentz equation for each diamondoid molecule. The results are shown in Fig. 3.39, which shows the variation in R_m and V_m as a function of polymantane order. Observe that both molar refractivity and molar volume increase with increasing polymantane order. Indeed, a twofold increase is observed between the lowest order (adamantane) and the highest order (hexamantane) diamondoid. It is worth to be mentioned that the R_m for each diamondoid and the observed trend of molecular refractivity as a function of polymantane order provide fundamental experimental data that may aid the further development of theoretical models (Richardson *et al.*, 2005; McIntosh *et al.*, 2004; Lu *et al.*, 2005; Drummond *et al.*, 2005; Willey *et al.*, 2005) aimed at understanding the basic atomic and electronic structures of bulk and individual isolated diamondoids, see Chapter 5. Thus it is concluded that the observed relatively mild increase in refractive index with increasing polymantane order is a result of an increase in the ratio between R_m and V_m (Choi *et al.*, 2008).

The variations of refractive index, n_D, at 589 nm (D-line), density and Kitaigorodskii packaging coefficient as a function of polymantane order are presented in Fig. 3.40. The Kitaigorodskii packaging coefficient is defined as the volume of molecules in a unit cell divided by the unit cell volume. The volume of molecules in the unit cell and the unit cell volume

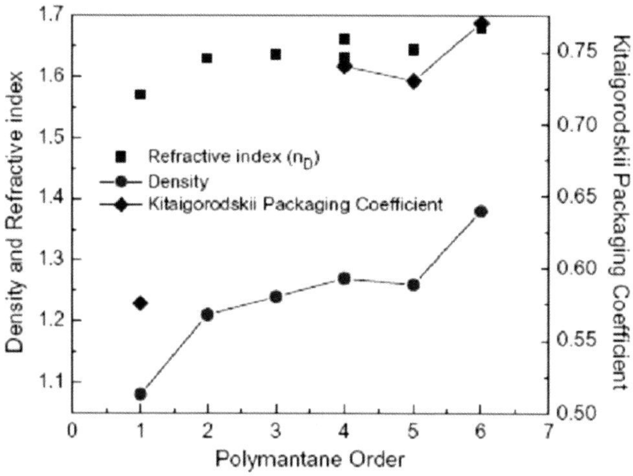

Fig. 3.40. Variations of refractive index [-] at 589 nm, density [g/cc] and Kitaigorodskii packing coefficient [-] as functions of polymantane order (Choi *et al.*, 2008).

for these diamondoids are obtained from X-ray crystal structure analysis as reported in Table 3.4 (Dahl *et al.*, 2003; Reynolds, 1978; Karle and Karle, 1965; Cernik *et al.*, 1978; Roberts and Ferguson, 1977; Dahl *et al.*, 2003a). The Kitaigorodskii packaging coefficient for diamondoids was also reported by Gavezzotti (1989), Gavezzotti (1991), Gavezzotti and Filippini (1992).

From Fig. 3.38, it can be observe that the refractive index for each wavelength exhibits two significant increases as the polymantane order increases. The first significant increase is from adamantane to diamantane and the second large increase is from [1(2,3)4]pentamantane to [12321] hexamantane, while the changes in refractive index are relatively small between diamantane, triamantane, [121]tetramantane, and [1(2,3)4]pentamantane. Those significant increases can be explained in terms of density or the Kitaigorodskii packaging coefficient. This is because it is obvious that the variation in the Kitaigorodskii packaging coefficient is proportional to the increase in density. It is observed that there is a mild decrease from [121]tetramantane to [1(2,3)**4**]pentamantane in both density and Kitaigorodskii packaging coefficient that cannot be clearly explained as the polymantane order increases. Nevertheless, the density and Kitaigorodskii packaging coefficient in [121]tetramantane are higher than

Table 3.4. Crystal structure data for six different diamondoid molecules (Choi *et al.*, 2008).

Diamondoid	Molecular weight	Molecular volume [Å^3]	Volume of molec. in unit cell [Å^3]	Volume of unit cell [Å^3]	Kitaigorodskii packing coef.	Density [g cc^{-1}]	References
Adamantane	136.125	N/A	N/A	839	N/A	1.08	[1]
Diamantane	188.314	N/A	N/A	1033	N/A	1.21	[2]
Triamantane	240.390	N/A	N/A	5171	N/A	1.24	[3]
[121]tetramantane	292.466	284	567	768	0.74	1.27	[4]
[1(2,3)4]pentamantane	344.542	329	1316	1815	0.73	1.26	[5]
[12312]hexamantane	342.526	319	957	1239	0.77	1.38	[6]

[1] Reynolds (1978).
[2] Karle and Karle (1965).
[3] Cernik *et al.* (1978).
[4] Roberts and Ferguson (1977).
[5] Dahl *et al.* (2003).
[6] Dahl *et al.* (2003).

those of [1(2,3)4]pentamantane, which means that the refractive index of [121]tetramantane should be higher than that of [1(2,3)4]pentamantane. In addition, an interesting phenomenon of birefringence is observed along with ordinary and extraordinary axes in [121]tetramantane, which means that the refractive index along the extraordinary axis (optic axis) is higher than that along the ordinary axis in [121]tetramantane, since its four cages are linearly connected in a row. Experimental data from Figs. 3.39 and 3.40 show the refractive index along the ordinary axis in [121]tetramantane to be lower than that of [1(2,3)]pentamantane. In this case, the density is obtained from X-ray structure analysis of the diamondoids and the dependence of optical properties on crystal structures is not considered.

Here, the birefringence observed in the refractive index of [121]tetramantane is successfully characterized using the Becke line method along with the ordinary and extraordinary axes. Finally, it is worth to note that the refractive index and dispersion parameters, such as Abbe number, of diamondoid molecules, presented in this section, may aid the further development of photonic technology.

3.4. Thermodynamic Properties of Diamondoid Molecules and their Derivatives

This section will be focused into two main subjects: Thermodynamic properties of diamondoid molecules and their derivatives, and solubilities of diamondoids and phase behavior of binary systems containing diamondoids.

Some physical aspects should be point out about diamondoids. Adamantane, in particular, is a cage hydrocarbon with a white or almost white crystalline solid nature, like solid wax, at normal conditions. Its odor resembles that of camphor. It is a stable and nonbiodegradable compound, which is combustible due to its hydrocarbon nature. It has not been found to be hazardous or toxic to living entities (Hedberg, 1948; Fort, 1976). In addition, adamantane can exist in gas, liquid and two solid crystalline states. On the other hand, diamantane can exist in gas, liquid and three different solid crystalline states, whereas higher diamondoids (tetramantane and upper) possess two or more solid crystalline states (Kabo *et al.*, 2000).

3.4.1. *Enthalpies and entropies*

Figure 3.41 shows the temperature dependence of the heat capacity of adamantane in condensed state between 5 and 600 K as reported both by Chang and Westrum (1960) and Kabo *et al.* (2000). The solid and liquid vapor pressures for adamantane and diamantane have been determined between ambient temperature and their estimated critical points using various measurement techniques by several researchers (Reiser *et al.*, 1996; Boyd *et al.*, 1971; Cullick and Magouirk, 1994). For instance, Table 3.5 summarizes equations representing the natural logarithm of adamantane and diamantane vapor pressures, *ln* P, as a function of absolute temperature, reported along with the temperature ranges of their validity. Figures 3.42 and 3.43 show the smoothed vapor–liquid–solid phase transition lines (vapor pressures) for adamantane and diamantane, which data are based on the vapor pressure data reported in Table 3.5 (Mansoori, 2007). On the other hand, there is a limited amount of data for vapor pressures of binary mixtures of adamantane and diamantane with

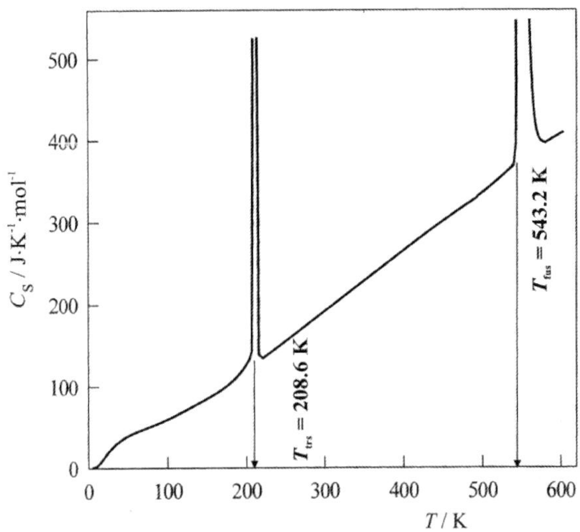

Fig. 3.41. The temperature dependence of the heat capacity in the condensed state for adamantane as measured by a scanning calorimeter (Kabo *et al.*, 2000). T_{trs} stands for temperature of transition from rigid crystal (fcc)-to-plastic crystal (cubic) state of adamantane and T_{fus} stands for fusion temperature.

Table 3.5. Vapor pressure equations of adamantane and diamantane for liquid–vapor and solid–vapor phase transitions.

Diamondoid	Phase transition kind	ln P [kPa] =	Temp. Range (K)	References
Adamantane	Liquid–vapor	–4670/T + 14.75	T > 543 K	[1]
	Solid–vapor	–6570/T + 18.18	483–543	[1]
		–6324.7/T + 17.827	366–443	[2]
		–9335.6/T + 65.206–15.349 *log*T	313–443	[2]
		–7300/T + 31.583–4.376 *log*T	333–499	[3]
Diamantane	Liquid–vapor	–5680/T + 14.858	516–716	[1]
	Solid–vapor	–7330/T + 18.00	498–516	[1]
		–7632.5/T + 18.333	353–493	[3]
		–18981.3/T + 190.735–55.4418 *log*T	332–423	[3]

[1] Reiser *et al.* (1996).
[2] Boyd *et al.* (1971).
[3] Cullick and Magouirk (1994).

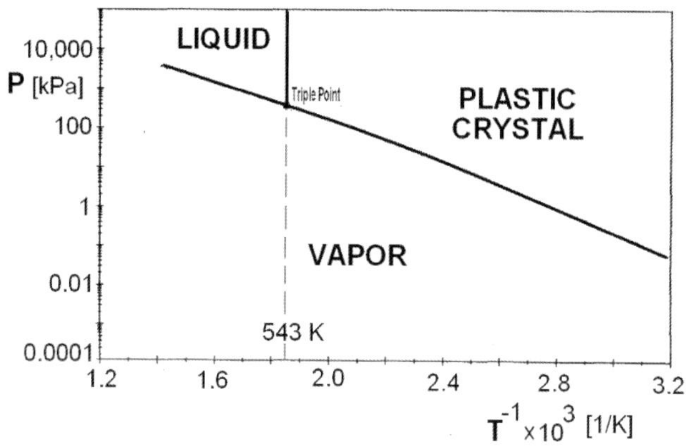

Fig. 3.42. Vapor–Liquid–Solid (plastic crystal) phase diagram of adamantane. The phase transition from plastic crystal to rigid crystal phase occurs at 208.6 K (1/T = 0.004794 K^{-1}). This diagram is based on the data of Table 3.5 (Mansoori, 2007).

other hydrocarbons available (van Miltenburg *et al.*, 2000; Poot *et al.*, 2003; Poot *et al.*, 2004).

In Table 3.6, we report the thermodynamic properties of adamantane and diamantane including various kinds of enthalpies and heat capacities

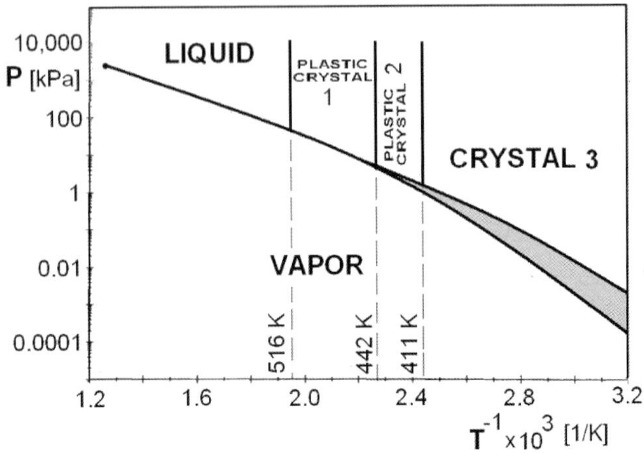

Fig. 3.43. Vapor–Liquid–Solids (plastic crystal 1, plastic crystal 2, crystal 3) phase diagram of diamantane. This diagram is based on the data of Table 3.5. The shaded area between vapor and plastic crystal 2 and crystal 3 phase transitions is indicative of the error range of the available data (Mansoori, 2007).

as reported by various researchers (Boyd *et al.*, 1971; Bratton *et al.*, 1967; Buttler *et al.*, 1971; Clark *et al.*, 1975; Clark *et al.*, 1979; Jochems *et al.*, 1982; Mansson *et al.*, 1970; Spinella *et al.*, 1978; Steele and Watt, 1977; Westrum, 1961; Westrum *et al.*, 1978).

3.4.2. *Other thermodynamic properties*

Some other thermodynamic properties of adamantane and diamantane in different phases are also available as follows:

(i) The standard molar thermodynamic functions for adamantane in the ideal gas state as calculated by statistical thermodynamics methods. These data are reported in Table 3.7.

(ii) The temperature dependence of the heat capacities of adamantane in the condensed state between 340 and 600 K as measured by a differential scanning calorimeter (DSC) and is shown in Fig. 3.41 (Kabo, 2000). According to this figure, liquid adamantane converts to a solid plastic with simple cubic crystal structure upon freezing. After further cooling, it moves into another solid state, an *fcc* crystalline phase.

Table 3.6. Thermodynamic properties of adamantane and diamantane. In this table, the average of the values reported by various investigators are reported (Mansoori, 2007).

Property	Diamondoid	Value	Units	T[K]	References
ΔH^{of}_{gas}	Adamantane	−133.60	kJ/gmol		[1,2–4]
	Diamantane	−145.90	kJ/gmol		[1]
ΔH^{of}_{solid}	Adamantane	−191.10	kJ/gmol		[1,2–4]
	Diamantane	−241.90	kJ/gmol		[2]
S^{o}_{solid}	Adamantane (crystalline phase II)	195.83	J/gmol●K		[5,6]
	Diamantane (crystalline phase III)	200.16	J/gmol●K		[7]
C_P^{solid}	Adamantane	189.74	J/gmol●K	298.15	[5]
(solid phase I)	Diamantane	220.20	J/gmol●K	295.56	[8]
		223.22		298.15	[7]
$\Delta H^{o}_{sublimation}$	Adamantane	59.90	kJ/gmol	323.00	[1,2,4,9–11]
	Diamantane	95.85	kJ/gmol	313.00	[11,12]

(*Continued*)

Table 3.6. (*Continued*)

Property	Diamondoid		Value	Units	T[K]	References
$\Delta H_{phase\ transition}$	Adam.	Solid I–Solid II	3.376	kJ/gmol	208.62	[5,6]
		Solid I–Liq	54.81		543.00	[11]
		Liq–Vapor	38.83			
	Diam.	Solid I–Solid III	4.445	kJ/gmol	407.22	[8]
		Solid I–Solid II	8.960	kJ/gmol	440.43	
		Solid I–Liq	8.646	kJ/gmol	517.92	
		Liq–Vapor	48.116		543.00	[11]
$\Delta S_{phase\ transition}$	Adam.	Solid I–Solid II	16.18	J/mol•K	208.62	[8]
	Diam.	Solid I–Solid III	10.92	J/mol•K	407.22	
		Solid I–Solid II	20.34	J/mol•K	440.43	
		Solid I–Liq	16.69	J/mol•K	517.92	
$\Delta H^{o}_{combustion}$	Adam. (solid phase)		−6030.04	kJ/gmol		[1–4]
	Diam. (solid phase)		−8125.58	kJ/gmol		[12]

Note: Solid adamantane possesses two crystalline phases and diamantane exists in three crystalline phases.

[1] Clark *et al.* (1979); [2] Boyd *et al.* (1971); [3] Butter *et al.* (1971); [4] Mansson *et al.* (1970); [5] Westrum (1961); [6] Shang and Westrum (1960); [7] Westrum *et al.* (1978); [8] Clark *et al.* (1979); [9] Jochems *et al.* (1982); [10] Bratton *et al.* (1967); [11] Reiser *et al.* (1996); [12] Clark *et al.* (1975).

Table 3.7. Temperatures, the molar enthalpies and entropies of phase transitions of adamantane (Kabo *et al.*, 2000).

Transition	T_{trs}(K)	ΔH_m (J mol^{-1})	ΔS_m (J K^{-1} mol^{-1})	Reference
CrII–CrI	208.60	3376	16.19	[1]
CrI–Liquid	543.20	13958 ± 279	25.70 ± 0.5	[2]

[1] Kabo *et al.* (1998).
[2] Kabo *et al.* (2000).

(iii) Table 3.8 contains the data of temperatures, molar enthalpies and molar entropies of phase transitions for adamantane and diamantane and other compounds of same cage hydrocarbon in the condensed state.
(iv) Table 3.9 contains the available enthalpies of formation, sublimation and combustion of adamantanes derivatives (methyl-adamantanes, dimenthyl-adamantane, trimethyl-adamantane and tetramethyl-adamantane) and diamantane derivatives (1-methyl-diamantane, 3-methyl-diamantane and 4-methyl-diamantane (Clark *et al.*, 1975; Steele and Watt, 1977). These data are compared with both adamantane and diamantane data (also included into Table 3.9).

3.4.3. *Thermodynamic properties for biomedical industry*

Adamantane derivatives have found widespread applications in medicine due to their diverse pharmacological action (see Sec. 6.2.1.1 for further detail). Their pharmacological action is likely to be a result of their specific molecular structure. Such derivatives contain the hydrophobic adamantane cage (which provides penetration of these molecules straight into a cell through the cell lipid membrane) with various hydrophilic substituents (which are responsible for the biological activity as well as relatively rapid assimilation and transportation of these substances to the specific body sites).

The spectrum of the biological activity of these medications is rather wide (Baisini *et al.*, 2003; Kin and Sohn, 2003; Monto, 2003; Morozov *et al.*, 2001). They possess neuroprotective and neuromodulating action, they are used for treatment of numerous viral diseases (e.g., influenza and

Table 3.8. Temperatures, the molar enthalpies and entropies of phase transitions of adamantane.

Formula	Compound	Transition type	T(K)	ΔH_m [J mol^{-1}]	ΔS_m [J K^{-1} mol^{-1}]	References
C_8H_{14}	Bicyclo[2.2.2]octane	CrII–CrI	164.25	4.59	27.9	[1]
		CrI–liq	447.48	8.35	18.7	
$C_{10}H_{16}$	Adamantane	CrII–CrI	208.60	3.38	16.2	[2]
		CrI–liq	543.20	13.80	25.4	[3]
$C_{11}H_{14}$	Pentacycloundecane	CrII–CrI	164.40	4.86	29.6	[4]
		CrI–liq	475.80	6.38	13.5	
$C_{14}H_{16}$	Heptacyclotetradecane	CrII–CrI	355.00	14.67	41.4	[5]
		CrI–liq	440.00	5.57	12.6	
$C_{14}H_{20}$	Diamantane	CrIII–CrII	407.22	4.46	10.8	
		CrII–CrI	440.43	8.96	20.2	
		CrI–liq	517.92	8.64	16.8	[6]

[1] Wong and Westrum (1970).
[2] Chang and Westrum (1960).
[3] Kabo *et al.* (2000).
[4] Kabo *et al.* (1995).
[5]Kabo *et al.* (1994).
[6] Westrum *et al.* (1978).

Table 3.9. Thermodynamic properties of methyl-derivatives of adamantane and diamantane. In this table, the average of the values reported by various investigators are reported (Mansoori, 2007).

Diamondoid	ΔH *of gas* (kJ/gmol)	ΔH *of solid* (kJ/gmol)	$\Delta H^{\circ}_{sublimation}$ (kJ/gmol)	$\Delta H^{\circ}_{combustion}$ (kJ/gmol) (solid phase)	References
Adamantane	−133.6	−191.1	59.9	−6030.04	See Table 4.7
1-Methyl adamantane	−171.6	−240.1	67.7	−6661.10	[1,2]
1,3-Dimethyl adamantane	−219.0	−287.3	67.8	−7294.00	[2]
1,3,5-Trimethyl adamantane	−255.0	−333.0	77.8	−7927.40	[2]
1,3,5,7-Tetramethyl adamantane	−283.3	−370.7	82.4	−8568.70	[1,2]
2-Methyl adamantane	−151.7	−220.8	67.7	−6680.40	[1,2]
Diamantane	−145.9	−241.9	96.	−8125.60	See Table 4.7
1-Methyl diamantane	−166.7	−247.4	80.6	−8799.40	[1]
3-Methyl diamantane	−157.3	−260.4	103.1	−8786.36	[1]
4-Methyl diamantane	−182.1	−261.5	79.4	−8786.20	[1]

[1] Clark *et al*. (1979).
[2] Steele and Watt (1977).

hepatitis), for stimulation of the immune system, and so on. For example, amantadine and the structurally similar derivatives of 1-aminoadamantane are able to prevent disfunction and death of nerve cells under a number of different cerebral afflictions such as Parkinson's and Alzheimer's diseases, hypoxic brain damage, neuroinfections and stroke (Morozov *et al.*, 2001). It should be pointed out that the syntheses of the pharmaceutically active adamantane derivatives are characterized by low yields of the target products (less than 10%). Therefore, the optimization of the conditions of their production requires detailed thermodynamic information. For this reason, Bazyleva *et al.* (2008, 2005) have been investigating thermodynamic properties of 1-aminoadamantane and 1-bromoadamantene in both crystalline and gaseous states. For 1-aminoadamantane, they obtained the following values for solid-to-solid phase transitions enthalpies through calorimetry,

$$\Delta_{crIII-crII} H_m^0 (241.4K) = 1716 \pm 10 (J/mol) \tag{3.3}$$

and

$$\Delta_{crII-crI} H_m^0 (284.6K) = 5309 \pm 5 (J/mol). \tag{3.4}$$

They also obtained values for sublimation, combustion, and formation enthalpies and the gaseous 1-aminoadamantane standard molar entropy (Bazyleva *et al.*, 2008):

$$\Delta_{sub} H_m^0 (298.15K) = 61.65 \pm 0.63 (kJ/mol), \tag{3.5}$$

$$\Delta_{comb} H_m^0 (cr, 298.15K) = -6169.2 \pm 1.9 (kJ/mol), \tag{3.6}$$

$$\Delta_{form} H_m^0 (cr, 298.15K) = -195.4 \pm 2.3 (kJ/mol), \tag{3.7}$$

$$\Delta_0^T S_{m,\exp}^0 (g, 298.15K) = 366.6 \pm 2.4 (J \cdot K^{-1} \cdot mol^{-1}), \tag{3.8}$$

For 1-bromoadamantane, the thermodynamic properties are as the following (Bazyleva *et al.*, 2005):

$$\Delta_{sub} H_m^0 (303.0K) = 71.77 \pm 0.31 (kJ \cdot mol^{-1}), \tag{3.9}$$

$$\Delta_{form} H_m^0(g, 298.15K)_{crII} = -132.8(kJ \cdot mol^{-1}), \tag{3.10}$$

$$\Delta_0^T S_{m,\exp}^0(g, 303.0K) = 377.2 \pm 1.5(J \cdot K^{-1} \cdot mol^{-1}), \tag{3.11}$$

$$\Delta_0^T S_m^0(cr, 303.0K) = 220.3 \pm 1.1(J \cdot K^{-1} \cdot mol^{-1}). \tag{3.12}$$

However, for polymantanes with order above 2, very limited data are available in the literature. For instance, for triamantane, the enthalpy and entropy of transition from crystalline phase II to crystalline phase I are reported (Jenkins and O'Brien, 1981) as

$$\Delta H_{\text{phase II-I}} \text{ (at 293.65 K)} = 1.06 \text{ kJ/mol}, \tag{3.13}$$

and

$$\Delta S_{\text{phase II-I}} \text{ (at 293.65 K)} = 3.77 \text{ kJ/mol}, \tag{3.14}$$

respectively. While for tetra-, penta-, and hexamantane and other higher diamondoids little data is reported in the literature. This is possibly due to the fact that of these compounds only anti-tetramantane has been successfully synthesized in the laboratory in small quantities (McKervey, 1980; Schleyer, 1969).

3.4.4. *Solubilities*

In this section, we report the available solubility data for diamondoids in liquid solvents (Table 3.10, Fig. 3.44, Figs. 3.45–3.47), as well as in supercritical (dense) (methane, ethane, and carbon dioxide) gases at a few temperatures and at various pressures and solvent densities (Figs. 3.48–3.51).

In Table 3.10, the solubility limits of adamantane and diamantane in various liquid solvents at normal conditions are reported (Reiser *et al.*, 1996). In producing these data, a known amount of diamondoid was titrated with various liquids at 25°C. Continuous stirring was used until all the diamondoid was dissolved, which defined the solubility limit. The solubility of adamantane in tetrahydrofuran (THF) is higher than in other organic solvents reported here. Overall cyclohexane is a better solvent for

Table 3.10. Solubilities of diamondoids in liquid solvents at 25°C (Reiser *et al.*, 1996).

Solvent	Adamantane (wt%)	Diamantane (wt%)
Carbon tetrachloride	7.0	5.0
Pentane	11.6	4.0
Hexane	10.8	3.9
Heptane	10.4	3.7
Octane	10.0	3.9
Decane	8.9	3.5
Undecane	7.9	3.2
Tridecane	7.3	2.7
Tetradecane	7.5	2.3
Pentadecane	7.1	2.2
Cyclohexane	11.1	6.3
Benzene	10.9	4.3
Toluene	9.9	4.5
m-Xylene	9.8	4.5
p-Xylene	9.6	4.5
o-Xylene	9.6	4.1
THF	12.0	4.0
Diesel oil	7.5	2.7
1,3, Dimethyl-adamantane	6.0	2.0

diamondoids among the liquids tested due to the similarities in the molecular structure of cyclohexane to diamondoids. It should be also pointed out that since diamondoid molecules are substantially hydrophobic their solubility in organic solvents is a function of their hydrophobicity (Reiser *et al.*, 1996). On the other hand, derivatization strategies (Cahill, 1990; Liu *et al.*, 1990) on diamondoids could be used in order to enhance their solubility and hence their potential applications in various separation schemes, chromatography and pharmaceuticals.

Recently, solubilities of adamantane and diamantane in pressurized hot water (PHW) have been reported by Karásek *et al.* (2008). In this study,

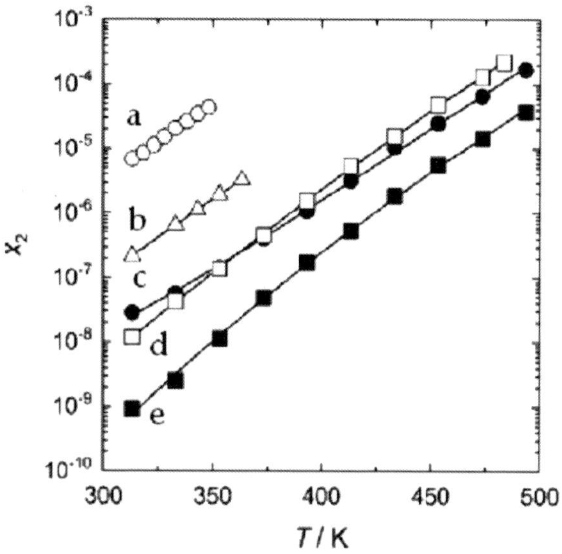

Fig. 3.44. Aqueous solubilities (equilibrium mole fractions, X_2) of diamondoids and selected PAHs: (a) naphthalene ($C_{10}H_8$); (b) phenanthrene ($C_{14}H_{10}$); (c) adamantane ($C_{10}H_{16}$); (d) anthracene ($C_{14}H_{10}$); (e) diamantane ($C_{14}H_{20}$) (Karásek *et al.*, 2008).

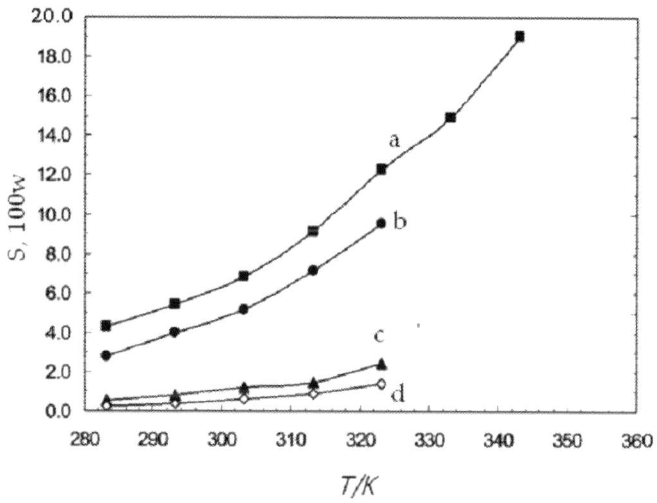

Fig. 3.45. Mass fraction solubility, *S*, of diamantane in four different organic solvents at various temperatures: (a) cyclohexane; (b) toluene; (c) ethyl acetate; (d) acetone (Chang *et al.*, 2008).

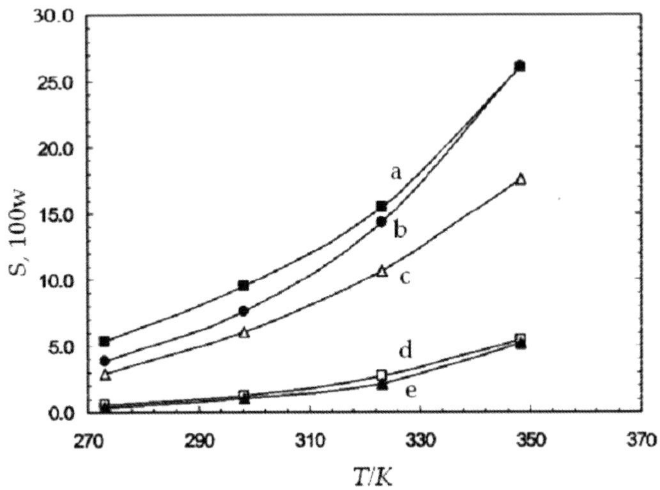

Fig. 3.46. Mass fraction solubility, *S*, of triamantane in five different organic solvents at various temperatures: (a) cyclohexane; (b) toluene; (c) heptane; (d) 1-pentanol, (e) ethyl acetate (Chang *et al.*, 2008).

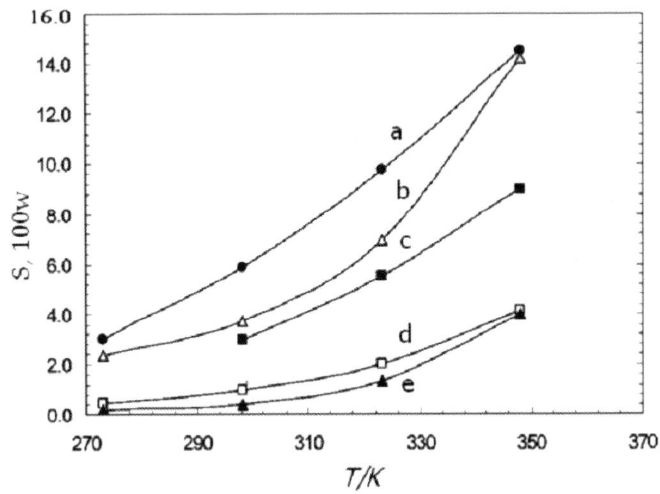

Fig. 3.47. Mass fraction solubility, *S*, of tetramantane in five different organic solvents at various temperatures: (a) toluene; (b) heptane; (c) cyclohexane; (d) 1-pentanol, (e) ethyl acetate (Chang *et al.*, 2008).

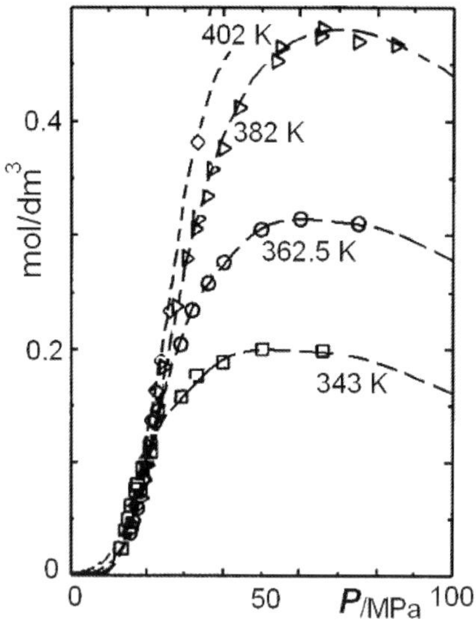

Fig. 3.48. Effect of temperature and pressure on solubility (in units of mol/dm^3) of adamantane in dense (supercritical) carbon dioxide gas (Kraska *et al.*, 2002).

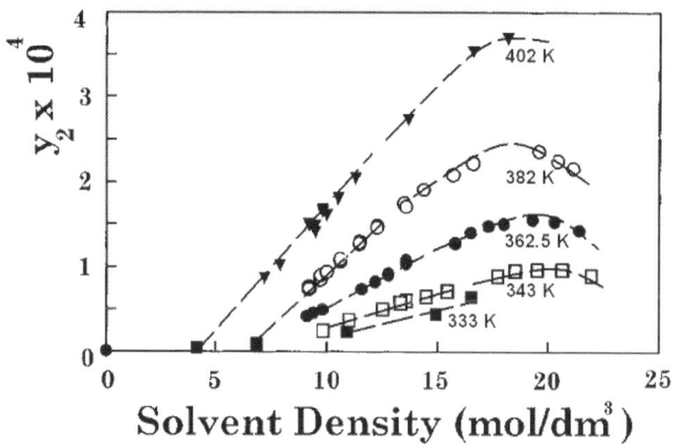

Fig. 3.49. Effect of temperature and supercritical solvent density on solubility of adamantane (in units of mole fraction) in dense (supercritical) carbon dioxide. Data of isotherm at 333 K is from (Smith and Teja, 1996). Data of isotherms at 343 K, 362.5 K, 382 K and 402 K are from (Swaid *et al.*, 1985).

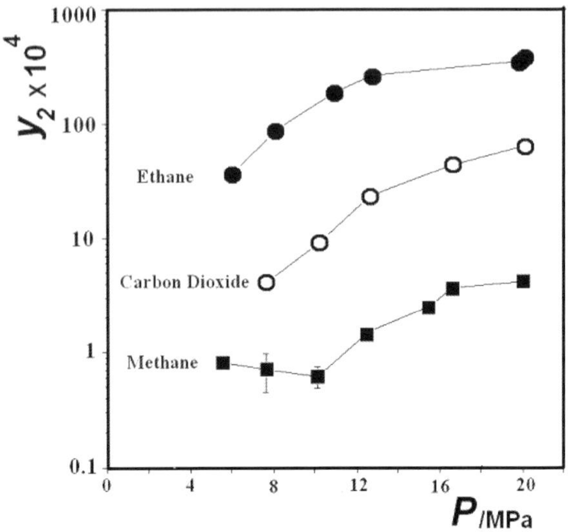

Fig. 3.50. Effect of pressure on the solubility (in units of mole fraction) of adamantane in dense (supercritical) carbon dioxide, methane, and ethane gases at 333 K (Smith and Teja, 1996).

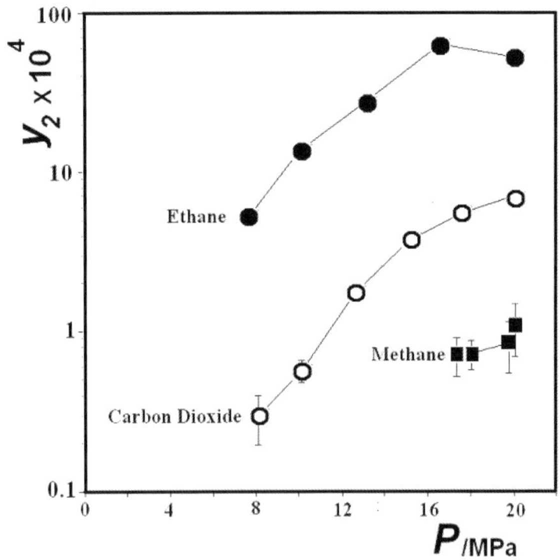

Fig. 3.51. Effect of pressure on solubility (in units of mole fraction) of diamantane in dense (supercritical) gases at 333 K (for carbon dioxide and ethane) and at 353 K (for methane) (Smith and Teja, 1996).

aqueous solubilities of lower diamondoid hydrocarbons as adamantane and diamantane were measured along the 5 MPa isobar at temperatures between 313 K and the triple-point (solid–liquid–vapor) temperature of the solute (see Fig. 3.44).

In addition, the activity coefficients of the individual diamondoids in saturated aqueous solutions were estimated from the solubility data employing two different approximation of the pure-solute heat capacity difference.

In general, characterization of solvent properties of PHW requires experimental data such as solubilities of several types of solutes in PHW. To date, this kind of experimental data have mostly been available for polycyclic aromatic hydrocarbons (PAHs) because of their environmental pollution concerns and their carcinogenic effects (Rössling and Franck, 1983; Karásek *et al.*, 2006; Karásek *et al.*, 2008; Miller *et al.*, 1998; Sanders, 1986).

PAHs, from a systematic point of view, can be pictured as hydrogen-terminated fragments of graphite, the most abundant allotrope of pure carbon. It should be noted that diamondoid molecules are, in fact, hydrogen-terminated fragments of another well-known allotrope of carbon, diamond. Figure 3.44 shows that, depending on temperature, the aqueous solubility of adamantane is lower than that of naphthalene by a factor of 240 to 400. The solubility of diamantane is 9- to 17-times lower than that of anthracene and 170- to 260-times lower than that of phenanthrene. Therefore, these data confirm the expectation that solubility of a diamondoid should be lower than solubility of an aromatic hydrocarbon of the same carbon number. The apparent similarity between the aqueous solubilities of adamantane ($C_{10}H_{16}$) and anthracene ($C_{14}H_{10}$) results from a fortuitous tradeoff between the solute–water interactions and the pure-solute properties (Karásek *et al.*, 2008).

Chang *et al.* (2008) recently measured the solubility of diamantane, triamantane, tetramantane, and their derivatives in acetone, cyclohexane, ethyl acetate, toluene, heptane, and 1-pentanol using the solid disappearance method for temperatures ranging from 273.15 to 348.15 K. In this investigation, the experimental solubility data is correlated using different thermodynamic models. Figures 3.45–3.47 show experimental data of mass fraction solubility, *S*, of diamantane, triamantane, and tetramantane, as a function of temperature in various organic solvents. According to this data, the solubility of diamondoids is higher in nonhydrogen bonding solvents than hydrogen-bonding solvents tried. Moreover, the dependence on diamondoid molecular size is not obvious perhaps because the difference in size of the three diamondoids reported is not quite significant. With the exception of tetramantane, both diamantane and triamantane exhibit a higher solubility in cyclohexane than in toluene. It can be said that cyclohexane is a good solvent for lower diamondoids. Note that the solubility for tetramantane represents the collective outcome of four tetramantane isomers since no attempt was made to separate them for this study. On the other hand, 1-hydroxyl diamantane and 1,6-dihydroxyl diamantane exhibit high solubility in hydrogen-bonding solvents (Chang *et al.*, 2008).

Solubilities of adamantane and diamantane in supercritical solvents (methane, ethane, and carbon dioxide dense gases) have been measured by various investigators (Kraska *et al.*, 2002; Smith and Teja, 1996; Swaid *et al.*, 1985). Experimental data (Kraska *et al.*, 2002) of the effect of temperature and pressure on the supercritical solubility of adamantane in dense (supercritical) carbon dioxide gas is shown in Fig. 3.48. On the other hand, Fig. 3.49 depicts the experimental data (Smith and Teja, 1996; Swaid *et al.*, 1985) of the effect of temperature and supercritical solvent density on the solubility of adamantane. Figure 3.50 shows the effect of pressure on the solubility of adamantane in various supercritical solvents at 300 K (Smith and Teja, 1996); whereas Fig. 3.51 shows the diamantane solubility data (Kraska *et al.*, 2002) in various supercritical solvents (carbon dioxide and ethane at 333 K and methane at 353 K).

Trends of solubility enhancement for each diamondoid follow regular behavior like other heavy hydrocarbon solutes in supercritical solvents with respect to variations in pressure and density (Park *et al.*, 1987; Hartono *et al.*, 1999). Supercritical solubilities of these lower diamondoids have been successfully correlated through cubic equations of state (Smith and Teja, 1996). Note that the supercritical fluid and liquid solubilities presented in Figs. 3.48–3.51 suggest that diamondoid molecules will preferentially partition themselves into the high pressure, high temperature and rather low-boiling fraction of any mixture including petroleum crude oil.

3.5. Concluding Remarks

As we have shown in this chapter, diamondoid molecules have unique physical properties with potential applications in electronics, medicine, molecular design, and especially in nanotechnology. The physical property data of diamondoids which are reported here included spectroscopic, optical, and thermodynamic properties.

The reported spectrometric properties of diamondoids included SXE data, XAS spectroscopy data, ionization potential data, photoion yield spectra data, vibrational (Raman) spectroscopy data and infrared spectroscopic data. Most of the data reported are for lower diamondoids specifically for adamantane, diamantine, triamantane.

The reported optical properties of diamondoids include their electron photoemission and their refractive indices. Since diamondoid surfaces have NEA as a result of which diamondoids are considered as electron photoemitters. NEA is useful for the development of electron emitters like the design of efficient cathodes that can supply electrons to the vacuum with little energy loss. Also selective substitution of H atoms on the singly or doubly hydrogenated sites of diamondoids may lead to very different functionalities for electronic devices. Such novel diamondoid properties as negative electron affinity, tuned electrical conductivity, optical absorption and heat flow could have variety of interesting applications. A limited amount of refractive index data of lower diamondoids are available. It has been rather difficult to perform conventional optical characterization of diamondoids, which normally exist in powder forms.

The reported thermodynamic properties of diamondoid molecules and their derivatives include latent heats and entropies of phase transition, phase behavior of binary systems containing diamondoids and solubilities of diamondoids in liquid solvents and in supercritical fluids.

The possibilities to selectively functionalize diamondoids open new ways for surface modification, mainly for high-efficiency field emitters in molecular electronics, as seed crystals for diamond growth, as robust mechanical coatings, in design of new medicine, drug delivery, combinatorial chemistry and host–guest chemistry. The properties of SAMs of diamondoids are of fundamental interest for a variety of emerging nanotechnology applications.

Bibliography

Bain, C. D., Troughton, E. B., Tao, Y. T., Evall, J., Whitesides, G. M. and Nuzzo, R. G. (1989). *J. Am. Che. Soc.*, **111**: 321–335.

Baisini, O., Pigozzi, M. G., Benini, F., Stellini, R., Reggiani, A., Quattrocchi, D., Salmi, A., Andri, G., Cominotti, A., Favret, M., Garciulo, F. and Lanzini A. (2003). *Hepatol. Res.*, **26**: 167–173.

Bauschlicher, C. W. and Langhoff, S. R. (1997). *Spectrochim. Acta A.*, **53**: 1225–1240.

Bazyleva, A. B., Blokhim, A. V., Kabo, A. G., Kabo, G. J., Emel'yanenko, V. N. and Verevkin, S. P. (2008). *J. Chem. Thermodyn.*, **40**: 509–522.

Bazyleva, A. B., Kabo, G. J., Paulechka, Y. U., Zaitsau, D. H., Blokhim, A. V. and Sevruk, V. M. (2005). *Thermochim. Acta.*, **436**: 56–67.

Bell, R. L. (1973). Negative Electron Affinity Devices, Clarendon, Oxford.

Bistricic, L., Pejovb, L. and Baranovic, G. (2002). *J. Mol. Struct. — Theochem.* **594**: 79–88.

Born, M. and Wolf, E. (1999) Principles of Optics: Electromagnetic Theory of Propagation, Interference and Diffraction of Light, Pergamon, Oxford.

Bostedt, C., van Buuren, T., Willey, T.M., Franco, N., Terminello, L. J., Heske, C. and Möller, T. (2004). *Appl. Phys. Lett.*, **84**: 4056–4058.

Boyd, R. H., Sanwal, S.N., Shary-Tehrany, S. and McNally, D. (1971). *J. Phys. Chem.*, **75**: 1264–1271.

Bratton, W. K., Szilard, I. and Cupas, C. A. (1967). *J. Org. Chem.* **32**: 2019–2021.

Butler, R. S., Carson, A. S., Laye, P. G. and Steele, W. V. (1971). *J. Phys. Chem.*, **3**: 277–280.

Cahill, P. A. (1990). *Tetrahedron Lett.* **31**: 5417–5420.

Castner, D. G., Hinds, K. and Grainger, D. W. (1996). *Langmuir,* **12**: 5083–5086.

Cernik, R., Evans, E. H. M., Hine, R. and Richards, J. P. G. (1978). *Solid State Commu.*, **27**: 1017–1019.

Chang, S. S. and Westrum , E. F., (1960). *J. Phys. Chem.*, **64**: 1547–1551.

Choi, J. H., Eichele, C., Lin, Y. C., Shi, F. G., Carlson, B., Sciamanna, S. (2008). *Scripta Materialia*, **58**: 5, 413–416.

Clark, T., Knox, T. M., McKervey, M. A., Mackle, H. and Rooney, J. J. (1975). *J. Am. Chem. Soc.*, **97**: 3835–3836.

Clark, T., Knox, T. M., McKervey, M. A., Mackle, H. and Rooney, J. J. (1979). *J. American Chem. Soc.*, **101**: 2404–2420.

Cooper, P. R. (1983). *Appl. Opt.*, **22**: 3070–3072.

Cullick, A. S., Magouirk, J. L. and Ng, H. J. (1994) Paper # P1994.01, Proceed. of the 73rd Annual Convention of the GPA (Gas Processors Association), Tulsa (OK), USA.

Dahl, J. E., Liu, S. G. and Carlson, R. M. K. (2003). *Science*, **299**: 96–99.

Dahl, J. E. P., Moldowan, J. M., Peakan, T. M., Clardy, J., Lobkovsky, E., Olmstead, M. M., May, P. W., Davis, T. J., Steeds, J. W., Peters, K. E., Pepper, A., Ekuan, A. and Carlson, R. M. K., (2003a). *Angew. Chem. Int. Ed.*, **42**: 2040–2044.

Dameron, A. A., Mllen, T. J., Hengstebeck, R. W., Saavedra, H. M. and Weiss, P. S. (2007). *J. Phys. Chem. C.*, **111**: 6747–6752.

deHeer, W. A. (1993). *Rev. Mod. Phys.* **65**: 611–676.

Drummond, N. D., Williamson, A. J., Needs, R. J. and Galli, G. (2005). *Phys. Rev. Lett.*, **95**: 096801.

Eden, R. C., Mol, J. L. and Spicer, W. E. (1967). *Phys. Rev. Lett.* **18**: 597–599.

Endo, K., Koizumi, S., Otsuka, T., Ida, T., Morohashi, T., Onoe, J., Nakao, A., Kurmaev, Z., Moewes, A. and Chong, D. P. (2003). *J. Phys. Chem. A.*, **107**: 9403–9408.

Filik, J., Harvey, J. N., Allan, N. L., May, P. W., Dahl, J. E. P., Liu, S. and Carlson, R. M. K. (2006). *Spectrochim. Acta A.*, **64**: 681–692.

Fokin, A. A., Tkachenko, B.A., Gunchenko, P.A., Gusev, D. V., Schreiner, P. R. (2005). *Chem. Eur.*, **11**: 7091–7101.

Fort, R. C. (1976). *Adamantane, The Chemistry of Diamond Molecules*. Marcell Dekker, New York, NY.

Fort, R. C. and Schleyer, P. R. (1964). *Chemical Rev.*, **64**: 277–300.

Garside, D. H., Molgaard, H. V., Smith, B. L. (1968). *J. Phys. B.*, **1**: 449–457.

Gavezzotti, A. (1989). *J. Am. Chem. Soc.*, **111**: 1835–1843.

Gavezzotti, A. (1991). *J. Phys. Chem.*, **95**: 8948–8955.

Gavezzotti, A. and Filippini, G. (1992). *Acta Crystallogr. B.*, **48**: 537–545.

Graupner, R., Ristein, J., Ley, L. and Jung, C. (1999). *Phys. Rev. B.*, **60**: 17023–17029.

Hähner, G. (2006). *Chem. Soc. Rev.*, **35**: 1244–1255.

Hartono, R., Mansoori, G. A. and Suwono, A. (1999). *Chem. Eng. Commum.*, **173**: 23–42.

Hedberg, K. W. (1948) *Part II. Determination of some molecular structures by the method of electron diffraction. A. Adamantane*, PhD Dissertation, Chemistry, Cal Tech.

Hermann, K. (2004). *Nanotech. L & Bus.*, **1**: 242–260.

Hertel, I. V., Steger, H., de Vries, J., Weisser, B., Menzel, C. and Kamke, W. *Phys. Rev. Lett.*, (1992). **68**: 784–787.

Himpsel, F. J., Knapp, J.A., van Vechten, J. A. and Eastman, D. E. (1979). *Phys. Rev. B.*, **20**: 624–627.

Hoffman, A., Comtet, G., Hellner, L., Dujardin, G. and Petravic, M. (1998). *Appl. Phys. Lett.*, **73**: 1152–1155.

Hybertsen, M. S. and Louise, S. G. (1986). *Phys. Rev. B.*, **34**: 5390–5413.

James, L. W. and Moll, J. L. (1969). *Phys. Rev.*, **183**: 740–753.

Janssens, E., Neukermans, S., Vanhoutte, F., Silverans, R. E., Lievens, P., Navarro-Vazquez, A. and Schleyer, P. v. R. (2003). *J. Chem. Phys.*, **118**: 5862–5871.

Jenkins, T. E. and Lewis, J. (1980). *Spectrochim. Acta A.*, **36**: 259–264.

Jenkins, T. E. and O'Brien, P. (1981). *Phys. Stat. Solid A.*, **67**: K161–K162.

Jensen, J. O. (2004). *Spectrochim. Acta A.*, **60**: 1895–1905.

Joblin, C., d'Hendecourt, L., Léger, A. and Défourneau, D. (1994). *Astron. Astrophys.*, **281**: 923–936.

Jochems, R., Dekker, H., Mosselman, C. and Somesen, G. (1982). *J. Chem. Thermodyn.*, **14**: 395–398.

Jorio, A.; Saito, R., Hafner, J. H., Leber, C. M., Hunter, M., McClure, T., Dresselhaus, G. and Dresselhaus, M. S. (2001). *Phys. Rev. Lett.*, **86**: 1118–1121.

Kabo, G. J., Blokhin, A. V., Kozyro, A. A., Diky, V. V., Simirsky, V. V., Ivanshkevich, L. S., Krasulin, A. P., Sevruk, V. M. and Frenkel, M. (1994). *J. Chem. Thermodyn.*, **26**: 129–142.

Kabo, G. J., Blokhin, A. V., Kozyro, A. A., Diky, V. V., Simirsky, V. V., Ivanshkevich, L. S., Krasulin, A. P., Sevruk, V. M., Marchand, A. P. and Frenkel, M. (1995). *J. Chem. Thermodyn.*, **27**: 707–720.

Kabo, G. J., Blokhin, A. V., Kozyro, A. A., Diky, V. V., Simirsky, V. V., Ivanshkevich, L. S., Krasulin, A. P., Sevruk, V. M. and Frenkel, M. (1998). *Thermochim. Acta.*, **313**: 111–124.

Kabo, G. J., Blokhin, A. V., Charapennikau, M. B., Kabo, A. G. and Sevruk, V. M. (2000). *Thermochimica Acta.*, **345**: 125–133.

Karásek, P., Planeta, J. and Roth, M. (2006). *J. Chem. Eng. Data.*, **51**: 616–622.

Karásek, P. Planeta, J. and Roth, M. (2008a). *J. Chem. Eng. Data.*, **53**: 160–164.

Karásek, P. Planeta, J. and Roth, M. (2008b). *J. Chem. Eng. Data.*, **53**: 816–819.

Karle, I. L. and Karle, J. (1965). *J. Am. Chem. Soc.*, **87**: 918–920.

Kim, J. S. and Sohn, Y. H. (2003). *Parkinsonism Relat. Disord.*, **9**: S99–S104.

Kittel, C. (2005). *Introduction to Solid State Physics*, John Wiley & Sons, Inc. New Caledonia, USA.

Klünder, K. Photoelectron Spectroscopy on Molecular Diamond. Diploma Thesis. Technical University of Berlin, 2007.

Kong, L.G., tan, Z. C., Xu, J., Meng, S. H. and Bao, X. H. (2003). *J. Chem. Thermodydnamics*, **35**: 1897–1904.

Konstantinova, K. K. and Kurbatova., S. V. (2006). *Russian J. Appl. Chem.*, **79**: 1251–1259.

Kraska, T., Leonhard, K. O., Tuma, D. and Schneider, G. M. (2002). *J. Supercritical Fluids*, **23**: 209–224.

Landa, S., Machacek, V. and Mzourek, V. (1933) *Collect. Czech. Chem. Commum.*, **5**: 1.

Lenzke, K, Landt, L., Hoener, M, Thomas, H., Dahl, J. E., Liu, S. G., Carlson, M. K., Möller, T. and Bosedt, C. (2007). *J. Chem. Phys.*, **127**: 084320.

Liu, S-G, Dahl, J. E. and Carlson, R. M. K. (2004) *Nanotech Conf. Technical Program Abstract*, NSTI, Cambridge, MA.

Love, J. C., Love J. C., Estroff, L.A., Kriebel, J. K., Nuzzo, R. G. and Whitesides, G. M. (2005). *Chem. Rev.*, **105**: 1103–1169.

Lu, A. J., Pan, B. C. and Han, J. G. (2005). *Phys. Rev. B.*, **72**: 35447.

Mansson, M., Rapport, N. and Westrum , E. F. (1970). *J. Am. Chem. Soc.*, **92**: 7296–7299.

Mansoori, G. A. (2007). *Advances in Chemical Physics*, **136**: 207–258.

Marsusi, F., Mirabbaszadeh, K. and Mansoori, G. A. (2009). *Physica E.*, **41**: 1151–1156.

McIntosh, G. C., Yoon, M., Berber, S. and Tománek, D. (2004). *Phys. Rev. B.*, **70**: 045401–045402.

McKervey, M. A. (1980). *Tetrahedron*, **36**: 971–992.

Miller, D. J., Hawthorne, S. B., Gizir, A. M. and Clifford, A. A. (1998). *J. Chem. Eng. Data.*, **43**: 1043–1047.

Monto, A. S. (2003). *Vacine*, **21**: 1796–1800.

Morar, J. F., Himpsel, F. J., Hollinger, G., Hughes, G. and Jordan, J. L. (1985). *Phys. Rev. Lett.*, **54**: 1960–1963.

Morozov, I. S., Ivanova I. A. and Lukicheva, T. A. (2001). *Pharm. Chem. J.*, **35**: 235–238.

Nunes, N., Moreira, L., Leitão, R. E. and Martins, F. (2007). *J. Chem. Thermodynamics*, **39**: 1201–1205.

Oomens, J., Polfer, N., Prali, O., Ueno, Y., Maboudian, R., May, P.W, Filik, J., Dahl., J. E., Liu, S. and Carlson, R. M. K. (2006). *J. Mol. Spectrosc.*, **238**: 158–167.

Paoletti, A. and Tucciarone, A., Eds. (1997). The Physics of Diamond, Proceedings of the International School of Physics Enrico Fermi, IOS Press, Amsterdam.

Park, S. J., Kwak, T. Y. and Mansoori, G. A. (1987). *Int. J. Thermophysics*, **8**: 449–471.

Pate, B. B. (1986). *Surf. Sci.* **165**: 83–142.

Poot, W., Kruger, K. M. and de Loos, T. W. (2003). *J. Chem. Thermod.*, **35**: 591–604.

Poot, W. and de Loos, T. W. (2004). *Fluid Phase Equil.*, **221**: 165–174.

Ramachandran, G. and Manogaran, S. (2007). *Theochem.*, **816**: 31–41.

Reiser, J., McGregor, E., Jones, J., Enick, R. and Holder, G. (1996). *Fluid Phase Equil.*, **117**: 160–167.

Reynolds, P. A. (1978). *Acta Cryst., A* **34**: 242–249.

Richardson, S. L., Baruah, T., Mehl, M. J. and Pederson, M. R. (2005). *Chem. Phys. Lett.*, **403**: 83–88.

Roberts, P. J. and Ferguson, G. (1977). *Acta Cryst. B.*, **33**: 2335–2337.

Rössling, G. L. and Franck, E. U. (1983). *Ber. Bunsen-Ges. Phys. Chem.*, **87**: 882–890.

Sanders, N. D. (1986). *Ind. Eng. Chem.*, **25**: 169–171.

Sandrock, R. (1982). *Rev. Sci. Instrum.*, **53**: 1079–1081.

Schleyer P. V. (1969). *Angewandte Chemie* Int'l Ed'n in English, **8**: 529–529.

Schreiner, P. R., Fokin A. A., Fokina, N. A., Tkachenko, B. A, Hausmann, H., Serafin, M., Dahl, J. E. P., Liu, S. G. and Carlson, R. M. K. (2006). *J. Org. Chem.*, **71**: 8532–8540.

Shaporenko, A., Brunnbauer M., Terfort A, Johansson L. S. O., Grunze M. and Zharnikov M. (2005). *Langmuir*, **21**: 4370–4375.

Smith, V. S. and Teja, A. S. (1996). *J. Chem. Eng. Data* **41**: 923–925.

Spinella, G. M., Andrews, J. T. S. and Bacon, W. E. (1978). *J. Chem. Thermodyn.*,**10**: 1023–1032.

Steele, W. V. and Watt, I. (1977). *J. Chem. Thermodyn.*, **9**: 843–849.

Stöhr, J. (1996). NEAFS Spectroscopy, Springer, Berlin.

Swaid, I., Nickel, D. and Schneider, G. M. (1985). *Fluid Phase Equilib.*, **21**: 95–112.

Timmermans, J. (1961). *J. Phys. Chem. Solids.*, **18**: 1–8.

Tkachenko, B. A., Fokina, N. A., Chernish, L. V., Dahl, J. E. P., Liu, S. G., Carlson, R. M. K., Fokin, A. A. and Schreiner, P. R. (2006). *Org. Lett.*, **8**: 1767–1770.

Ulman, A. (1996). *Chem. Rev.* **96**: 1533–1554.

van Buuren, T., Dinh, L. N., Chase, L.L., Sickhaus, W. J. and Terminello, L. J. (1998). *Phys. Rev. Lett.*, **80**: 3803–3806.

van der Weide, J., Nemanich, R. J. (1993). *Appl. Phys. Lett.*, **62**: 1878–1880.

van Ekeren, P. J., van Genderen, A. C. G. and van den Berg G. J. K. (2006). *Thermochimica Acta.*, **446**: 33–35.

van Miltenburg, A., Poot, W. and de Loos, T. W. (2000). *J. Chem. Eng. Data.*, **45**: 977–979.

Wang, Y, Kioupakis, E.; Lu, X, Wegner, D.; Yamachika, R., Dahl, J. E., Carlson, R. M. K., Louise, S.G. and Crommie, M. F. (2008). *Nat. Mater.*, **7**: 38–42.

Westrum, E. F. (1961). *J. Phys. Chem. Solids*, **18**: 83–85.

Westrum, E. F., McKervey, M. A., Andrews, J. T. S., Fort Jr., R. C. and Clark, T. (1978). *J. Chem. Thermod.*, **10**: 959–965.

Willey, T. M., Bostedt, C., van Buuren, T., Dahl, D. E., Liu, S. G., Carlson, R. M. K., Meulenberg, R. W., Nelson, E. J., Terminello, L. J. and Möller, T. (2005). *Phys. Rev. Lett.*, **95**: 113401.

Willey, T. M., Bostedt, C., van Buuren, T., Dahl, D. E., Liu, S. G., Carlson, R. M. K., Meulenberg, R. W., Nelson, E. J. and Terminello, L. J. (2006). *Phys. Rev. B.*, **74**: 205432.

Willey, T. M., Fabbri, J. D., Lee, J. R. I., Schreiner, P. R., Fokin, A. a., Tkachenko, B. a., Fokina, N. A., Dahl, J. E. P., Carlson, R. M. K., Valence, A. L., Yang, W., Terminello, L. J., van Buuren, T. and Melosh, N. A. (2008). *J. Am. Chem. Soc.*, **130**: 10536–10544.

Williamson, A. J., Bostedt, C., van Buren, T., Willey, T. M, Terminello, L. J. and Galli, G. (2004). *Nano Lett.*, **4**: 1041–1045.

Wong, W. K. and Westrum, E. F. (1970). *J. Phys. Chem.*, **74**: 1303–1308.

Xue, Y. and Mansoori, G. A. (2008). *Int'l J. Nanoscience*, **7**: 63–72.

Yang, W. L., Brouet, V., Zhou, X. J., Choi, H. J., Louie, S. G., Cohen, M. L., Kellar, S. A., Bogdanov, P. V., Lanzara, A., Goldoni, A., Parmigiani, F., Hussain, Z. and Shen, Z. X. (2003). *Science*, **300**: 303–307.

Yang, W. L., Fabbri, J. D., Willey, T. M., Lee, J. R. I., Dahl, E., Carlson, R. M. K., Schreiner, P. R., Fokin, A. A., Tkachenko, B. A., Fokina, N. A., Meevasana, W., Mannella, N., Tanaka, K., Zhou, X. J., van Buuren, T., Kelly, M. A., Hussain, Z., Melosh, N. A. and Shen, Z. X. (2007). *Science*, **316**: 1460–1462.

Yashkin, S. N., Novoroselova, O. V. and Svetlov, D. A. (2008). *Russian J. Phys. Chem. A.*, **82**: 849–854.

Zharnikov, M. and Grunze, M. (2002). *J. Vac. Sci. Technol. B.*, **20**: 1793–1807.

4

Diamondoids as Nanoscale Building Blocks

4.1. Introduction

A short and precise definition of nanotechnology is the statement by the US National Science and Technology Council (NSTC, 2000) which states:

> "The essence of nanotechnology is the ability to work at the molecular level, atom by atom, to create large structures with fundamentally new molecular organization. The aim is to exploit these properties by gaining control of structures and devices at atomic, molecular, and supramolecular levels and to learn the efficient manufacturing and use of these devices".

In short, nanotechnology is the ability to build micro and macro materials and products with atomic precision (NSTC, 2000).

Nanotechnology may also be defined as the practical applications of nanoscience to fulfill a need or bring about a new possibility. Nanoscience is the knowledge about the collective behavior and nature of a group of electrons, ions, atoms, molecules and their interactions and properties when their number is limited and they occupy a space of up to 200 nanometers in at least one dimension. Science, in general, is the pursuit and accumulation of knowledge in order to understand and orient nature and human activities. To establish a simple distinction between nanoscience and nanotechnology is not an easy task, as both are intimately related.

The promise and essence of the nanoscale science and technology is based on the demonstrated fact that materials at the nanoscale have properties (i.e., chemical, electrical, magnetic, mechanical, and optical) quite different from the bulk materials. Some of such properties are, somehow, intermediate between properties of the smallest elements (atoms and molecules) from which they can be composed of, and those of the macroscopic materials (Mansoori, 2005).

Two different methods are envisioned for nanotechnology to build nanostructured systems, components and materials: One method is the "top-down" approach, and the other is the "bottom-up" approach (Siegel *et al.*, 1999). In the top-down approach, the idea is to miniaturize macroscopic structures, components and systems towards a nanoscale of the same. In the bottom-up approach, the atoms and molecules constituting the building blocks are the starting point to build the desired nanostructure.

In the top-down approach, a macro-sized material is reduced in size to reach the nanoscale dimensions. The photolithography used in the semiconductor industry is one example of the top-down approach. In the bottom-up strategy, we need to start with molecular building blocks (MBBs) and assemble them to build a nanostructured material (Mansoori *et al.*, 2007). Diamondoids are one of the best MBBs for nanotechnology.

The most fundamentally important aspect of the bottom-up approach is that the nanoscale building blocks, because of their sizes of a few nanometers, impart to the nanostructures created from them new and possibly preferred properties and characteristics heretofore unavailable in conventional materials and devices. For example, metals and ceramics produced by consolidating nanoparticles with controlled nanostructures are shown to possess properties substantially different from materials with coarse microstructures. Such differences in properties include greater hardness, higher yield strength, and ductility in ceramic materials. The band gap of nanometer-scale semiconductor structures increases as the size of the microstructure decreases, raising expectations for many possible optical and photonic applications. Considering that nanoparticles have much higher specific surface areas, in their assembled forms there are large areas of interfaces. One needs to know in detail not only the structures of these interfaces, but also their local chemistries and the effects of segregation and interaction among MBBs, and also between MBBs and their surroundings. Nanostructure sizes, size distributions, compositions and assemblies are key aspects of nanoscience and nanotechnology, and it is important to understand these aspects as well as possible.

Nanotechnology MBBs are distinguished for their unique properties. Some examples are graphite, fullerene, carbon nanotube, diamondoids, nanowires, nanocrystals, and amino acids. All these molecular building blocks, and more, are candidates for various applications in nanotechnology. These building blocks have quite unique properties which are not found in small molecules. Some of these MBBs are electrical conductors, some are semiconductors, some are photonic, and the characteristic dimension of each is a few nanometers. For example, carbon nanotubes are about five times lighter and five times stronger than steel. Many nanocrystals are photonic, and they guide light through air since their spacing of the crystal pattern is much smaller than the wavelength of light being controlled. Nanowires can be made of metals, semiconductors, or even different types of semiconductors within a single wire. They are upwards of ten nanometers and can be made into a conductor or semiconductor. Aminoacids and DNA, the basis for life, can also be used to build nanomachines. Adamantane (a diamondoid) is a tetrahedrally symmetric stiff hydrocarbon that provides an excellent building block for positional (or robotic) assembly as well as for self-assembly. In fact, over 20,000 variants of adamantane have been identified and synthesized, and even more are possible (Mansoori, 2005, 2008), providing a rich and well-studied set of MBBs.

The applications of MBBs would enable the practitioner of nanotechnology to design and build systems on a nanometer scale. The controlled synthesis of MBBs and their subsequent assembly (self-assembly, self-replication, or positional-assembly) into nanostructures is a fundamental theme of nanotechnology. These promising nanotechnology concepts with far-reaching implications (from mechanical to chemical processes; from electronic components to ultra-sensitive sensors; from medical applications to energy systems; and from pharmaceuticals to agricultural and food chains) will impact every aspect of our future.

In this chapter, we present applications of diamondoids in nanotechnology. This will include a discussion of advantages of diamondoids as nanotechnology MBBs; their applications in mechanosynthesis, in host–guest chemistry and in inclusion compounds formation. We also present a detailed discussion about the formation and applications of cyclodextrin–adamantane inclusion complex (CAIC) which is as a supramolecular inclusion compound with several unique applications in medicine.

4.2. Diamondoids as Molecular Building Blocks for Nanotechnology

Amino acids, diamondoids, carbon nanotubes, DNA, fullerenes, functionalized polymer nanofibers, Ge/Si/SiO_2 nanostructures, gold and silver nanoparticles, graphite, nanodiamond powders, polyhedral heteroborane clusters, proteins, titanium dioxide, and zinc oxide nanorods are among the more important MMBs for nanotechnology (Mansoori *et al.*, 2007). As fundamental construction parts for bottom-up approach, MMBs are ultimately responsible for unique, sometimes unprecedented nanosystems characteristics and properties, quite different from those exhibited by bulky materials of the same chemical nature.

As diamondoids are almost entirely constituted by sp^3 bonded carbons arranged in a three-dimensional network, these stiff, dense, covalent solids present many advantages over other types of MMBs:

(a) With some restrictions, diamondoids are reasonably well described by molecular mechanics models. Thus, investigations at computational level can be performed in atomic details, in a fast and affordable way (Drexler, 1992).
(b) Diamondoid specificity towards receptors have advantages over proteins, because diamondoid structures exhibit greater stiffness, higher thermal resistance, and a greater atom number density than proteins, enabling stronger van der Waals interactions, leading to higher binding energies and possible higher affinities (Drexler, 1992).
(c) In spite of being a small molecule, adamantane possesses six primary and four secondary linking sites (Fig. 4.1). Thus, it is possible to build complex, but orderly, three-dimensional structures with adamantane and higher order diamondoids, but only tubular or planar structures can be constructed with graphite or other *quasi*-two-dimensional crystalline substances.

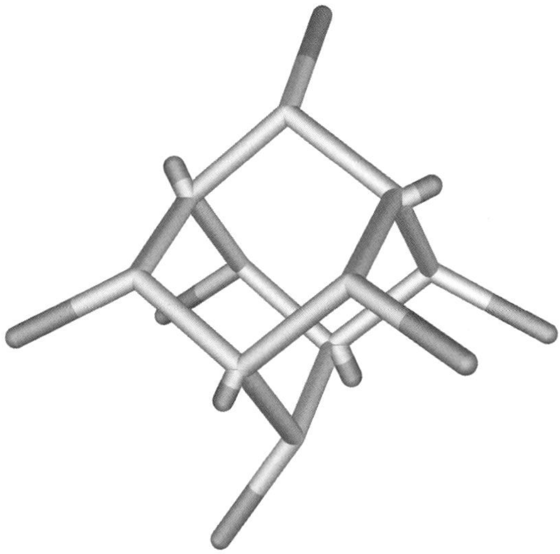

Fig. 4.1. Six primary and four secondary linking sites of adamantane.

(d) The root-mean square amplitude of vibration for diamond at room temperature is 0.002 nm. In comparison, the same property of for a 1 N/m AFM (atomic force microscope) cantilever at room temperature is 0.06 nm (Rugar and Ransma, 1990). Such low vibrational amplitudes are very desirable for precision construction and operation of nano-electro-mechanical systems (NEMS). Moreover, micro-electro-mechanical systems (MEMS) constructed with diamondoids exhibit expected operational lifetimes 10,000 times higher than MEMS made out of other materials, such as polysilicon (Carlson *et al.*, 2007).

4.2.1. *Diamondoids in mechanosynthesis*

Diamondoids may become a fundamental material for construction of nanometric mechanical systems. Their conformationally rigid cages, allied to their strength and stiffness have made these MMBs a target for theoretical studies on molecular manufacturing since the middle 1980's. Molecular manufacturing, according to Drexler (1992), is the construction of objects to complex atomic specifications using sequences of chemical reactions directed by nonbiological molecular machinery. Drexler (1986) proposed a machine-directed bottom-up approach starting with atoms to form diamond flexible fibers. Such fine fibers would exhibit over 50 times more strength than the same weight of aluminum.

Mechanosynthesis is considered as a fundamental strategy to molecular manufacturing. The main idea is to guide chemical reactions at atomic or molecular level through precisely applied mechanical forces, instead of using local steric and electronic properties of the reagents as driving forces (Drexler, 1992). Such level of manipulation is presently unavailable as an experimental procedure. An initial demonstration of mechanosynthesis was performed with Si individual atoms by vertical manipulation, using a near-contact atomic force microscope (NC-AFM) operated at low temperature (Fig. 4.2). As neither bias voltage nor voltage pulse was applied between the probe and the sample, this experiment is considered the first demonstration of exclusively mechanical positional synthesis or mechanosynthesis (Oyabu *et al.*, 2003).

Diamond mechanosynthesis tools have been studied theoretically by various investigators (Walch and Merkle, 1998; Merkle and Freitas, 2003; Peng *et al.*, 2003; Mann *et al.*, 2004; Freitas, 2004a, 2004b; Peng *et al.*, 2006; Freitas *et al.*, 2007; Freitas and Merkle, 2008; Freitas, 2009; Allis *et al.*, 2011). Experimental methodologies of mechanosynthesis are still to come.

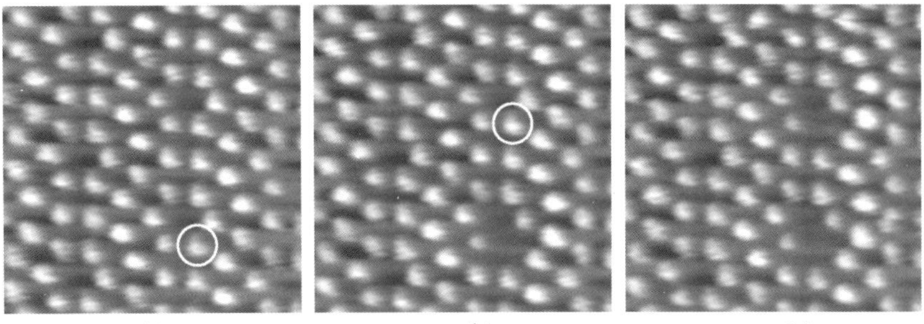

(a) (b) (c)

Fig. 4.2. Sequence showing purely mechanical manipulation of two single Si atoms, marked with a circle in (a) and (b), performed successively over the selected atomic positions of a Si(111)-(7 × 7) surface, by using near contact atomic force microscopy (NC-AFM). As a result, (c) two vacancies were created at the selected atomic positions (Oyabu *et al.*, 2003).

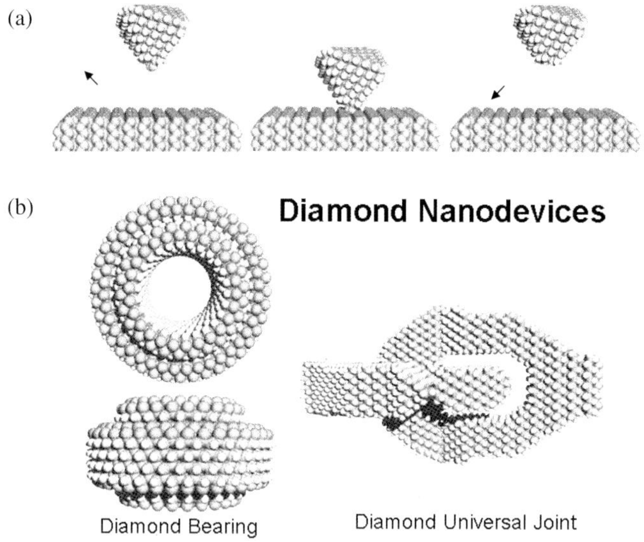

Fig. 4.3. (a) Model shows mechanosynthetic tool positioning a carbon dimmer (pointed out by the black arrow) on a flat diamond surface. Repetition of precise movements like this would supposedly result in positional assembly of diamonds to form complex nanostructures as (b) a bearing or a universal joint made of diamond (white spheres represent H atoms). Adapted from Freitas, 2004a.

Figure 4.3a shows a simulation of the steps necessary to place a carbon dimer on a flat diamond surface with a mechanosynthetic tool. Once the tool is lifted away from the surface, two new carbon atoms would become covalently bonded to the diamond surface. Precise repetition with successive dimers would supposedly yield functional three-dimensional arrangements, as a diamond bearing or a universal joint (Fig. 4.3b).

Some of the envisioned applications of molecular manufacturing based on diamondoids are: The design of an artificial red blood cell called respirocyte, nanomotors, nanogears, molecular machines, and nanorobots (Merkle, 1999; Drexler, 1999; Freitas, 1998). The other potential application of molecular manufacturing of diamondoids is in the design of molecular capsules and cages for various applications including drug delivery.

In 2004, Freitas Jr. filled the first known patent on fabrication of a carbon dimer placement tool tip for positional diamond mechanosynthesis

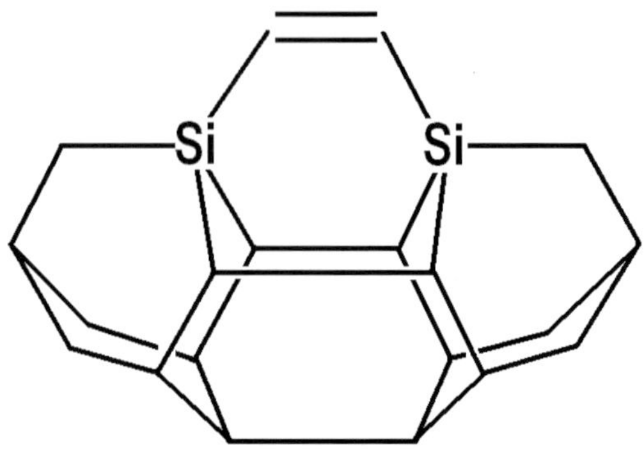

Fig. 4.4. DCB6-Si dimer placement tool tip molecule. One of the members of the DCB6-X family (X = Si, Ge, Sn, Pb) analyzed by Density Functional Theory (DFT) (Sec. 5.2.1) and found to be capable of holding and positioning a CC dimer in a manner suitable for positionally controlled diamond mechanosynthesis under vacuum, at room temperature (Merkle and Freitas, 2003).

(Freitas, 2004b). This triamantane-like molecule, called DCB6-Si (6-membered dicarbon brigde) (Fig. 4.4) was first presented by Merkle and Freitas Jr. (2003) as a member of the DCB6-X family (X = Si, Ge, Sn, or Pb). DCB6-Si exhibits a highly strained ethynylene group bonded to two Si atoms as the dicarbon bridge. According to theoretical computational analysis, this group would work as the source for the carbon dimer positioned on the diamond surface, as DCB6-Si and the other members of the DCB6-X family are stable under vacuum at room temperature or even in higher temperatures. In addition, ethynylene group would remain in the horizontal position, with low probability of rearrangements. When Si atoms are successively replaced by heavier members of IV A group, dimer discharge is facilitated because of the weakening of the X–C≡C bond. In the same 2004 patent, Freitas Jr. (2004b) presented a process to deposit a large diamond handle attached to the tool tip molecule to complete the tool set and proposed to extend the use of this process to other mechanosynthetic reactions involving nondiamond substances.

For the concept of molecular manufacturing to become successful, a systematic study of the fundamental theory of the molecular processes involved and the possible technological and product capabilities are needed (Herman, 1999).

4.2.2. *Naturally occurring diamondoids as molecular components of nanosystems*

In this section, we focus on the diamondoid molecules which are found in petroleum fluids. Lower diamondoids (adamantane, diamantane, triamantane, and their alkylated derivatives) have been found in naturally occurring petroleum fluids since the 1930's while natural occurrence of higher diamondoids (tetramantane, pentamantane, and hexamantane) was first reported in 1995 (Lin and Wilk, 1995). These and other naturally-occurring diamondoids may also be used as structural components of nanosystems. There have been a number of patents on purification of higher diamondoids from petroleum and their subsequent use in molecular electronics and nanotechnology in recent years (Dahl and Carlson, 2002, 2003a–d; Dahl *et al.*, 2004; Carlson *et al.*, 2007).

As we discussed in Sec. 1.2, the number of possible isomers for higher diamondoids increases rapidly with the number of cages. While there are four tetramantane structures: [1(2)3], [121] and two enantiomeric [123] tetramantanes, (see Fig. 1.4, where one of the two [123]tetramantane enantiomers is shown), there are nine isomer pentamantanes with molecular formula $C_{26}H_{32}$ and one pentamantane with molecular formula $C_{25}H_{30}$. There are 39 possible structures for hexamantanes and 160 postulated structures for heptamantanes (Carlson *et al.*, 2007). Hence, a great variety of shapes and lengths can be found within a relatively narrow range of carbon atoms constituents. Naturally-occurring diamondoids may also be chemically modified to bear alkyl groups, heteroatoms and to interconnect with other diamondoids to form even more complex structures that may be used as functional parts in nanosized systems.

Figure 4.5 presents an overview of a proposed process, starting from the purification of the diamondoids from petroleum, followed by the cataloging and indexation of the isolated molecules, their use as structural components and their assembly to achieve functional subsystems. This cataloging and indexation system proposed in the patent by Carlson *et al.* (2007) is based on the projection of atoms from the three-dimensional molecular structure onto the most appropriate plane, which are, according to the inventors, the (110) crystallographic plane for axial rods, the (111) and the (100) planes for screws, and the (100) plane for gears.

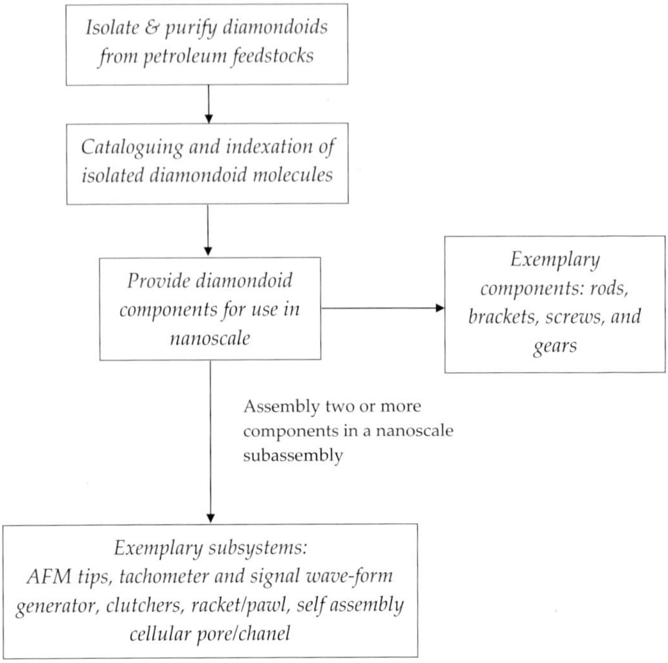

Fig. 4.5. An overview of converting petroleum diamondoids in molecular components of nanosystems (Carlson and Dahl, 2007).

Figure 4.6 (a to f) exhibits possible other petroleum higher diamondoids and derivatives proposed by Carlson *et al.* (2007) as having suitable sizes and shapes to be used as molecular components for nanometric systems. [12121]hexamantane presented in Fig. 4.6a, is one of many axial (rod-shape) diamondoids which may be found in petroleum. Other proposed examples are [121]tetramantane, [1212]pentamantane, [121212] heptamantane, [1212121]octamantane and even larger ones if they could be found in petroleum.

Molecules presenting other forms, such as "**L**", "**Y**", "**X**" and "+" shapes are also required for components of nanosystems, thus, structures like [1213]pentamantane in Fig. 4.6b, the smallest possible "**L**"- shaped diamondoid, would be a useful building block. In addition, due to its asymmetry, such component may impart structural specificity, which is vital in biological applications. Other chiral structures are shown in Fig. 4.6c: A pair of 30-carbon atom enantiomeric [12341]hexamantanes

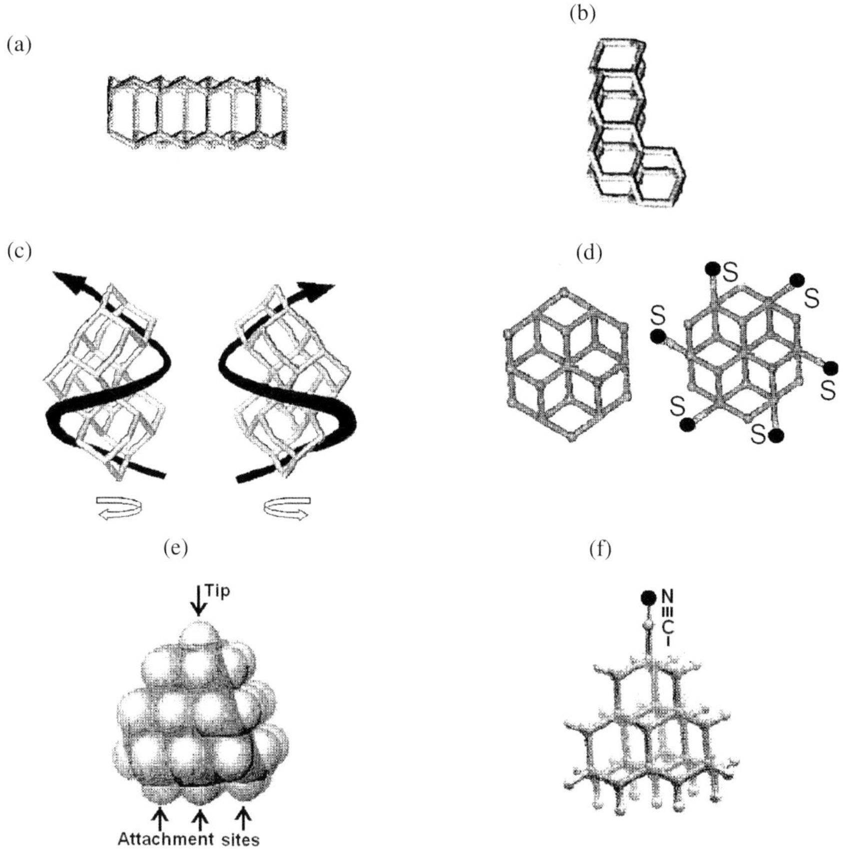

Fig. 4.6. Higher diamondoids which may be used as building blocks in nanoscale construction:

(a) [12121]hexamantane with 1.138 nm in length suggested as a rigid diamondoid rod. (b) Chiral "L"-shaped [1213]pentamantane. (c) A pair of enantiomeric [12341]hexamantanes. Black spirals show the right- and left-handed grooves of these 30-carbon-atom molecular screws. White curved arrows indicate the direction of twist needed to screw into a material. (d) *Peri*-condensed hexamantane (see also Fig. 1.3) may be used as a molecular rotor when six thiol (–SH) groups are bonded to tertiary carbons on its periphery. (e) [1(2,3)4]pentamantane as an atomic force microscope (AFM) probe. (f) Nitrile derivative of E is suggested as a possible way of improving probe performance. Adapted from Carlson *et al.*, 2007.

molecules exhibiting opposite helicities may be in hand to hold pieces together, when used as screws.

Flat diamondoid structures, if found or if they could be synthesized, are important in nanoscale construction as gears, rotors, and impellers. For example, *Peri*-condensed hexamantane ([12312]hexamantane) is a basically disc-shaped molecule having "teeth" or "blades" as extensions which fits this description. In principle, extensions may be attached to it by chemical functionalization to generate the necessary other structures. Figure 4.6D shows the molecular structure of *Peri*-condensed [12312] hexamantane (see also Fig. 1.4). As an example of its nanoscale application is its use as a molecular rotor when six thiol (–SH) groups are bonded to tertiary carbons on its periphery, forming clockwise-biased blades.

Compact three-dimensional diamondoids may be used as AFM tips. For example, [1(2,3)4]pentamantane (Fig. 4.6e) is a pyramidal shaped molecule suitable for such application. The tip of the "pyramid" comprises a tertiary carbon bonded to a hydrogen atom while seven to ten atoms at the attachment site are available to be coupled to an assembler arm. The main advantages of applying compact pyramidal diamondoids as AFM tips, in place of other advanced technologies, are related to the sensibility imparted by the presence of a single atom at the end of the probe which is considered as "the sharpest possible tip", instead of a large number of atoms found in round tips made of nanotubes, for example. Moreover, the ease of functionalization may allow the tailoring of the probe for very specific applications. For example, a thiol group would allow efficient probing and manipulation of certain single metal atoms. Similarly, the linear nitrile (–C≡N) group attached to [1(2,3)4]pentamantane (Fig. 4.6f) would provide a chemically-modified AFM tip presenting increased sensibility.

Diamondoids, in general, are amenable for the design of self-assembled nanoscale structures. Self-assembly is defined as the spontaneous assembly of molecules into structured, stable, noncovalently joined aggregates (Whiteside *et al.*, 1991; Mansoori, 2005). Diamondoids of different shapes are super-imposable by nature, as they generally fit the diamond carbon lattice. When arranged in a proper manner, they tend to mimic this strong lattice to build specific structures. In Chapter 7, we report the results of self-assembly simulation of some diamondoids and derivatives.

Alternatively, a given nanostructure design might be planned to have different diamondoid components as almost perfectly-fitted building blocks, leaving only short segments between the parts. Linking chemical sites can be used to interconnect these short segments and to finish the design keeping the whole framework together and as close as possible to the diamond structure. Disulfide (–S-S–) bonds are preferable as linkers because the chemical processes involved generate side products and waste materials recoverable by hydrocracking (Maesen *et al.*, 2007). Figure 4.7(a and b) is an example of a pore-like superstructure which is claimed resulting from interconnection of six [12312]hexamantanes due to two thiol groups disulfide bond (not shown) each in opposite sites of every hexamantane

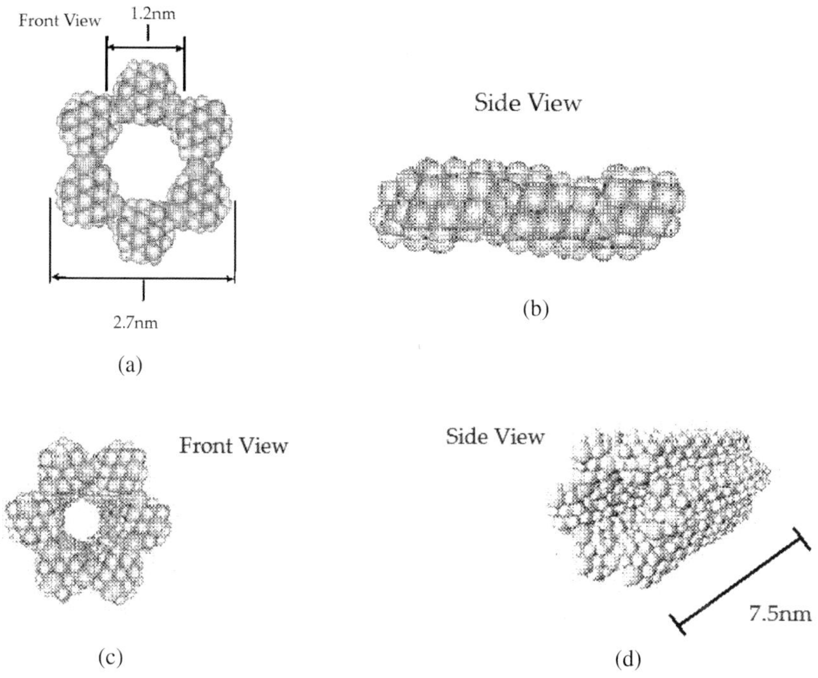

Fig. 4.7. (a) Membrane pore flat structure containing six *peri*-condensed hexamantane moieties linked by disulfide bridges (hexamantane hexamer) (front view) (b) Side view of "A". (c) Transmembrane channel formed by the stacking of eight disulfide-bonded layers of the hexamantane hexamer presented in "A" (side view). (d) Front view of "C". Adapted from Carlson *et al.*, 2007.

outerring (Carlson *et al.*, 2007). It is said to achieve this structure, the oxidation process was used to link every two thiol groups from distinct thiolated molecules together as a disulfide bridge (–S-S–) to yield a hexamer. Size-exclusion chromatography separated target hexamers, then, remaining thiol intermediates were removed using other chromatography methods. HPLC (high performance liquid chromatography) equipped with Hypercarb™ columns which is a highly shape-selective column yield the purified disc-shaped product as shown by Fig. 4.7(a and b).

Tubular nanostructures may be obtained by taking advantage of stacking tendency (self-assembly or coacervation) of flat pore-like nanostructures. However, more rigid nano-arrangements might be accomplished by having extra disulfide bridges attaching additional layers of hexamers. Thus, complex, stronger nanostructures may be constructed. For example, a tubular nanostructure formed by two disulfide-linked layers of hexamers would be 1.7 nm long, which is equivalent to the thickness of a lipid monolayer. In contrast, a tubular nanostructure containing eight covalently-bonded layers of hexamers would be 7.5 nm in length, Fig. 4.7(c and d), long enough to span a lipid bilayer similar to a cellular membrane. Both tubular structures may have interesting biological applications as artificial membrane channels.

4.3. Diamondoids for Host–Guest Chemistry

The science of host–guest chemistry was born out of a class of macrocyclic compounds which include crown ethers, calixarenes, cyclobisamides, cyclodepsipeptides, norbornene-constrained cyclic peptides, cyclo (Adm-Cyst)3, and cyclodextrins. The main aim in host–guest chemistry is to construct molecular receptors by a self-assembly process so that such receptors could, to some extent, gain molecular recognition capability. The goal of such molecular recognition capability is to either mimic or block a biological effect caused by molecular interactions (Chadwick and Widdows, 1991; Mansoori, 2005; Mansoori, 2007; Ramezani and Mansoori, 2007; Moghaddam *et al.*, 2011).

Crown Ethers: The first family of these host compounds is macrocyclic polyethers called "crown ethers" as shown in Fig. 4.8 and it was first

Fig. 4.8. Molecular structure of ether-crown host ligands: (1) 18-Crown-6 ether ($C_{12}H_{24}O_6$); (2)–(7)-Various adamantane-containing crown ethers (named Diamond Crowns) synthesized by Ranganathan *et al.* (1999a).

discovered in 1967 by C.J. Pedersen (1967). Crown ethers are oligomers made from a quite broad family of monomers and they strongly bind with certain cations forming three-dimensional complexes. Ranganathan *et al.* (1999a) produced the first examples of a novel family of adamantane-containing crown ethers (named Diamond Crowns) as shown in Fig. 4.8. Cage-functionalized crown ethers have also been reported with oxa-adamantyl moieties as a part of the cyclic framework to provide relatively rigid crown ethers with modified ion-complexing properties and improved solubility in nonpolar solvents — a property particularly useful for studying molecular recognition and inclusion phenomena (Ranganathan *et al.*, 1999a).

The outstanding feature of adamantane-bearing crown ethers (which are also called "Diamond Crowns") is that α-amino acids can be incorporated to the adamantane-crown backbone (Ranganathan *et al.*, 1999a). This family of compounds provides the valuable models for studying selective host–guest chemistry, ion transportations and ion-complexation (Ranganathan *et al.*, 1999b).

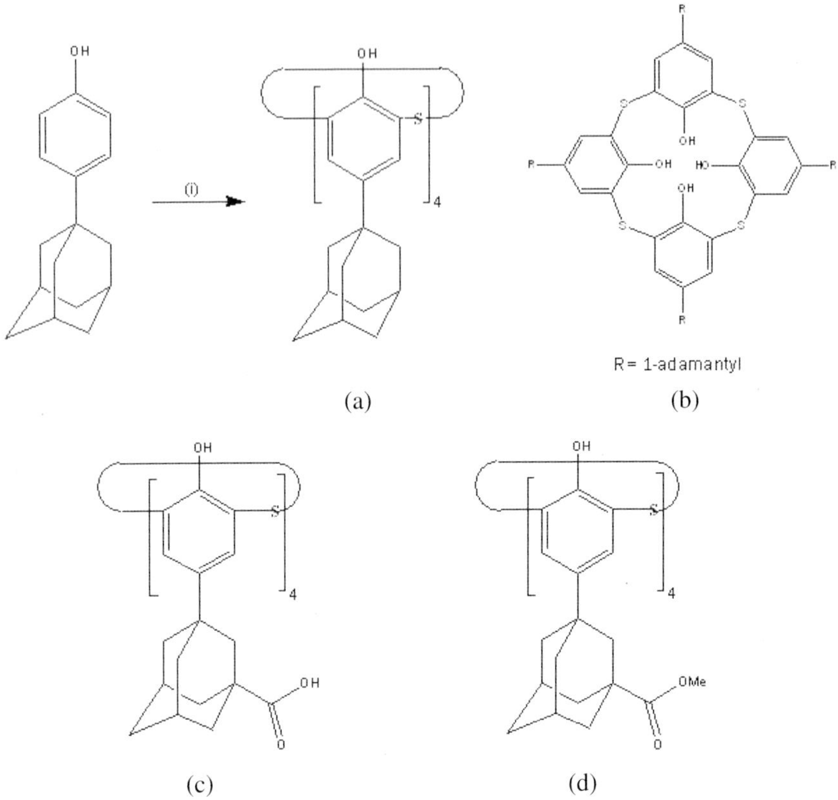

Fig. 4.9. (a) Synthesis route of the molecule (b): (i) S8, NaOH, tetraethyleneglycol dimethyl ether, heat, (28%). (b) Adamantane Upper rim derivative based on the thiacalix[4] arene platform. (c, d) The carboxylic acid and ester derivative of adamantane can be also used as substituent's (Shokova *et al.*, 2002).

Calixarenes: Calixarenes are also macrocyclic compounds and they are considered as one of the best building blocks to design molecular hosts in supramolecular chemistry (Mandolini and Ungaro, 2000). Synthesis of Calix[4]arenes which have been adamantylated has been reported (Kovalev *et al.*, 1996; Hirsch *et al.*, 1997; Shokova *et al.*, 2002). In Calix[4]arenes, adamantane or its ester/carboxylic acid derivatives were introduced as substituents (Fig. 4.9).

The purpose of this synthesis was to learn how to employ the flexible chemistry of adamantane in order to construct different kinds of molecular

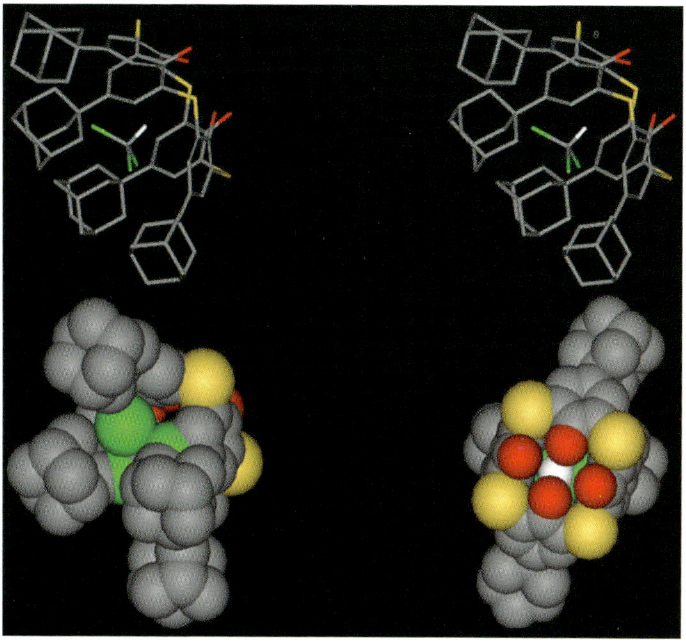

Fig. 4.10. Lateral stereo views of adamantane-derivative thiacalix[4]arene (top) presented in Fig. 4.9. A $CHCl_3$ molecule has been entrapped inside the inclusion compound. The bottom view (left bottom) and top view (right bottom) have been also shown. H atoms have been removed from the inclusion compound for more clarity. (Cl, OH, S, H, and C atoms have been colored green, red, yellow, white, and gray, respectively) (Ramezani and Mansoori, 2007).

hosts. The X-ray structure analysis of p-(1-adamantyl) thiacalix[4]arene (Hirsch *et al.*, 1997; Shokova *et al.*, 2002) demonstrated that it contained four $CHCl_3$ molecules, one of which was located inside the host molecule cavity, and the host molecule assumed the cone-like conformational shape (Fig. 4.10).

Cyclobisamides: Other types of macrocycle compounds have been synthesized using adamantane and its derivatives. Recently, a new class of cyclobisamides has been synthesized using adamantane derivatives, which shows the general profiles of amino acid (serine or cystine)-ether composites as depicted, for example by Fig. 4.11. They were shown to be efficient ion transporters (especially for Na^+ ions) in the model membranes (Ranganathan *et al.*, 2002).

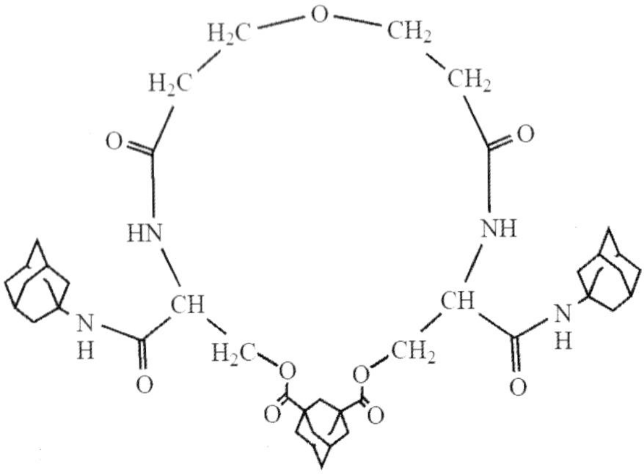

Fig. 4.11. A cyclobisamide containing adamantane (Redrawn from Ranganathan *et al.*, 2002).

Cyclodepsipeptides: Adamantane has been also used as a cage-like alicyclic (both aliphatic and cyclic) bridge to construct a new class of tyrosine-based cyclodepsipeptides (tyrosinophanes) (Ranganathan *et al.*, 1999b). Macrocyclic peptides composed of an even number of D and L amino acids can self-assemble to form a tube through which ions and molecules can be transported across the lipid bilayers. Although they rarely exist in nature, they are synthesized to be employed in the host–guest studies (Fig. 4.12a) and to act as ion transporters in the model membranes (Fig. 4.12(b and c)) (Ranganathan *et al.*, 1999b). The adamantane-bridged, leucine-containing macrocycle 4.12b shows a modest ability to transport Na^+/K^+ ions across the model membranes (Ranganathan *et al.*, 1999b). The adamantane-constrained macrocycle 4.11c is also suitable for attachment of different functional groups to design artificial proteins (Ranganathan *et al.*, 1999b).

Norbornene-constrained cyclic peptides: The adamantane-containing cyclic peptides are efficient metal ion transporters and utilization of adamantane in such compounds improves their lipophilicity and thus membrane permeability (Ranganathan *et al.*, 2000a). A new class of norbornene-constrained cyclic peptides has been synthesized using adamantane as a second bridging ligand (Fig. 4.13). The macrocycle 4.13a is a specific ion transporter for monovalent cations while the cyclic peptide 4.13b is able to transport both mono- and divalent cations across the model membranes (Ranganathan *et al.*, 2000a).

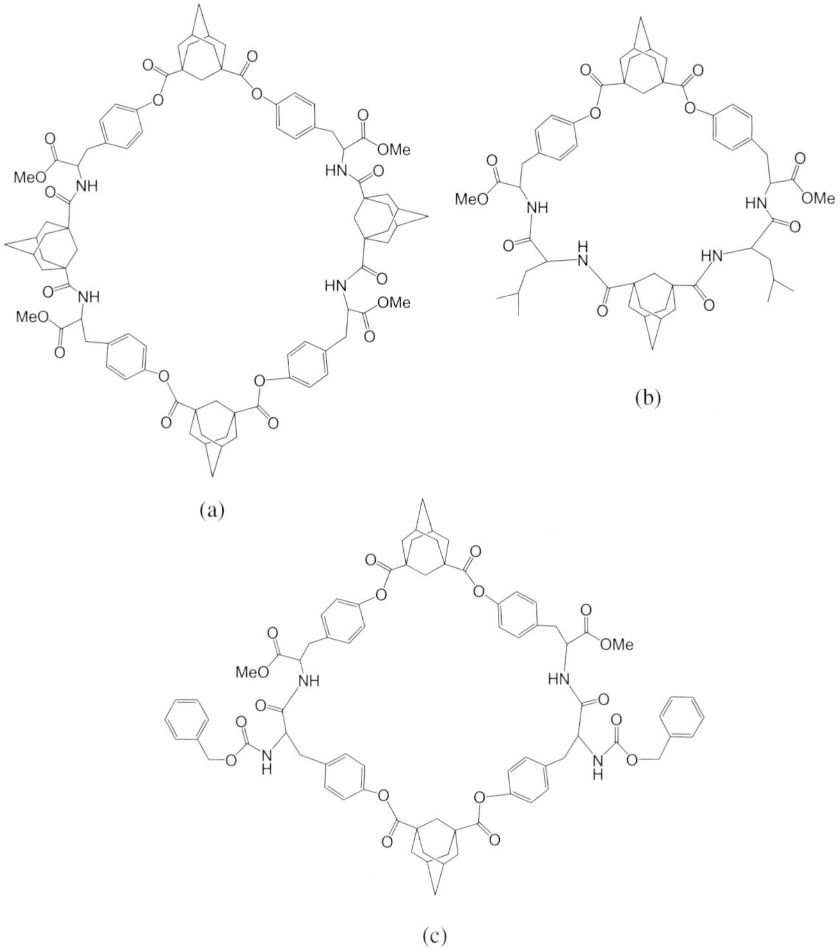

Fig. 4.12. Adamantane-bridged tyrosine-based cyclodepsipeptides are suitable models for host–guest studies and they are also able to act as ion transporters (Ranganathan *et al.*, 1999b).

Cyclo (Adm-Cyst)3: Peptidic macrocycles are especially useful models for discovering protein folding mechanisms and designing novel peptide-made nanotubes as well as other biologically important molecules. These large cyclic peptides tend to fold in such a way that they can adopt a secondary structure like β-turns, β-sheets and helical motifs. A new series of double-helical cyclic peptides have been synthesized

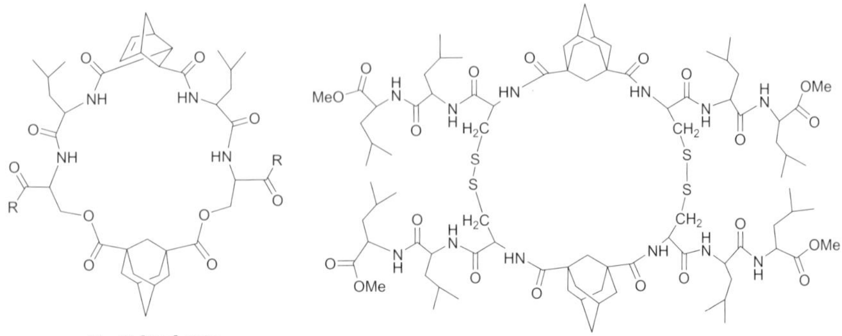

R = Val-Val-OMe

(a) (b)

Fig. 4.13. Adamantane-containing norbornene ( )-constrained cyclic peptides possess the ability to transport ions across the model membranes in both specific and nonspecific ways (Ranganathan *et al.*, 2000a).

among them are the adamantane-constrained cystine-based cyclic trimers [cyclo (Adm-Cyst)3]. They have attracted a great deal of attention due to their figure "8"-like helical topologies and special way of hydrogen binding and symmetries (Ranganathan *et al.*, 2000b; Karle, 1999) (Fig. 4.14). The cyclo (Adm-Cyst)3 molecule was able to transport K^+ ions through the model membranes and it was a valuable model to study the mechanism of secondary structure formation in proteins (Karle and Ranganathan, 2003).

Cyclodextrins: The cyclic oligosaccharides known as cyclodextrins (CDxs) with several D-glucopyranosyl residues linked by α-1, 4-glucosidic bonds as shown in Fig. 4.15. They are inclusion compounds formed by enzymatic decomposition of starch during intramolecular transglycosylation reaction (Szetjtli, 1998) and they contain six, seven, or eight glucopyranosyl units, and are called α-, β-, and γ-CDx, respectively. Depending on the number of glucose units, there are three types of natural CDxs, namely α, β, and γ, consisting of six, seven, and eight glucoses, respectively. The interior lining of the parent CDxs cavities is somehow hydrophobic. Table 4.1 shows some physical properties of α-, β-, and γ-CDx.

Adamantane is one of the best guests entrapped within the CDxs cavities (Jaime *et al.*, 1991; Fujita *et al.*, 1999; Krois and Brinker, 1998; Ogoshi *et al.*, 2007; Strenalyuk and Haaland, 2008; Ren *et al.*, 2009; Granadero *et al.*, 2010). Its noticeable association constant with

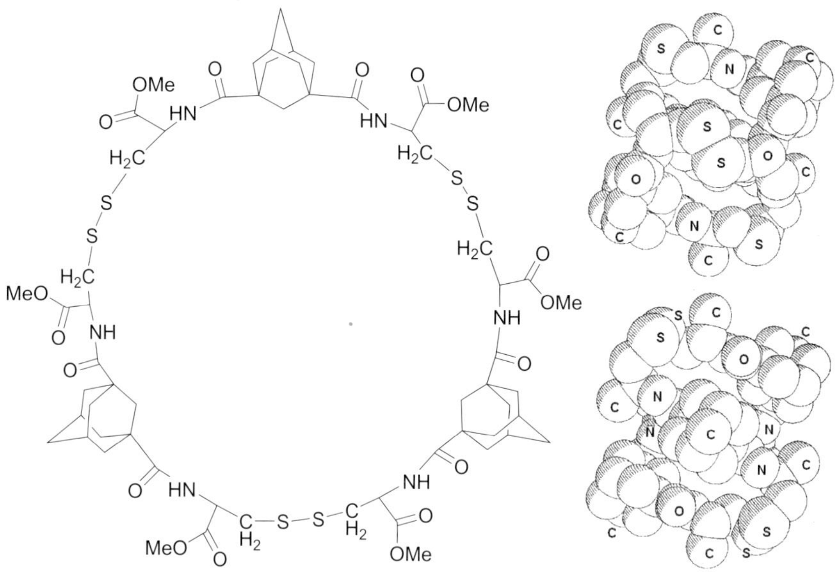

Fig. 4.14. The cyclo (Adm-Cyst)3 as adopts a "figure 8"-like helical structure. The chiral amino acid, cystine, configuration determines the helix disposition (right-handed or left handed helix). Adamantane plays an important role as a ring size controlling agent (Ranganathan *et al.*, 2000b).

CDxs (~10^4 to 10^5 M^{-1}) as reported in Table 4.2 denotes a high affinity to interact with the hydrophobic pocket of CDxs, which is a valuable linking system to join different molecules together (see Fig. 4.15).

Interestingly, this system adsorbs and immobilizes molecules on a solid support and has been exploited to immobilize an adamantane-bearing polymer onto the surface of a β-CDx-incorporated silica support. In this case, adamantane acts as a linker to attach a dextran-adamantane-COOH polymer to a solid support through a physical entrapment mechanism and thus contributes to the formation of a stationary phase for chromatographic purposes. The aforementioned stationary phase could be readily prepared under mild conditions and is stable in aqueous media. It has revealed some cation-exchange properties suggesting its application to the chromatography of proteins (Karakasyan *et al.*, 2004).

Poly-adamantane molecular rods: Covalent attachment between adamantane molecules is a key strategy to string several adamantanes

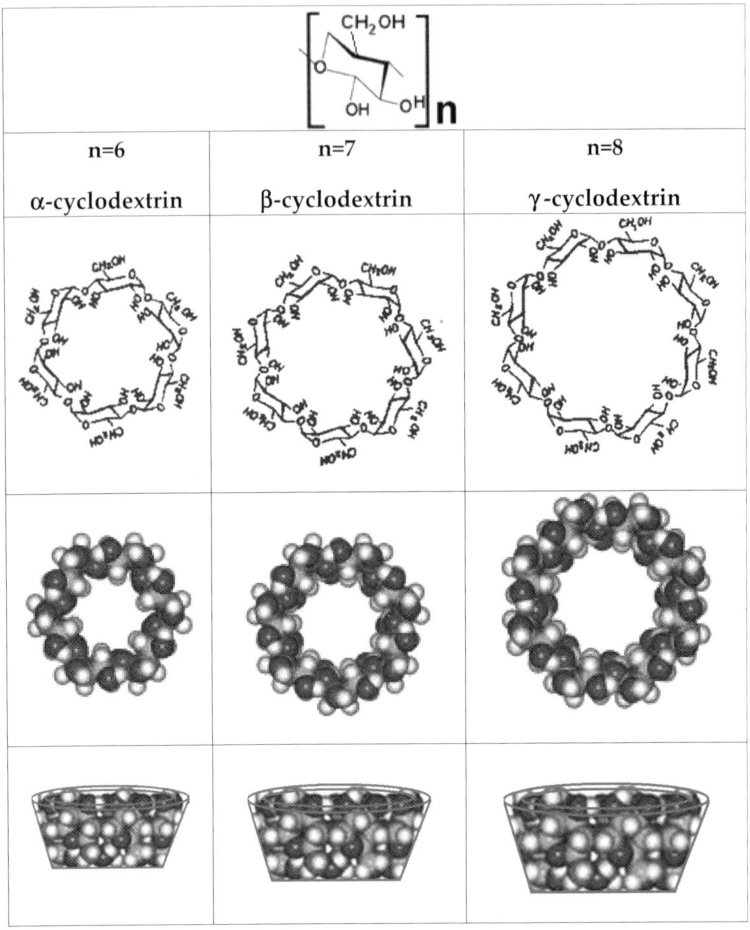

Fig. 4.15. Chemical formula and structure of Cyclodextrins. Their shapes are like truncated cones and they have relatively hydrophobic interiors. They have the ability to form inclusion complexes with a wide range of substrates in aqueous solution. This property has led to their application for encapsulation of drugs in drug delivery (Mansoori, 2005).

together and construct molecular rods. The McMurry coupling reaction was employed to obtain poly-adamantane molecular rods (see Fig. 4.16) (Ayres *et al.*, 1994). Another example of poly-adamantane molecular rods is the synthesized tetrameric 1,3-adamantane and its butyl derivative as shown in Fig. 4.17 (Ishizone *et al.*, 2001).

Table 4.1. Cyclodextrin properties (Del Valle, 2004).

Property	**α-CDx**	**β-CDx**	**γ-CDx**
Number of glucopyranose units	6	7	8
Molecular mass (g/mol)	972	1135	1297
Solubility in water (%, w/v)	14.5	1.85	23.2
Outer diameter (nm)	1.46	1.54	1.75
Cavity diameter (nm)	0.47–0.53 (0.6)[a]	6.0–6.5 (0.75)[a]	0.75–0.83 (0.9-1)[a]
Height of torus (nm)	0.79	0.79	0.79
Cavity volume (nm^3)	0.17	0.26	0.43

[a]Data in parentheses are from Cramer *et al.* (1967).

Table 4.2. Association constant (K), standard Gibbs free energy (ΔG°), enthalpy (ΔH°) and entropy (ΔS°) of formation for the interaction between cyclodextrins and adamantane-1-carboxylic acid (Cromwell *et al.*, 1985).

Cyclodextrin	**pH**[a]	**K (M^{-1})**	**ΔG⁰ (kcal/mol)**	**ΔH⁰ (kcal/mol)**	**ΔS⁰ (kcal/mol.K)**
α	4.08	1.3×10^2	-2.9 ± 0.1	-3.2 ± 0.1	-1.0 ± 0.5
α		5.0×10^2	-3.6 ± 0.1[b]	-8.8 ± 0.1[b]	-17 ± 0.5[b]
α	8.50	1.4×10^2	-2.94 ± 0.04	-3.22 ± 0.06	-0.3 ± 0.3
β	4.08	2.9×10^5	-7.45 ± 0.15	-7.53 ± 0.02	-0.1 ± 0.5
β	8.50	1.9×10^4	-5.85 ± 0.04	-4.85 ± 0.02	3.4 ± 0.2
γ	4.08	2.4×104[d]	-5.9 ± 0.1[c]	-0.1 ± 0.1	22.0 ± 0.5
γ	8.50	3.3×103	-4.81 ± 0.05	1.20 ± 0.05	20.2 ± 0.3

4.4. Adamantane in Inclusion Compounds

Supramolecular are complex and ordered aggregates composed of numerous molecules, ions and/or coordination compounds formed through attractive intermolecular forces. The construction of supermolecules has created a field of chemistry called "supramolecular chemistry" and involves two important processes: Molecular recognition and self-assembly. Details of supramolecular chemistry are discussed well by Lehn (1988).

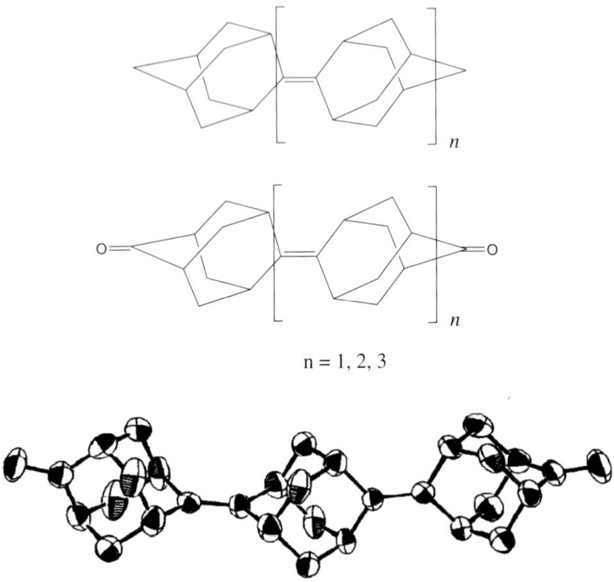

Fig. 4.16. Poly-adamantane molecular rods (Ayres *et al.*, 1994).

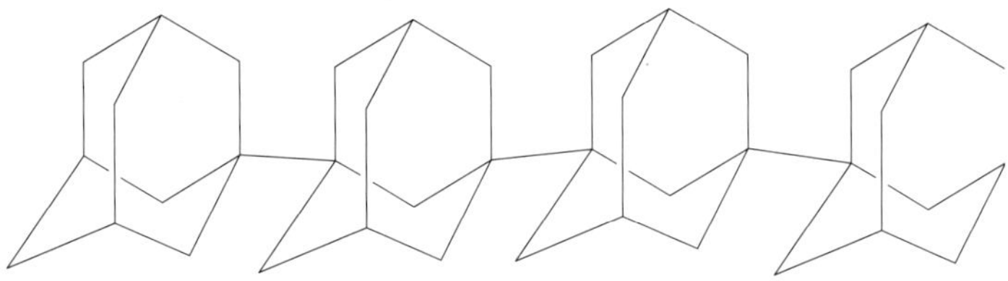

Fig. 4.17. Synthetic design of a molecular rod made of adamantanes: The tetrameric 1,3-adamantane (Ishizone *et al.*, 2001).

Molecular recognition is characterized by specific binding of molecules in structurally well-defined patterns due to intermolecular forces. Some examples of molecular recognition are enzyme–substrate interaction and DNA base pairings. Generally, the molecule that bears the binding site is called "host" or "receptor" and the "recognized" molecule is called "guest" or "substrate".

The major studied host molecules in supramolecular chemistry are crown ethers, calixarenes, cyclodextrins (CDx), criptands, cyclophanes, spherands and porphyrins (Lehn, 1988; Vogtle, 1992). Among these, CDxs are, by far, the most important and their chemistry is one of the more active fields in supramolecular chemistry. Similarly, CDxs are the main host in nanotechnology applications of adamantane inclusion compounds, although adamantane inclusion compounds with a few other hosts, e.g., calix[4] arenes (Chapman and Sherman, 1997), curcubiturils (Hettiarachchi and Macartney, 2006; Liu *et al.*, 2005) and derivatives of resorcinarene octols cavitands (Amrhein *et al.*, 2002) have been also reported over the years.

4.4.1. *Cyclodextrin (CDx)-adamantane inclusion complexation (CAIC) supramolecules*

Natural or chemically modified CDxs form inclusion complexes with a large number of hosts including adamantanes, as reported by Cromwell *et al.* (1985); Palepu and Reinsborough (1990); Shortreed *et al.* (1993); Brown *et al.* (2003); Poon and Cheng (2008), and Haddad *et al.* (2011).

Adamantane derivative β–CAICs show very high association constants, comparable to values found for protein–ligand complexes and they are reversible. For example, when β–CAICs are exposed to ligand-free solvents (de Jong *et al.*, 2001; Brown *et al.*, 2003), they can be modulated and controlled by the presence of co-solutes (ions or polar small molecules) in the solution (Harries *et al.*, 2005). Such characteristics make β–CAICs quite interesting models for selectivity studies in protein–ligand complexes. In addition, selective, strong, and yet reversible interactions exhibited by these host–guest pairs may turn them into useful connecting points in molecular system design, especially when some degree of flexibility or dynamic features are required.

Table 4.2 shows the association constant (K), formation Gibbs free energy (ΔG°), formation enthalpy (ΔH°) and formation entropy (ΔS°), all at standard state, for α–, β–, and γ–CAICs-1-carboxylic acid in acidic (pH 4.08) and basic (pH 8.50) media. At the lower (acidic) pH value, carboxylic acid neutral form predominates over the anionic carboxylate form. Thus, differences in the formation thermodynamic properties can be associated with the electrical charge of the ligand. In contrast, under the same pH condition, differences between formation thermodynamic properties are related to the diameter and volume of the host cavity (see Table 4.1 for CDxs properties).

Most frequently host:guest ratio between CDxs and ligands is 1:1. However, 2:1, 1:2, 2:2, or even more complicated associations are possible. (Szetjtli, 1998). 1:1 complexes are predominant in β– and γ–CDx-adamantane-1-carboxylic acid complexes, whereas 2:1 ratios appear to be the preferential binding mode of the neutral form of this ligand and α–CDx. The association constant for the 2:1 complex is $K = 5.0 \times 10^2\ M^{-1}$. This value is almost 4 times higher than K values for 1:1 complexes, both in neutral or anionic form. In contrast, there is no preferential binding of the ligand with α–CDx in the 1:1 complex. This suggests that the carboxylic group, in either neutral or anionic form, is directed away from the host cavity. The large ΔH° value for the 2:1 complex (8.8 ± 0.1 kcal/mol) suggests that the binding of the second host molecule involves a significant improvement in hydrogen bonding or van der Waals interactions.

Figure 4.18 presents a possible structure for the 2:1 complex with hydrogen bonds formed between the hydroxyl groups at the secondary face of the two α-CDx molecules.

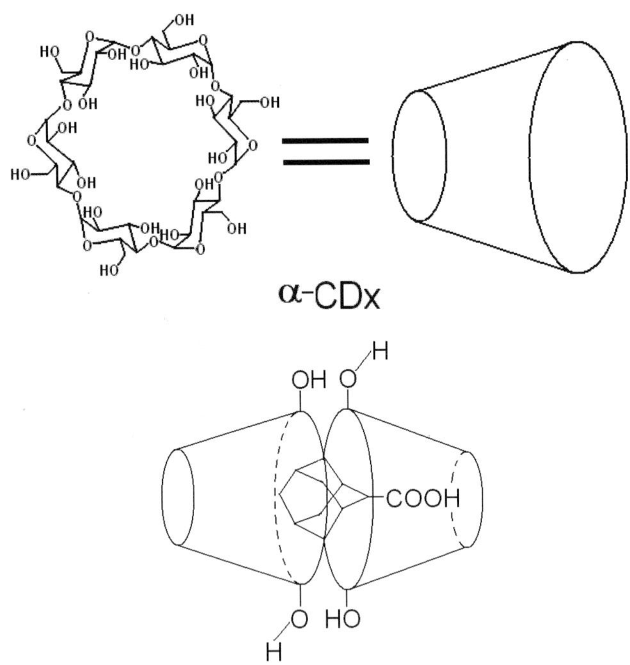

Fig. 4.18. Proposed structure for the 2:1 α-CAICs-1-carboxylic acid (α-cyclodextrin-adamantane-1-carboxylic acid complex) (Redrawn from Cromwell *et al.*, 1985).

The β-CDx-adamantane-1-carboxylic acid interactions are the most exergonic (DG° = −7.45 ± 0.15 and -5.85 ± 0.04 kcal/mol, for neutral and anionic forms, respectively) and the most exothermic of the 1:1 complexes (ΔH° = -7.53 ± 0.02 and -4.85 ± 0.02 kcal/mol for neutral and anionic forms, respectively), with the preference for the neutral form of the ligand. *K* values place the stability of the complexes in the order β-CDx > γ-CDx > α-CDx, with *K* value for the β-CDx-ligand complex approximately 140-fold higher than the *K* value for α-CDx-ligand complex. Considering similarities between the chemical natures of all the three host–guest systems, the high stability of the β-CDx-adamantane-1-carboxylic acid inclusion complex may be attributed to the close-fitting of the adamantane moiety into the β-CDx cavity. The estimated diameter and volume for the adamantanyl group are 7 Å (Shortreed *et al.*, 1993) and 147 $Å^3$ (Rudyak *et al.*, 2009), respectively. CDx diameter data given in Table 4.1 suggest that adamantane cage guest should loosely fit into the γ-CDx cavity, but it should tightly fit into the β-CDx cavity. However, α-CDx cavity is too small to allow full penetration of adamantane. ΔG° values (Table 4.2) also confirm the preferential behavior of adamantane-1-carboxylic acid for the β-CDx cavity, which is large enough to allow maximum penetration, and yet small enough to enable short-range interactions, such as van der Waals forces, between the hydrophobic interior of the CDx cavity and the apolar adamantane cage (Cromwell *et al.*, 1985).

Figures 4.19(a, b, c) show adamantane-1-carboxylate guest and models for its 1:1 complex with α- and β-CDx, based on NMR experiments and computer-aided molecular modeling (Rudiger *et al.*, 1996). Collected data

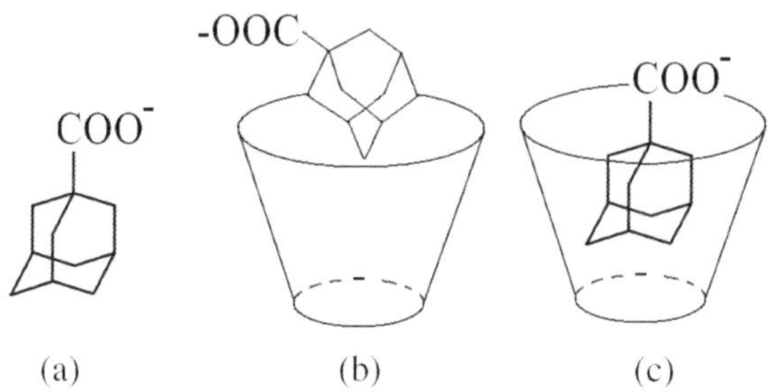

Fig. 4.19. (a) Adamantane-1-carboxylate and its inclusion complexes with (b) α- and (c) β-cyclodextrin.

indicated that the smaller cavity in α-CDx leads to a partial immersion with the symmetry axis tilted towards the CDx axis (Fig. 4.19b) and support the deep immersion mode for β-cyclodextrin-adamantane-1-carboxylate complex (Fig. 4.19c). It is important to notice that in the β-CDx-ligand complex there is no room for water molecules inside the cavity. This complex also exhibits an optimal contact between the C–H bonds of the host and guest molecules. In contrast, α-cyclodextrin-adamantane-1-carboxylate complex allows interactions only at the secondary face of the host. The tilted orientation of the carboxylate group is due to an additional stabilization by hydrogen bonding with the α-CDx secondary hydroxyl groups (Fig. 4.19b).

4.4.2. *Nanotechnology applications of CAICs (CDx-adamantane inclusion complexes)*

Nanotechnology applications of CAICs reach many fields of research and applications. The major recent nanotechnology applications of CAICs which will be further discussed in this book include the following three subjects: (1) Nonviral gene therapy (Bernal *et al.*, 2008; Pun and Davis, 2002; Pun *et al.*, 2004; Park *et al.*, 2006; Bartlett *et al.*, 2007). (2) Enzyme immobilization (Fragoso *et al.*, 2002; Villalonga *et al.*, 2006, 2007(a and b); Valdivia *et al.*, 2007; Camacho *et al.*, 2007, 2009; Holzinger *et al.*, 2009). (3) Physical crosslinking of polymers (Auzely-Velty and Rinaudo, 2002; Kretschmann *et al.*, 2006; Munteanu *et al.*, 2009).

Other recent and less explored nanotechnology applications of CAICs are in cancer drug delivery (Schluep *et al.*, 2009), in enzyme simulation/modeling (Li *et al.*, 2008), in chemical sensors (Chen *et al.*, 2007) and in photoactive supramolecular systems (Faiz *et al.*, 2006; Wen *et al.*, 2006).

4.4.2.1. *Application of CAICs in nonviral gene therapy*

Gene therapy is a new process, which plays an important role in genetic science and engineering. In this process, DNA must be delivered into the cell and, simultaneously, it must be protected from enzymatic degradation. For this purpose, various nanodelivery systems are suggested (Pun and Davis, 2002; Nikakhtar *et al.*, 2007; Cheng *et al.*, 2011). CAICs have been applied in modified-polycation-based nonviral gene delivery. Pun and Davis (2002) developed a new method for modifying polyplexes using β-CDx-containing polymers (β-CDxPs) in inclusion complexes with adamantine–PEG conjugate. This method was applied to β-CDxP-based polyplexes to enhance particles stability in salt solutions and to mitigate undesirable interactions after intravenous administration. Figure 4.20 shows the structure of polycation β-CDxP6-PEG_{3400} and its guest which is an adamantane-1-PEGylated molecule (Ada-PEG_{5000}). A schematic representation of the methodology for inclusion of Ada-PEG_{5000} into β-CDx on the surface of preformed polyplexes is presented in Fig. 4.21. This approach relies on the addition of Ada-PEG_{5000} (and other Ada-PEG conjugates) to solutions of condensed plasmid DNA-(β-CDx) particles (1:1 mol/mol). Phosphate-buffered saline (PBS) with a concentration of 150 mM was then added to the mixture. The average diameter of β-CDxP6 polyplexes increases from 58 to 272 nm within ten minutes after PBS addition, evidencing high level of aggregation in the absence of the Ada-PEG host. However, PEGylation *via* inclusion complexation of the β-CDxP6 polyplexes with Ada-PEG molecules reduces particle aggregation. Apparently, such improvement is also related to the length of the PEG chain: ten minutes after PBS addition, PEGylated particles with Ada-PEG_{3400} form aggregates of 204 nm in diameter, while particles PEGylated with Ada-PEG_{5000} only increase in diameter to 95 nm. Additionally, the particles PEGylated with Ada-PEG_{5000} demonstrate sustained stability in physiological PBS concentration of 150 mM. PEGylated polyplexes are also stable in culture media containing 10% serum, while unmodified polyplexes provoke fast aggregation and precipitation of serum proteins. A clear evidence of PEGylated polyplexes stability in such conditions arises from the fact that around only 18% of the DNA was recovered in the aggregates of PEGylated polyplexes. Thus, most DNA was still suspended in the PBS-serum mixture.

Special ligands, designed to target specifically the asialoglycoprotein receptor were conjugated with Ada-PEG particles (Fig. 4.21), in order to assess cell entry *via* this receptor. The strategy was to galactosylate β-CDxP-based particles and measure transfection efficiency in hepatoma cells. The obtained galactosylated β-CDxP-based particles were able to transfect hepatoma cells with 10-fold higher efficiency than control

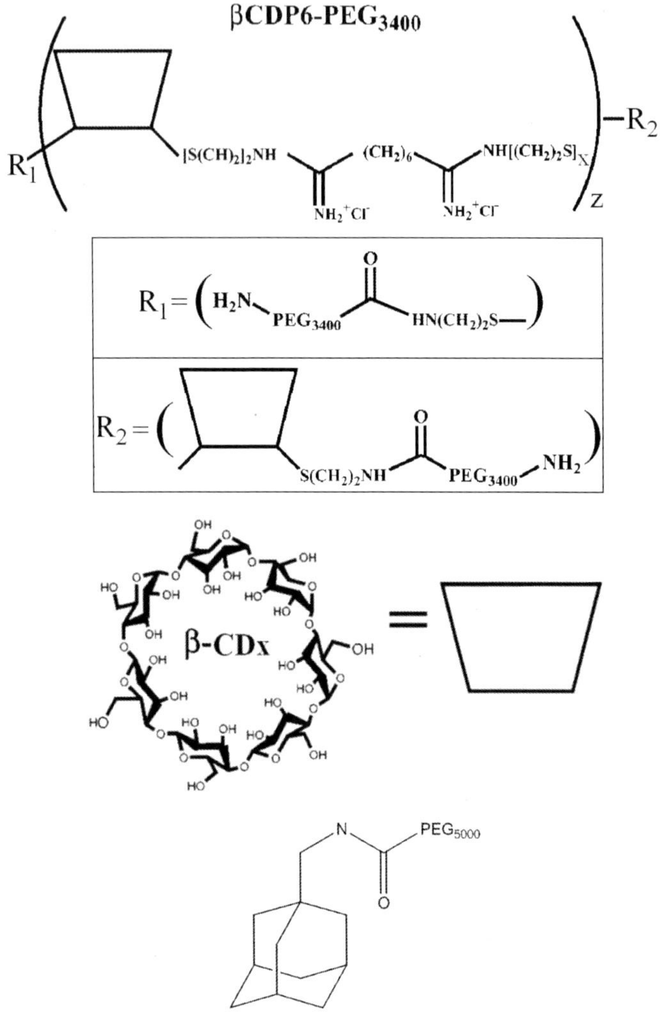

Fig. 4.20. Structure of polyplex-modifier polycation β-β-CD×P6-PEG3400 and guest molecule Ada-PEG5000 (adamantanyl group coupled with polyethylene glycol). β-CD×P6-PEG3400 contains β-cyclodextrins in the polymer backbone and is cationic due to repeating amidine groups. Number 6 denotes the number of methylene groups separating the charges. The average molecular weight of the polycation corresponds to a "z" of approximately 4 (Pun and Davis, 2002).

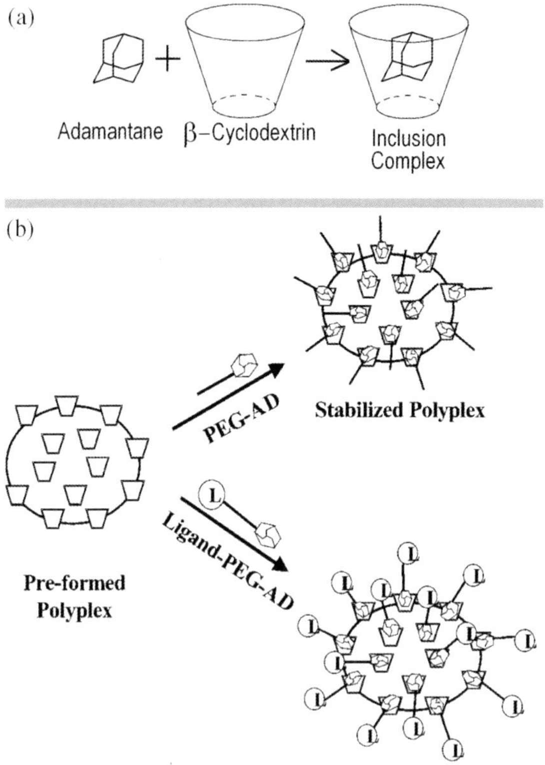

Fig. 4.21. (a) Inclusion compound formation stage. (b) Post-DNA-complexation PEGylation by inclusion compound formation by PEG-Ada and ligand-PEG-Ada. PEG is conjugated to adamantane. Adamantane forms an inclusion complex with β-cyclodextrin on the polyplex surface and brings PEG or PEG-L with it to decorate the polyplex and provide steric stabilization and targeting. Special ligands designed to target a particular receptor are conjugated with Ada-PEG particles (Ligand-PEG-Ada) in order to assess cell entry *via* receptor. (Redrawn from Pun and Davis, 2002).

(glucosylated particles), but showed no preferential transfection in a cell line lacking the asialoglycoprotein receptor. Thus, PEGylation *via* inclusion complexation with Ada-PEG molecules are shown to be amenable to chemical modifications designed to allow gene transfer to a specific cell type, in this case, hepatoma cells.

More recently, a different modified polyplex, based on β-CDx-adamantane inclusion complex (β-CDx/Ada), was presented by Pun *et al.* (2004) and tested as *in vitro* and *in vivo* gene delivery systems. In this work, both branched and linear polyethylenimine (i.e., BPEI and LPEI, respectively) were grafted with β-CDx *via* reaction of mono-tosylated cyclodextrins with the primary and secondary amines of PEI. Levels of CDx grafting varied from 5 to 16% of β-CDx for BPEI and 12% of β-CDx for LPEI. Self-assembly of β-CDx-BPEI and β-CDx-LPEI polymers with plasmid DNA yielded small spherical polyplex particles with average diameters ranging from 100 to 160 nm. Figure 4.22 shows schematics of β-CDx-BPEI (a) and PEGylated β-CDx-BPEI (b) polyplexes formation (Pun *et al.*, 2004). Ada-PEG_{5000} inclusion complex of both β-CDx-BPEI and β-CDx-LPEI grafted polyplexes were also tested regarding their size, morphology, and salt stability. Complexation with Ada-PEG_{5000} resulted in PEGylated nanoparticles having practically the same size and shape of noncomplexed polyplexes. But the increase in stability was remarkable with the addition of Ada-PEG_{5000}. When added in β-CDx/Ada (2:1 ratio), complete nanoparticle stabilization was achieved by forming a protective hydrophilic PEG brush layer on the nanoparticle surface, while all unmodified β-CDx-PEI polyplexes aggregated in salt solutions.

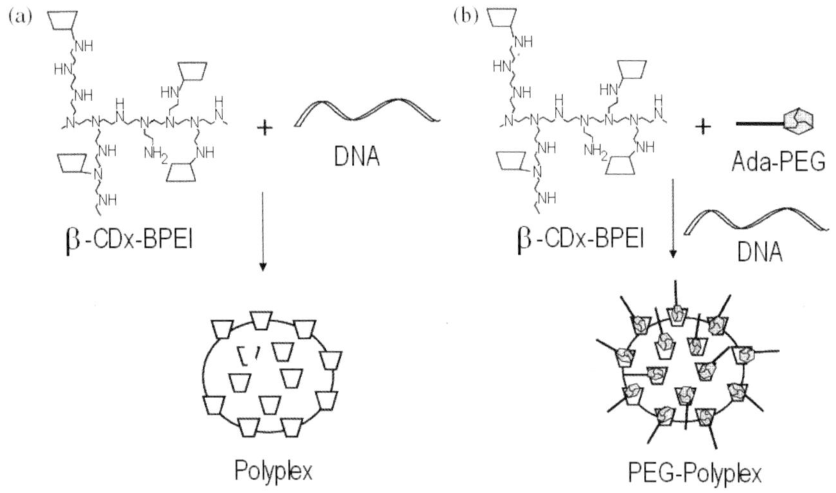

Fig. 4.22. Schematics of β-Cyclodextrin-grafted branched polyethylenimine (β-CDx-BPEI) and its process of (a) DNA complexation to give a polyplex, and (b) PEGylated adamantane (Ada-PEG)/DNA complexation to give a PEGylated polyplex (PEG-Polyplex) (Adapted from Pun *et al.*, 2004).

Cyclodextrin grafting to BPEI (β-CDx-BPEI) reduced polymer toxicity, as complexes. Specifically, 5% and 8% of graftings had IC_{50}'s (half maximal inhibitory concentrations) of 0.64 and 1.45 mM, respectively. These are much higher than the IC_{50} of noncomplexed BPEI which was of 0.28 mM.

Moreover, higher levels of cyclodextrin graftings (i.e., 12% and 16%), increased IC_{50} values to 4.8 and 6.7 mM, respectively. Possible reasons for improved tolerability are: Increased polymer solubility, capping of primary amines, or reduced binding affinity exhibited by grafted polycations.

In contrast, increased β-CDx grafting reduced *in vitro* transfection efficiency. The best CDx-BPEI for plasmid delivery to PC3 (human prostatic carcinoma) cells was determined to have 8% PEI grafting available for inclusion complex formation. Cyclodextrin modification of LPEI reduced polymer toxicity, as well. The IC_{50} for LPEI (with 12% of CDx grafting) was increased from 0.38 to 0.92 mM for β-CDx-LPEI. Inclusion of Ada-PEG_{5000} did not practically lower transfection with CDx-LPEI but significantly lowered the number of expressing cells with CDx-bPEI.

Low toxicity of the CDx-PEI delivery particles determined for *in vitro* experiments was also confirmed with *in vivo* experiments. Inclusion complexes of β-CDx-lPEI polyplexes and Ada-PEG_{5000} did not produce any acute toxicity when intravenously administrated to Balb/c mice (laboratory-bred strain of albino "House Mice"), in doses up to 120 μg. Internal organs of Balb/c mice did not show any adverse symptoms generally associated with LPEI polyplex intravenous administration at the similar dose levels, e.g., liver necrosis, red blood cell adhesion in the lung, which are frequently lethal to the animals (Chollet *et al.*, 2001). The liver was the primary accumulating organ and the main site of delivered DNA expression, whilst no other organs exhibited measurable gene expression. As unmodified PEI polyplexes are known to accumulate DNA to the lungs (Chollet *et al.*, 2001), the use of β-CDx-LPEI-Ada-PEG_{5000} polyplex particles may provide a good delivery vehicle for liver gene expression from a systemic administration.

Supramolecular assembly of β-CDx and adamantane was suggested as an approach for localized delivery of nucleic acids (Park *et al.*, 2006). Figure 4.23 shows a representation of polyplex immobilization process based on the formation of β-CDx-Ada inclusion complex. Nanoparticle complexes or polyplexes (50 to 100 nm in diameter) of β-CDx-LPEI and DNA were specifically immobilized on adamantane-functionalized gold

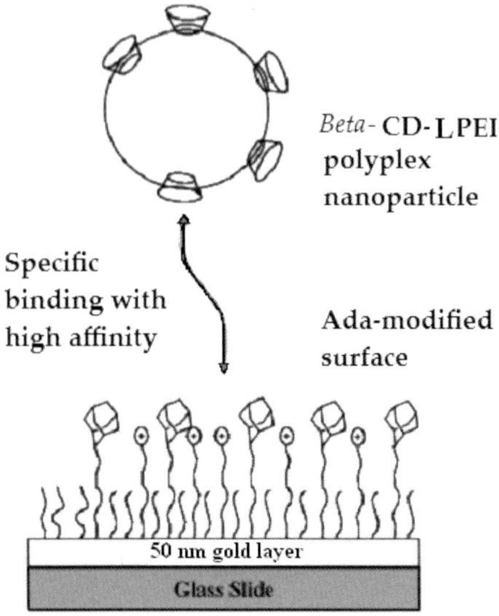

Fig. 4.23. Schematic of β-Cyclodextrin-linear polyethylenimine (β-CDx-LPEI) nanoparticle immobilization process on adamantane-modified (Ada-modified) surfaces by inclusion complex formation (Redrawn from Park *et al.*, 2006).

surfaces by inclusion complex formation. This method takes advantage of all the amenities of self-assembly inclusion complexation: Minimal changes in guest surface properties, no need for additional chemical bonding, stability, and reversibility, to selectively immobilize gene delivery nanoparticles in substantially higher density (1289 pg of DNA/mm^2 area). β-CDx-PEI nanoparticles demonstrated significantly higher adsorption on adamantane-modified surfaces than PEI nanoparticles. Multivalent host–guest interactions, i.e., supramolecular multipoint anchoring (Fulton and Stodart, 2001), increase binding affinity of β-CDx-PEI nanoparticles to the adamantane surfaces by several orders of magnitude when compared to the binding of single cyclodextrin molecules, with no evident interference with polymer/DNA interactions.

4.4.2.2. *Application of CAIC in enzyme immobilization*

Enzyme immobilization is attractive in a multitude of R&D fields and has found many applications in several analytical, medical, industrial, and biotechnology processes.

Cyclodextrin (CDxs)-adamantane inclusion complexations *(***CAICs)** are interesting candidates as attachment components for enzyme immobilization, due to their physical nature in their applications in host–guest interactions with enzymes. Although chemical modification of enzyme surface is still necessary, host or guest groups can be attached to the enzyme with suitable-sized linkers, thus, preventing excessive conformational distortions of the enzyme. Applications of such supramolecular devices, like CAICs**,** have already been reported as biosensors (Fragoso *et al.*, 2002; Villalonga, 2007(a,b); Camacho *et al.*, 2007, 2009; Holzinger *et al.*, 2009), nanocatalyst for enantioselective reactors (Villalonga *et al.*, 2006), and as protective environment to improve pharmacological activity of enzymes (Valdivia *et al.*, 2007).

Figure 4.24 depicts two different strategies for supramolecular enzyme immobilization on electrodes, in order to build amperometric biosensors. In Fig. 4.24a, L-phenylalanine dehydrogenase (biocatalyst) was tagged with 1-adamantanyl groups and supramolecularly assembled to β-CDx-coated-Au electrodes. The resulting device was used as a biosensor to quantify L-phenyalanine. Washing of the electrodes in aqueous (0.3%) solution of sodium dodecylsulfate (SDS) for 2 hours disrupted the hydrophobic interactions between electrode, β-CDx moieties and 1-adamantanyl residues bonded to the enzyme, allowing the release of the enzyme (biocatalyst) from the electrode surface with no detectable electrocatalytic activity after washing. Nevertheless, further dipping of the electrode in solutions containing the same tagged enzyme yielded electrodes able to give an amperometric response, roughly at the same level as that obtained in the first trial mentioned above. These trials are indicative of the reversible nature of the supramolecular attachment and suggested that the enzyme could be successfully re-immobilized on the CDx-coated electrode.

Figure 4.24(b) shows a poly-adamantane-pyrrole (PAP) derivative (adamantanyl-11-*N*-pyrrolyl-1-undecyl carboxylic acid amide) which is electropolymerized on platinum electrodes to act as an affinity-binding electrically-conducting polymer. Supramolecular complexation with the β-CDx-tagged enzyme named GOX (glucose oxidase), completed the elaboration of devices capable of detecting and quantitatively measuring redox events associated with the presence of the GOX substrate (glucose).

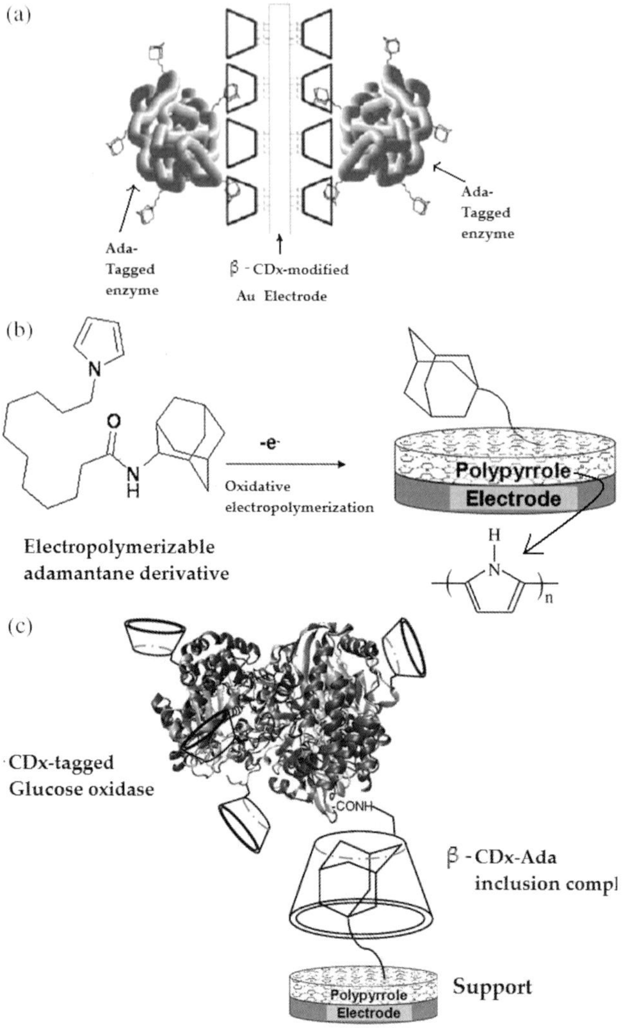

Fig. 4.24. (a) Supramolecularly-mediated immobilization of adamantane-tagged L-phenylalanine dehydrogenase (Ada-Tagged enzyme) on β-cyclodextrin (CDx)-modified Au electrodes (Villalonga *et al.* 2007a). (b) Schematic representation of oxidative polymerization of an electropolymerizable adamantane derivative (Adamantanyl-11-N-pyrrolyl-1-undecyl carboxylic acid amide) to give an electroconductive adamantane-tagged polypyrrole (Redrawn from Holzinger *et al.*, 2009). (c) Schematic presentation of β-cyclodextrin-adamantane (β-CDx-Ada) inclusion complex. The principle of the supramolecular assembly is sketched with β-CDx-tagged glucose oxidase and the poly (adamantanepyrrole-coated electrode (Redrawn from Holzinger *et al.*, 2009).

This is an example of amperometric glucose biosensors. The selective recognition of this analyte is electrochemically processed to produce a signal in the form of current, the magnitude of which is proportional to the quantity of glucose present. Since GOX catalyzes the aerobic oxidation of glucose in the presence of oxygen with the concomitant production of hydrogen peroxide (H_2O_2) then, these particular biosensors were set up to amperometrically detect generated H_2O_2. The sensitivity of the PAP-GOX electrode was 0.175 [$mAM^{-1}.cm^{-2}$] with a maximum current density, $J_{max} \cong 10$ [$\mu A.cm^{-2}$]. The J_{max} reflects the amount of immobilized GOX allowing the maximum reaction rate at saturating glucose conditions. Taking into account the specific activity of GOX prior to the immobilization, this sensitivity is low when compared with previous reports on compact GOX monolayer elaborated with other affinity polymers (Cosnier *et al.*, 1998; Cosnier and Lepellec, 1999) instead of PAP. Probably, only a small part of the PAP complexes with β-CDx at the polymer-solution interface, due to the high hydrophobicity of adamantane groups (Holzinger *et al.*, 2009).

More elaborated architectures for CAIC-based amperometric biosensors, including single-walled carbon nanotubes (SWNTs) as constructing parts, were also presented in the same report (Holzinger *et al.*, 2009). SWNTs were deposited on platinum electrodes (Fig. 4.25a) and coated with electrodeposited PAP film as affinity binding system. Afterwards, the electrodes were incubated with β-CDx modified GOX. Figure 4.25b presents the resulting biosensor response as a function of glucose concentration. The SWNT-PAP-based biosensor exhibits a sensitivity of 8.72 [$mAM^{-1}cm^{-2}$], with the $J_{max} = 105$ [$\mu A\ cm^{-2}$]. This suggests that the geometrical effect of SWNT leads to a larger area covered by the PAP film and hence an increase in the amount of attached GOX. In a more elaborated arrangement, β-CDx gold nanoparticles were deposited as intermediate layer for the specific anchoring of adamantane-tagged GOX (Fig. 4.25c). The resulting glucose sensor exhibits the highest sensitivity (31.02 [$mAM^{-1}cm^{-2}$]) with the $J_{max} = 350$ [$\mu A\ cm^{-2}$]. Comparison with the performance of the corresponding similar configuration without SWCNT (sensitivity 0.980 [$mAM^{-1}cm^{-2}$] and $J_{max} = 75$ [$\mu A\ cm^{-2}$]) evidenced improved performance due to the SWNT coating. In addition, a combined synergetic effect of SWCNTs and gold nanoparticles is likely to provide a higher amount of immobilized enzyme.

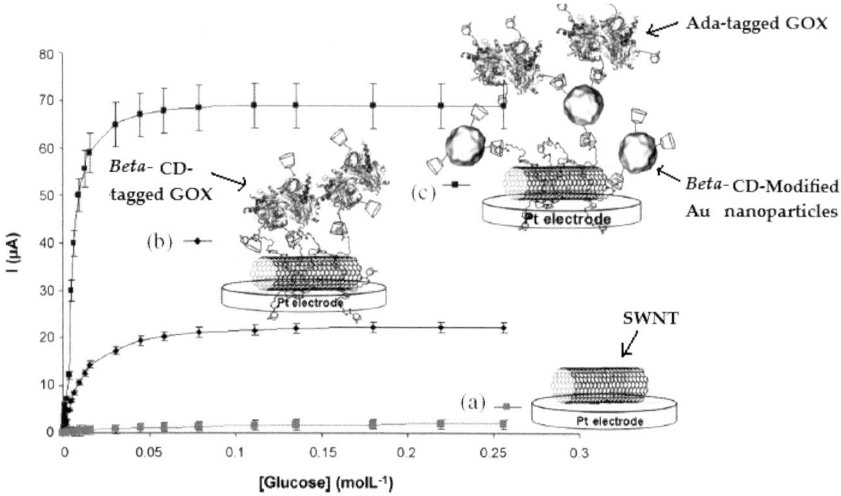

Fig. 4.25. Sketch of SWNT electrodes and related calibration curves for glucose. (a) Pure SWNT coating, (b) SWNTs covered by poly(adamantane-pyrrole) film and post-functionalized by β-cyclodextrin-tagged GOX and (c) SWNTs covered by a poly(adamantane-pyrrole) film, post-functionalized by adamantane-tagged (Ada-tagged) GOX with β-CD gold nanoparticles as intermediate layer. Adapted from (Holzinger *et al.*, 2009).

4.4.2.3. *Application of CAIC in physical crosslinking of polymers*

Polymer crosslinking is commonly defined as the formation of polymer networks by coupling of linear or branched polymer chains through covalent bonds. Such network materials are also called thermosets or thermosetting polymers. Crosslinked polymer chains lose their ability to flow past one another. As a consequence, the polymer will not melt or flow, thus, it cannot be molded after crosslinking. Moreover, thermosetting polymers are insoluble because crosslinking process enormously increases their molar masses. Actually scrap thermoset polymers are not recyclable as polymers. To overcome thermosetting polymers intractability, other alternatives to the classical chemical crosslinking have been proposed. An example is the introduction of strong secondary physical bonding between polymer chains to make them thermoplastics which are tractable. This process is called physical crosslinking (Stevens, 1999).

The β-CDx-Ada host–guest chemistry (see Sec. 4.3) may be used for polymer physical crosslinking to produce intelligent hydrogels presenting switchable transparency in a narrow temperature range. Intelligent hydrogels that are sensitive to external stimuli (like temperature, light, pH, or salinity) are of increasing interest because of their potential applications in sensing and actuating devices (Sangeetha and Maitra, 2005). Kretschmann and co-workers (2006) obtained noncovalent crosslinked hydrogel structures using adamantanyl-containing copolymers (Fig. 4.25/26(a, b, c)) and a β-CDx dimer (Fig. 4.26d). Guest copolymers A, B, and C are believed to complex with β-CDx dimer to form supramolecular networks as shown in Fig. 4.27. After interacting with the β-CDx dimer, copolymers A, B, and C were subjected to viscosity measurements at low concentrations (zero-shear viscosity, η) at 10°C. Viscosity data of copolymers A, B, and C solutions and of guest copolymer mixed with β-CDx solutions were also collected, for comparison (Fig. 4.28). As expected, the viscosity of the solutions of pure copolymers increased with increasing their molecular mass (molecular mass of A, B, and C copolymers are 50, 66, and 76 kg/mol, respectively). Nevertheless, the comparison reported by Fig. 4.28 can only be considered qualitative because of the different kinds of monomers used for copolymerization of each of the three copolymers. Addition of β-CDx to these copolymer solutions had no significant effect on the viscosity. The slight decrease in viscosity observed in the case of copolymer C after addition of β-CDx might have resulted from the disruption of some intermolecular assemblies when the hydrophobic side chains were included into the β-CDx cavity. When aqueous solutions of copolymers A, B, and C were mixed with the β-CDx dimer (Fig. 4.26d), the viscosity of the solutions increased dramatically with instant formation of stable gels. These data strongly suggest the formation of supramolecular crosslinking between guest copolymers A, B, and C and the β-CDx dimer D.

Changes in phase transition temperature of the guest copolymers A, B, and C were investigated by turbidity measurements of pure copolymer aqueous solutions, or in the presence of a defined amount of methyl-β-CDx (me-β-CDx) or the β-CDx dimer, respectively. No significant alterations were found in copolymer A. Nevertheless, the addition of me-β-CDx led to 35°C increase in the cloud points of copolymers B and C, (Fig. 4.29) which correlates to the lower critical solution temperature (LCST) behavior of poly(N-isopropylacrylamide), the homopolymers correspondent to $m = 0$ in copolymer structures presented in Fig. 4.26b and Fig. 4.26c. This

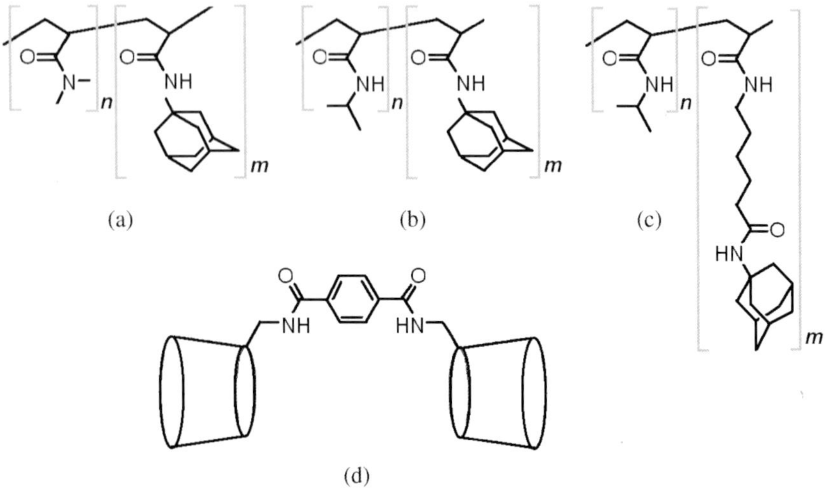

Fig. 4.26. Guest copolymers (a) (Mn 50 kg/mol), (b) (Mn 66 kg/mol), (c) (Mn 76 kg/mol), and β-cyclodextrin dimmer D. n/m ratio is 20:1 in all copolymers (Kretschmann *et al.*, 2006).

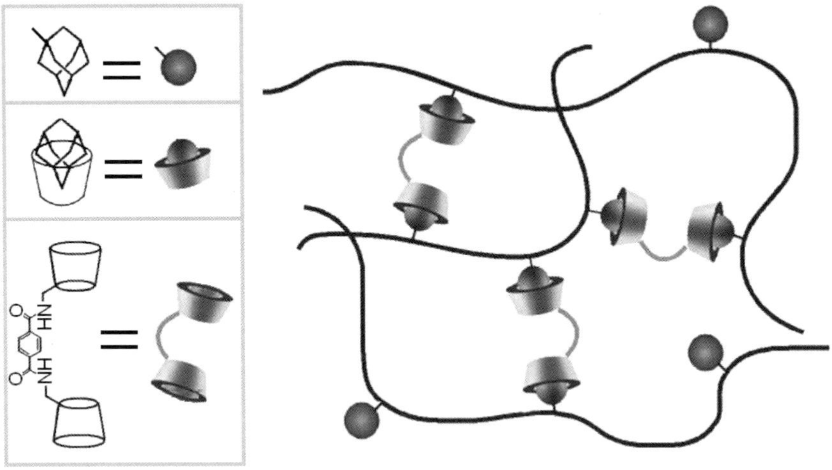

Fig. 4.27. Estimated structure of the supramolecular networks of "guest" copolymers A, B, and C (shown in Fig. 3.25) which are believed to join with β-CDx dimer "host" producing the intelligent hydrogel (Redrawn from Kretschmann *et al.* 2006).

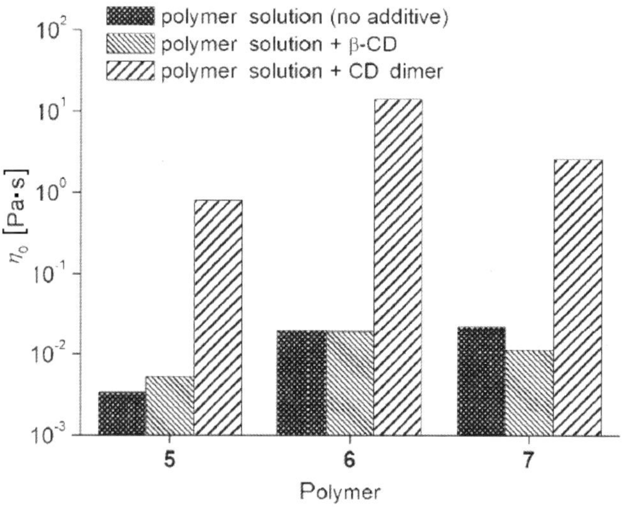

Fig. 4.28. Comparison of viscosities at 100°C of copolymer solutions without CDx linkers, with a mono-β-CDx functionalized linker, and β-CDx dimer D (Kretschmann *et al.*, 2006).

increase of the cloud points relative to the cloud points of pure B and C is a consequence of hydrophobic adamantanyl inclusion in me-β-CDx. When β-CDx dimer was added to the aqueous polymer solutions at half the molar ratio (relative to the amount of me-β-CDx), the cloud points decreased significantly. This observed effect again can be explained by the cross-linkage of single polymer chains upon complexation of the β-CDx dimer. Cloud points of 14.0°C for polymer B and 15.7°C for polymer C can be explained by the mobility and solubility restriction resulting from the new supramolecular interactions formed. As a consequence of LCST behavior alterations, supramolecular hydrogels present switchable transparency in a narrow temperature range: Gels formed from copolymer A are transparent at room temperature, while those from copolymers B and C are turbid at room temperature and transparent below their cloud points.

4.5. Concluding Remarks

This chapter was devoted to the role of diamondoids and derivatives as the MBBs for nanotechnology, since they exhibit a set of unique physico-chemical characteristics, which allow them to be utilized in bottom-up

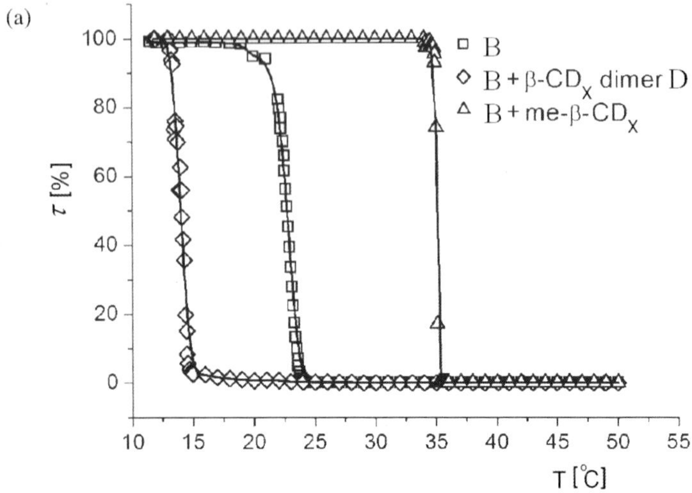

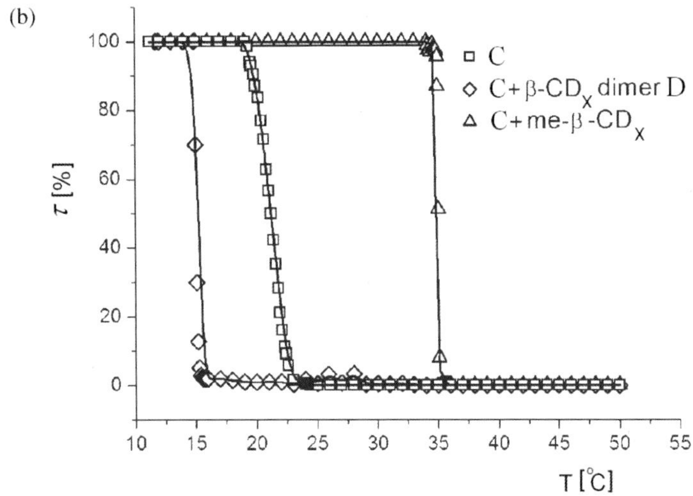

Fig. 4.29. Turbidity measurements of aqueous solutions of the copolymers (a) B and (b) (c) without addition of host molecules and measurements of the supramolecular complexes of the polymers with β-CDx dimer and methyl-β-CDx (me-β-CDx), respectively (Kretschmann et al., 2006).

strategies for the assembling of nanodevices. Diamondoids are stiff, dense, covalently/sp^3 bonded carbons in three-dimensional networks. Because of these characteristics, they have many advantages over other types of MMBs. Diamondoids possess six or more primary and four or more secondary symmetrically located linking sites which makes them excellent candidates for accurate and preconceived design of nanoassemblies. There exist naturally occurring diamondoids having such shapes, as "**L**", "**S**", "**X**", "**Y**", "**∇**", "**+**" "▽" and more which can be used as building components of nanosystems.

A number of unprecedented diamondoids are being recovered from petroleum sources. These new compounds present a variety of challenging sizes and shapes, which can be envisioned as nanoscopic mechanical parts for molecular engines.

Due to their varieties of shapes and derivatization, diamondoids are amenable for the design of self-assembled nanoscale structures. We have different diamondoid components as almost perfectly-fitted MBBs. Even tubular and spherical nanostructures may be obtained by taking advantage of self-assembly or coacervation of simpler diamondoids.

In this chapter, we also presented a detailed discussion about the important role of diamondoids and derivatives in host–guest chemistry, such as applications of adamantane-constrained cystine-based cyclic trimers [cyclo (Adm-Cyst)3] which is a double-helical cyclic peptide and poly-adamantane molecular rods. Adamantane is known as one of the best guests entrapped within cyclodextrin (CDx) cavities. This is due to its high energy affinity to interact with the hydrophobic pocket of CDxs.

Considering that CDxs are the most important host molecules in supramolecular chemistry and the main host in nanotechnology applications of adamantane inclusion compounds, CDxs-adamantane inclusion complexation (CAIC) has found many applications in supramolecular chemistry. This includes applications of CAIC in nonviral gene therapy, in enzymes immobilization, in physical crosslinking of polymers, just to name a few.

Adamantane inclusion compounds with β-CDx are the major force in the present practical nanotechnology applications of diamondoids. Most of the work done in this subject is related to medical and pharmacology areas. Nevertheless, supramolecular assembling directed by the strong affinity between adamantane and β-CDx show good prospects in polymer science, analytical chemistry and possibly other fields.

Overall diamondoids are one of the best candidates for use as MBBs in molecular nanotechnology to design nanostructures with predetermined physicochemical properties although; the tangible benefits of such structures are yet to emerge.

Bibliography

Allis, D. G., Helfrich, B., Freitas, R. A. and Merkle, R. C. (2011). *J. Comp'l and Theo'l Nanoscience*, **8**: 1139–1161.

Amrhein, P., Shivanyuk, A., Johnson, D. W. and Rebek Jr., J. (2002). *J. Am. Chem. Soc.*, **124**: 10349–10358.

Arun, R., Ashok, K. C. K. and Sravanthi, V. V. N. S. S. (2008). *Sci Pharm.*, **76**: 567–598.

Auzely-Velty, R. and Rinaudo, M. (2002). *Macromolecules* 2002, **35**: 7955–7962.

Ayres, F. D., Khan, S. I., Chapman, O. L. (1994). *Tetrahedron Letts.*, 35: 8561–8564.

Bartlett, D, W., Su, H., Hildebrandt, I. J., Weber, W. A. and Davis, M. E. (2007). *Proc Nat. Acad. Sci. USA*, **104**: 15549–15554.

Bernal, V., Arense, P., Blatz, V. *et al.* (2008). *J. Appl. Microbiology*, **105**(1): 42–50.

Bogunia-Kubik, K., Sugisaka, M. (2002). *Biosystems*, 65: 123–138.

Boussif, O., Lezoualch, F., Zanta, M. A., Mergny, M. D., Scherman, D., Demeneix, B. and Behr, J. P. (1995). *Proc. Nat. Acad. Sci. USA*, **92**: 7297–7301.

Brown, S. E., Easton, C. J., Kelly, J. B. (2003). *J. Incl. Phenom. Macrocycl. Chem.*, **46**: 167–173.

Camacho, C., Chico, B., Cao, R., Matías, J. C., Hernández, J., Palchetti, I., Simpson, B. J., Mascini, M. and Villalonga, R. (2009). *Biosens. Bioelectron.*, **24**: 2028–2033.

Camacho, C., Matias, J. C., Chico, B., Cao, R., Gomez, L., K., Simpson, B. K. and Villalonga, R. (2007). *Electroanalysis*, **19**: 2538–2542.

Carlson, R. M., Dahl, J. E. and Liu, S. (2007). US Patent No. US 7309476 B2.

Chadwick, D. J. and Widdows, K. (eds). (1991). Host-Guest Molecular Interactions, from Chemistry to Biology, CIBA Foundation Symposia Series, No. 158, J. Wiley & Sons, New York, NY.

Chapman, R. G. and Sherman, J. C. (1997). *Tetrahedron*, **53**: 15911–15945.

Chen, Y., Han, K.-Y. and Liu, Y. (2007). *Bioorg. Med. Chem.*, **15**: 4537–4542.

Cheng, R., Feng, F., Meng, F. *et al.* (2011). *J. Controlled Release*, **152**: 2–12.

Chirio, D., Cavalli, R., Trotta, F., Carlotti, M. E. and Trotta, M. (2007). *J. Inclusion Phenom. Macrocyclic. Chem.*, **57**: 645–649.

Chollet, P., Favrot, M., Hurbin, A. and Coll, J.-L. (2001). *J. Gene Med.*, **4:** 84–91.

Cosnier, S., Galland, B., Gondran, C. and Lepellec, A. (1998). *Electroanalysis*, **10**: 808–813.

Cosnier, S. and Lepellec, A. (1999). Poly(pyrrole–biotin): A new polymer for biomolecule grafting on electrode surfaces *Electrochim. Acta.*, **44**: 1833–1836.

Cramer, F., Saenger, W. and Spatz, H.-C. (1967). *J. Am. Chem. Soc.*, **89**: 14–20.

Cromwell, W. C., Bystrom, K. and Eftink, M. R. (1985). *J. Phys. Chem.*, **89**: 326–332.

Dahl, J. E. and Carlson, R. M. (2002). US Patent No. US2002147373-A1.

Dahl, J. E. and Carlson, R. M. (2003a). Patent No. WO200296804-A.

Dahl, J. E. and Carlson, R. M. (2003b). US Patent No. US2002143217-A1.

Dahl, J. E. and Carlson, R. M. (2003c). US Patent No. US2002143217-A1.

Dahl, J. E. and Carlson, R. M. (2003d). US Patent No. US2003100808-A1.

Dahl, J. E., Carlson, R. M. and Liu, S. (2004). US Patent No. US2004016397-A1.

de Jong, M., Huskens, J., Reinhoudt, D. (2001). *Chem. Eur. J.*, **7**: 4164–4170.

De Rosa, G. and La Rotonda, M. I. (2009). *Molecules*, **14**: 2801–2823.

Del Valle, E. M. M. (2004). *Process Biochem.*, **39**: 1033–1046.

Drexler, K. E. (1986). *Engines of Creation: the coming era of nanotechnology.* Anchor books: New York.

Drexler, K. E. (1992) *Nanosystems: Molecular Machinery, Manufacturing and Computation.* John Wiley & Sons, Inc. New York, NY.

Drexler, K. E. (1999). *Trends in Biotech.*, **17**: 5–7.

Faiz, J. A., Williams, R. M., Silva, M. J. J. P., De Cola, L., De Cola and Pikramenou, Z. (2006). *J. Am. Chem. Soc.*, **128**: 4520–4521.

Fatouros, D. G., Hatzidimitriu, K. and Antimisiaris, S. G. (2001). *Eur. J. Pharm. Sci.*, **13**: 287–296.

Felgner, P. L., Gadek, T. R., Holm, M., Roman, R., Chan, H. W., Wenz, M., Northrop, J. P., Ringold, G. M. and Danielsen, M. (1987). *Proc. Nat. Acad. Sci. USA*, **84**: 7413–7417.

Fragoso, A., Caballero, J., Almirall, E., Villalonga, R. and Cao, R. (2002). *Langmuir*, **18**: 5051–5054.

Freitas Jr., R. A. (1996). *Nanotechnology*, 2, 8–13.

Freitas Jr., R. A. (1998). *Artificial Cells, Blood Substitutes & Biotechnology*, **26**: 411–430.

Freitas Jr., R. A. (2004a). Pathway to diamond-based molecular manufacturing. *First Foresight Conference on Advanced Nanotechnology*, 22 October, Washington, DC.

Freitas Jr., R. A. (2004b). U.S. Provisional Patent Application No. 60/543,802, filed 11 February 2004; U.S. Patent Pending, 11 February 2005.

Freitas Jr., R. A. (2009). *Int. J. Robotics Res.*, **28**: 548–557.

Freitas Jr., R. A., Allis, D. G. and Merkle, R. C. (2007). *J. Comput. Theor. Nanosci.*, **4**: 433–442.

Freitas Jr., R. A. and Merkle, R. C. (2008). *J. Comput. Theor. Nanosci.*, **5**: 760–861.

Fujita, K., Chen, W-H., Yuan, D.-Q., Nogami, Y., Koga, T. (1999). *Tetrahedron, Asymmetry*, 10: 1689–1696.

Fulton, D. A.; and Stoddart, J. F. (2001). *Bioconjugate Chem.*, **12**: 655–672.

Granadero, D., Bordello, J., Pérez-Alvite, M. J. *et al.* (2010). *Int'l J. Mol. Sciences*, **11**: 173–188.

Grcic, F. J., Voinovich, D., Moneghini, M., Becirevic, L. M., Magarotto, L., Jalsenjk, I. (2000). *Eur. J. Pharm. Sci.*, **9**: 373–379.

Haddad, R., Holzinger, M., Villalonga, R. *et al.* (2011), *Carbon*, **49**(7), 2571–2578.

Hanania, E. G., Kavanagh, J., Hortobagyi, G., Giles, R. E., Champlin, R. and Deisseroth, A. B. (1995). *Am. J. Med.*, **99**: 537–552.

Harries, D., Rau, D. C. and Parsegian, V. A. (2005). *J. Am. Chem. Soc.*, **127**: 2184–2190.

Herman, A. (2010). *J. Comput'l and Theor'l Nanoscience*, **7**(11), 2360–2366.

Hettiarachchi, D. S. N. and Macartney, D. H. (2006). *Can. J. Chem.*, **84**: 905–914.

Hirayama, F. and Uekama, K. (1999). *Adv. Drug Delivery Rev.*, **36**: 125–141.

Hirsch, K. A., Wilson, S. R., Moore, J. S. (1997). *Chemistry-A European J.*, 765–771.

Holzinger, M., Bouffier, L., Villalonga, R. and Cosnier, S. (2009). *Biosens. Bioelectron.*, **24**: 1128–1134.

Ishizone, T., Tajima, H., Matsuoka, S., Nakahama, S. (2001). *Tetrahedron Letts.*, **42**: 8645–8647.

Jaime, C., Redondo, J., Sanchez-Ferrando, F., Virgili, A. (1991). *J. Molecular Structure*, **240, 8**: 317–329.

Karakasyan, C., Millot, M.-C., Vidal-Madjar, C. (2004). *J. Chromatography B.*, **808**: 63–67.

Karle, I. L. (1999). *J. Molecular Structure*, **474:** 103–112.

Karle, I. L., Ranganathan, D. (2003). *J. Molecular Structure*, **647**: 85–96.

Kovalev, V., Shokova, E., Khomich, A., Luzikov, Y. (1996). *New J. Chemistry*, April 1996, **20**: 483–492.

Kretschmann, O., Choi, S. W., Miyauchi, M., Tomatsu, M., Harada, A. and Ritter, H. (2006). *Angew. Chem. Int. Ed. Eng. Ed.*, **45**: 4361–4365.

Krois, D. and Brinker, U. H. (1998). *J. Amer. Chem. Soc.*, **120**: 11627–11632.

Lehn, J. M. (1988). *Angew. Chem. Int. Ed. Engl. Ed.*, **27**: 89–112.

Li, X, Qi, Z., Liang, K., Bai, X., Xu, J., Liu, J., Shen, J. (2008). *Catal. Lett.*, **124**: 413–417.

Lin, R. and Wilk, Z. A. (1995). *Fuel*, **74**: 1512–1521.

Liu, S., Zavalij, P. Y. and Isaacs, L. (2005). *J. Am. Chem. Soc.*, **127**: 16798–16799.

Maesen, T., Carlson, R. M., Dahl, J. E., Liu, S., Timken, H. K. C. and Qureshi, W. R. (2007). US Patent No. US7173160 B2.

Mandolini, L. and Ungaro, R. (eds). (2000). *Calixarenes in Action*, World Scientific Pub. Co., New York, NY.

Mann, D. J., Peng, J., Freitas Jr., R. A., Merkle, R. C. (2004). *J. Comput. Theor. Nanosci.*, **1**: 71–80.

Mansoori, G. A. (2005). *Principles of Nanotechnology (Molecular-Based Study of Condensed Matter in Small Systems)*. World Scientific. Pub. Co., Hackensack, NJ.

Mansoori, G. A., George, T. F., Assoufid, L. and Zhang, G. (2007). *Molecular Building Blocks for nanotechnology-from diamondoids to nanoscale materials and applications*. Topics in Applied Physics, **109,** Springer: New York.

Merkle, R. C. (1999). Trends in Biotechnology, **17**: 271–274; See also: *ibid.*, *Nanotechnology*, 2000, **11**: 89–99.

Merkle, R. C. and Freitas Jr. R. A. (2003). *J. Nanosci. Nanotechnol.*, **3**: 319–324.

Moghaddam, S., Yang, C., Rekharsky, M. *et al.* (2011). *J. Am. Chem. Soc.*, **133**(10): 3570–3581.

Munteanu, M., Choi, S. and Ritter, H. (2009). *Macromolecules*, **42**: 3887–3891.

Nikakhtar, A., Nasehzadeh, A., Mansoori, G. A. (2007). *J. Comput'l & Theor'l Nanoscience*, **4**: 1–8.

NSTC, (2000). *National Nanotechnology Initiative: Leading to the Next Industrial Revolution.* A Report by the Interagency Working Group on Nanoscience, Engineering and Technology Committee on Technology, National Science and Technology Council, Washington, D. C.

Ogoshi, T., Yamagishi, T., Nakamoto, Y. and Harada, A. (2007). *Chemistry Lett's*, **36**(8): 1026–1027.

Oyabu, N., Custance, O., Yi, I., Sugawara, Y. and Morita, S. (2003). *Phys. Rev. Lett.*, **90**: Article No. 176102.

Palepu, R. and Reinsborough, V. C. (1990). *Aust. J. Chem.*, **43**: 2119–2123.

Park, I.-K., Recum, H. A. v., Jiang, S. and Pun, S. H. (2006). *Bioconjugate Chem.*, **13**: 630–639.

Pedersen, C. J. (1967). *J. Am. Chem. Soc.*, **89**: 7017–7036.

Peng, J., Freitas Jr., R. A. and Merkle, R. C. (2003). *J. Comput. Theor. Nanosci.*, **1**: 62–70.

Peng, J., Freitas Jr., R. A., Merkle, R. C., von Ehr, J. R., Randall, J. N., Skidmore, G. D. (2006). *J. Comput. Theor. Nanosci.*, **3**: 28–41.

Poon, K. H.-N. and Cheng, Y.-L. (2008). *J. Incl Phenom. Macrocycl. Chem.*, **60**: 211–222.

Pun, S. H., Bellocq, N. C., Liu, A., Jensen, G., Machemer, T., Quijano, E., Schluep, T., Wen, S., Engler, H., Heidel, J. and Davis, M. E. (2004). *Bioconjugate Chem.*, **15**: 831–840.

Pun, S. H. and Davis, M. E. (2002). *Bioconjugate Chem.*, **13**: 630–639.

Ramezani, H. and Mansoori, G. A. (2007). *Diamondoids as Molecular Building Blocks for Nanotechnology*. In Molecular Building Blocks for Nanotechnology: From Diamondoids to Nanoscale Materials and Applications, (Springer, New York), *Topics in Applied Physics*, **109**: 44–71.

Ranganathan, D., Haridas, V. and Karle, I. L. (1999a) *Tetrahedron*, **55**: 6643–6656.

Ranganathan, D., Haridas, V., Kurur, S., Nagaraj, R., Bikshapathy, E., Kunwar, A. C., Sarma, A. V., Vairamani, M. (2000a). *J. Organic Chem.*, **65**: 365–374.

Ranganathan, D., Haridas, V., Nagaraj, R., Karle, I. L., Isabella, L. (2000b). *J. Organic Chem.*, **65:** 4415–4422.

Ranganathan, D., Samant, M. P., Nagaraj, R., Bikshapathy, E. (2002). *Tetrahedron Lett.s*, **43**: 5145–5147.

Ranganathan, D., Thomas, A., Haridas, V., Kurur, S., Madhusudanan, K. P., Roy, R., Kunwar, A. C., Sarma, A. V., Vairamani, M., Sarma, K. D. (1999b). *J. Organic Chem.*, **64:**, 3620–3629.

Ren, S., Chen, D. and Jiang, M. (2009). *J. Polymer Sci.*, **47**(17): 4267–4278.

Rudiger, V., Eliseev, A., Simova, S., Schneider, H.-J., Blandamer, M. J., Cullis, P. M. and Meyer, A. J. (1996). *J. Chem. Soc., Perkin Trans.*, **2**: 2119–2123.

Rudyak, V. Y., Avakyan, V. G., Nazarov, V. B. and Alfimov, M. V. (2009). *Nanotech. Russ.*, **4**: 27–37.

Rugar, D. and Hansma, P. (1990). *Physics Today*, Oct., 23–30.

Sangeetha, N. M. and Maitra, U. (2005). *Chem. Soc. Rev.*, **34**: 821–836.

Schluep, T., Hwang, J., Hildebrandt, I. J., Czernin, J., Choi, C. H. J., Alabi, C. A., Mack, B. C. and Davis, M. E. (2009). *Proc Nat. Acad. Sci. USA*, **106**: 11394–11399.

Seel, C and Vogtle, F. (1992). *Angew. Chem. Int. Ed. Engl. Ed.*, **31**: 528–549.

Segura, T. and Shea, L. D. (2001). *Annu. Rev. Mater. Res.*, **31**: 25–46.

Shokova, E., Tafeenko, V., Kovalev, V. (2002). *Tetrahedron Letts.*, **43:** 5153–5156.

Shortreed, M. E., Wylie, R. S. and Macartney, D. H. (1993). *Inorg. Chem.*, **32**: 1824–1829.

Siegel, R. W., Hu, E. and Roco, M. C., Editors (1999). *Nanostructure Scence and Technology — A Worldwide Study*. Prepared under the guidance of the IWGN, NSTC (WTEC, Loyola College), Maryland.

Smith, J., Zhang, Y. and Niven, R. (1997). *Adv. Drug Delivery Rev.*, **26**: 135–150.

Stevens, M. P. (1999). *Polymer Chemistry — An introduction*. Oxford University Press: New York.

Strenalyuk, T. and Haaland, A. (2008). *Chemistry — A Europ. J.*, **14**(33): 10223–10226.

Suh, J., Lee, S. H. and Zoh, K. D. (1992). *J. Am. Chem. Soc.*, **114**: 7916–7917.

Szetjtli, J. (1998). *Chem. Rev.*, **98**: 1743–1753.

Thomas, D. and Suthanthiran, M. (2003). *Immunol. Rev.*, **196**: 161–175.

Thomas, M. and Klibanov, A. M. (2003). *Appl. Microbiol. Biotechnol.*, **62**: 27–34.

Valdivia, A., Perez, Y., Cao, R., Banos, M., Garcıa, A. and Villalonga, R. (2007). *Macromol. Biosci.*, **7:** 70–75.

Villalonga, R, Camacho, C., Cao, R.,Hernandez, J., and. Matıas, J. C. (2007b). *Chem. Commun.*, 942–944.

Villalonga, R., Fujii, A., Shinohara, H., Asano, Y., Cao, R., Tachibana, S. and Ortiz, P. (2007a). *Biotechnol. Lett.*, **29**: 447–452.

Villalonga, R., S. Tachibana, S., Cao, Ortiz, P. D., Gomez, L. and Asano, Y. (2006). *J.Exp. Nanosci.*, **1**: 249–260.

Walch, S. P., Merkle, R. C. (1998). *Nanotechnology*, **9**: 285–296.

Wen, G.-T., Zhu, M.-Z., Wang, M.-X., Wang, Y.-H. and Guo, Q.-X. (2006). *J. Incl. Phenom. Macrocyc. Chem.*, **56**: 221–224.

Whitesides, G. M., Mathias, J. P. and Seto, C. T. (1991). *Science*, **254**: 1312–1319.

Zhao, H., Yang, K., Martinez, A., Basu, A., Chintala, R., Liu, H.-C., Janjua, A., Wang, M. and Filpula, D. (2006). *Bioconjugate Chem.*, **17**: 341–351.

5

Properties of Diamondoids through Quantum Calculations

5.1. Introduction

Quantum mechanical computations are advanced to a level so that we can predict the properties of atoms and molecules with unprecedented precision. Many advances have been made recently which focus on characterization of diamondoid molecules through quantum calculations due to their great potential applications in several segments of nanotechnology.

In this chapter, we present quantum mechanical investigations into properties of diamondoids showing the utility of such prediction schemes in characterization of diamondoid molecules. Initially, we briefly explain some frequently used terms in quantum calculations of nanosystems in order to provide the reader a familiarity with the principles involved including *ab initio* calculations, density functional theory (DFT), and the related commercial and developing scientific computer packages (or codes). Afterward, we present a number of case studies involving the quantum mechanical investigations of physical and structural properties of diamondoid molecules (Araujo *et al.*, 2011).

5.2. Schrödinger Equation and *ab initio* Calculations

Quantum mechanical calculations are based on the infamous Schrödinger wave equation, as is shown below.

$$i\hbar\frac{\partial \Psi}{\partial t} = -\frac{\hbar^2}{2m}\frac{\partial^2 \Psi}{\partial x^2} + \phi(x)\Psi(x,t) \equiv \tilde{H}\Psi(x,t), \tag{5.1}$$

In this equation $\hbar = h/(2\pi)$, h is the Planck's constant, ψ is wave function, t is time, m is mass, x is position, $\hat{\mathrm{H}}$ is Hamiltonian, and ϕ is the potential energy. This fundamental equation has a role in quantum mechanics analogous to the Newton's second law in classical mechanics. This equation describes the time-dependence of the wave function ψ in quantum mechanical systems.

In the mathematical solution of the Schrödinger equation and its application for characterization of atoms and molecules, inner system of each atom and molecule which is constituted of many fundamental particles is associated with the complex Hilbert space. Hilbert space is the generalization of the Euclidean vector algebra to a space with any finite or infinite number of dimensions. Each instantaneous state of the quantum mechanical system is described by a unit vector. Every such time-dependent unit vector encodes the probabilities for the outcomes of all possible measurements applied to the system.

The *ab initio* (meaning "from the first principles") calculations include the variety of computational methods used for characterization and prediction of atomic and molecular structures and properties based on the principles of quantum mechanics. *Ab initio* calculations may be used to find, for example, the interamolecular and intermolecular potential energies including the bond lengths, bond angles, and molecular geometries.

Many efforts have been made for the analytical as well as numerical solution of quantum mechanics for characterization of atoms and molecules. As a result, a variety of approximation techniques and numerical methods for calculations have been developed. Among those techniques the Hartree–Fock (HF) approximation and the DFT method are more widely developed and used. In what follows, we briefly introduce these approaches and their applications for diamondoids.

5.2.1. *The Hartree–Fock (HF) approximation*

The basic idea behind the Hartree–Fock approximation is that the motion of each electron is assumed to be described by a molecular orbital. Each molecular orbital is assumed to be made of a linear combination of atom-centered "basis set".

While molecular orbital concept is valid for hydrogen atom it is an approximation for other atoms and molecules. To reach to this approximation it is like assuming that each electron in an atom or molecule is only exposed to the average Coulomb repulsion of all the other electrons in that atom or molecule.

A "basis set" is actually a set of functions used to create the molecular orbitals. Each molecular orbital is expanded as a linear combination of such functions with the weights (or coefficients) to be determined. The Hartree–Fock iterative procedure then calculates the coefficients in such linear expansions. As a result, the Hartree–Fock approximation is a self-consistent field (SCF) interactive procedure to calculate the so-called "best possible" single determinant solution to the time-independent Schrödinger equation.

In a mathematical formalism, this approximation consists of finding the best set of one electron orbitals in the Slater determinant that represents the ground state. Therefore, according to the variational principle, the best set is the one which yields the lowest total energy for the interacting system. The resultant orbitals are given by a set of modified one-particle Schrödinger equations that now is named Hartree–Fock equation (Fock, 1930):

$$\begin{aligned}&\left[-\frac{h^2}{2m}\nabla^2+\phi(\mathrm{r})+\int dr^t\frac{e^2}{\left|\mathrm{r}-\mathrm{r}^t\right|}n(\mathrm{r}^t)\right]\psi_i(\mathrm{r})\\&-\int dr^t\frac{e^2}{\left|\mathrm{r}-\mathrm{r}^t\right|}n(\mathrm{r},\mathrm{r}^t)\psi_i(\mathrm{r}^t)=E\psi_i(\mathrm{r})\end{aligned} \tag{5.2}$$

This approximation assumes that each electron moves under the combined influence of the external potential $\phi(\mathbf{r})$ (Hartree and Fock-exchange potential). Here, the Hartree potential is expressed as the electrostatic potential due to the electron density distribution $n(\mathbf{r})$ via the Coulomb interaction $e^2/|\mathbf{r}-\mathbf{r}'|$. On the other hand, the nonlocal Fock exchange potential is due to the one-particle density matrix $n(\mathbf{r}, \mathbf{r}')$ as a consequence of the Pauli principle. Both, the electron density and density matrix are given by the N lowest energy orbitals of Eq. (5.2). The many-electron wave function Ψ is approximated as a product of single-particle functions ψi, i.e., ψ ($\mathbf{r}1$, $\mathbf{r}2$, ...) = $\psi_1(\mathbf{r}1)\cdot\ \psi_2(\mathbf{r}2)\cdot\ \psi_3(\mathbf{r}3)\cdot\ldots\cdot\ \psi_N(\mathbf{r}_N)$. Moreover, the equation yields orbitals which can be used to construct excited states beyond ground state.

5.2.2. *Density functional theory (DFT)*

Instead of concentrating on molecular orbitals, which is the case for the Hartree–Fock approximation, the DFT emphasis is on the density of electrons. DFT is now-a-days a general approach for the *ab initio* calculations used to solve the Schrödinger equation of quantum many-particle systems. In DFT calculations, the original many-body problem is rigorously recast in the form of an auxiliary single-particle problem (Hohenberg and Kohn, 1964; Kohn and Sham, 1965a; Kohn and Sham, 1965b). In fact, the Thomas–Fermi theory was the first theory of electronic energy in terms of the electron density distribution (Lieb, 1981).

DFT is based on two general theorems:

(i) Any physical property of an interacting electron gas, in its fundamental state, can be written as a unique functional of the electron density $\rho(\mathbf{r})$ and, in particular, its total energy $E(\rho)$. The total energy of an electron gas in the presence of a background potential may be written as a functional of the charge density.

(ii) The total energy $E(\rho)$ reaches its minimum for the true density $\rho(\mathbf{r})$, i.e., derived from the Schrödinger equation. Therefore, the true charge density of the system is that which minimizes this functional, subject to it having the correct normalization.

For electrons moving in a potential field due to ions, $E(\rho)$ is expressed as

$$E(\rho)=-e^2\int[\rho^-(\mathbf{r})\rho^+(\mathbf{r}')/|\mathrm{r}-\mathrm{r}'|]d^3\mathrm{r}d^3\mathrm{r}'+e^2/2\int[\rho^-(\mathrm{r})\rho^-(\mathrm{r}')/|\mathrm{r}-\mathrm{r}'|]$$
$$d^3rd^3r'+e^2/2\int[\rho^+(\mathrm{r})p^+(\mathrm{r}')/|\mathrm{r}-\mathrm{r}'|]d^3\mathrm{r}d^3\mathrm{r}'+T[\rho(\mathrm{r})]+Exc[\rho(\mathrm{r})],$$

(5.3)

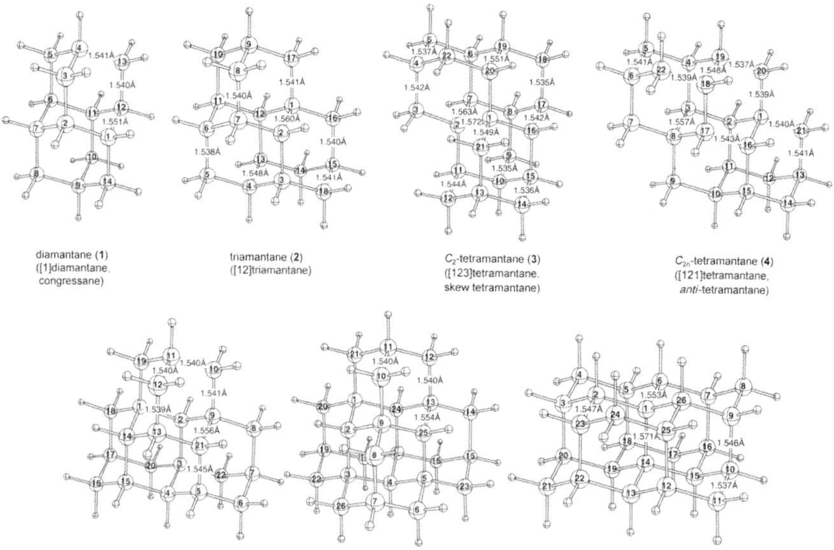

diamantane (1) ([1]diamantane, congressane)

triamantane (2) ([12]triamantane)

C_2-tetramantane (3) ([123]tetramantane, skew tetramantane)

C_{2h}-tetramantane (4) ([121]tetramantane, *anti*-tetramantane)

C_{3v}-tetramantane (5) ([1(2)3]tetramantane, isotetramantane)

T_d-pentamantane (6) ([1(2,3)4]pentamanate)

D_{3d}-hexamantane (7) (cyclohexamantane)

Fig. 5.1. The B3LYP/6-31G* optimized geometries of selected diamondoids with numbering of carbons (Fokin *et al.*, 2005). All of these structures are also shown on Fig. 1.4.

where $\rho^+(\mathbf{r})$ is the number of elementary positive charges per unit volume. The first three terms in the right-hand side of the above equation are due to the classical Coulomb electron–ion, electron–electron and ion–ion interactions, respectively. The fourth term is the kinetic energy of a noninteracting electron system of density $\rho(\mathbf{r})$. The last term is the exchange and correlation energy (Borisenko and Ossicini, 2004). DFT tends to give very accurate results for simulating fundamental electronic and structural properties of matter (metals, semiconductors, and insulators).

To illustrate the application of DFT for diamondoids, Fig. 5.1 shows optimized geometries of some diamondoid molecules calculated using the three-parameter hybrid functionals of Becke (Becke, 1988) and the correlation functional of Lee, Yang and Parr (Lee *et al.*, 1988).

These approximations are known as the B3LYP "level" of the DFT. The B3LYP/6-31G* optimized geometries were performed by Fokin *et al.* (2005) in investigation of properties such as enthalpies of formation, ionization potentials, strain energies, and functionalization of diamondoid molecules. In this section, some aspects of the mostly used theoretical approximations used to predict diamondoid properties will be discussed.

5.2.3. *Some basic ab initio computations for diamondoids*

Diamondoids, as the molecular building blocks for nanotechnology, are essentially diamond fragments with hydrogen termination. Thus, it is expected to present insulating and toughness properties as bulk diamond, due to structural similarities. *Ab initio* calculations can be a very useful tool in predicting and anticipating structural and electronic properties of diamondoid molecules.

In addition to the Hartree–Fock and DFT methods, we should also mention of the quantum Monte Carlo (QMC) calculation approach and the tight-binding (TB) approximation which also are used for diamondoids property calculations.

QMC is a stochastic computation/simulation technique based on random number generation of the variable to estimate the function statistically. Drummond *et al.* (2005) have used QMC and DFT for predicting of the properties of diamondoid molecules. Their QMC calculations were performed with CASINO code (Needs *et al.*, 2002) using Slater–Jastrow (Foulkes *et al.*, 2001) trial wave functions of the form $\psi_T = D^{\uparrow} D^{\downarrow} \exp[J]$. Here $D^{\uparrow}$ and $D^{\downarrow}$ are Slater determinants (Slater, 1929) of up and down spin orbitals taken from the DFT calculations and $\exp[J]$ is a Jastrow correlation factor, which includes electron–electron and electron–ion terms expanded in Chebyshev polynomials. Also Jensen *et al.* (2004) published a rather detailed assignment of the normal modes of vibration of adamantane and deuterated-adamantane, and produced the normal mode based empirical correction factors at the Hartree–Fock, DFT (B3LYP) and 2nd order Møller–Plesset perturbation theory (MP2) levels.

The TB approximation approach assumes that valence electrons are completely delocalized when a solid is formed. On the contrary, core electrons remain much localized and, thus, discrete core levels of the atoms are only very slightly broadened in the solid state. The corresponding

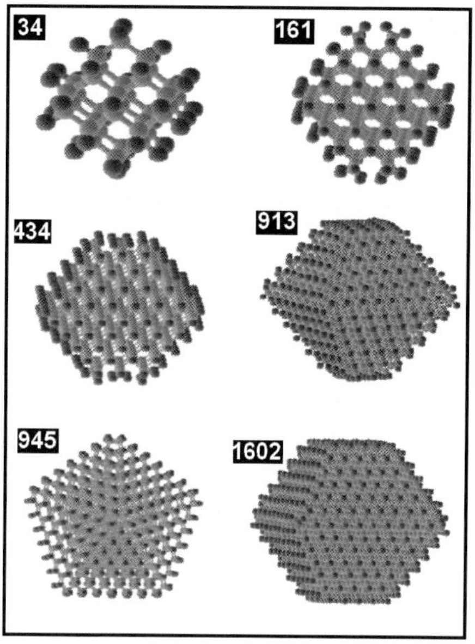

Fig. 5.2. Illustrations of the hydrogen passivated large nano-diamond clusters studied by Areshkin *et al.* (2004). The numbers of carbon atoms for the structures are also reported in this figure.

wave functions are not very different, in the vicinity of each atom, from the atomic wave functions (Horsfield and Bratkovsky, 2000).

A self-consistent environment-dependent tight binding calculation for very large diamondoid clusters (named as "hydrogen passivated nano-diamonds" as shown in Fig. 5.2) was performed by Areshkin *et al.* (2004) to examine their electron emission-related properties.

For sizes larger than 2.5 nm particle bandgap was found to be equal to the bandgap of bulk diamond. Coulomb potential distributions and electron affinities of clusters were found to be insensitive to the particle size if it exceeded 1.0 nm. Tunneling probabilities for homogeneous and inhomogeneous emission models were estimated. The simulation results indicate that the low emission threshold for hydrogen passivated diamond nano-clusters is due to hydrogen-assisted emission from the edges of small unpassivated islands. They predicted no LUMO energy shift down to 34 carbon atoms for such very large diamondoid clusters.

5.2.4. *Commercial and scientific computer codes (ab initio packages)*

A number of commercial and scientific *ab initio* computer codes have been developed in order to predict and simulate properties of different atomic, molecular, and nano systems. In what follows, we briefly present four such packages which are more widely used for diamondoids characterizations, and we present some of the results based on these packages in this chapter.

Gaussian 03 (www.gaussian.com) is a computational chemistry software program. The name originates from the use of Gaussian orbitals to speed up calculations compared to those using Slater-type orbitals. This improved performance on slower computer hardware and facilitated the growth of computational chemistry, particularly *ab initio* methods such as Hartree–Fock (Frisch, 2003).

FLAPW (iffwww.iff.kfa-juelich.de/~bluegel/flapw/flapw_home.html) stands for "Full-Potential Linearized Augmented Plane Wave". This package is an all-electron method which within DFT is universally applicable to all atoms of the periodic table and to systems with compact as well as open structures.

It is widely considered to be the most precise electronic structure method in solid state physics. It is considered among the most accurate methods for performing electronic structure calculations for crystals.

MOLPRO (www.molpro.net) package is a complete system of *ab initio* programs for molecular electronic structure calculations. As distinct from other commonly used quantum chemistry packages, the emphasis is on highly accurate computations. Using recently developed integral-direct local electron correlation methods, which significantly reduce the increase of the computational cost with molecular size, accurate *ab initio* calculations can be performed for much larger molecules than with most other programs.

Atomistix ToolKit (ATK) and Virtual NanoLab (VNL) (*www.atomistix.com*) combine *ab initio* approximations such as nonequilibrium Green's functions (NEGF) and DFT for simulating and modeling electrical properties of nanostructures where two electrodes are coupled with a nanostructure. The entire system is treated self-consistently under finite bias conditions. ATK whose predecessor is TranSIESTA-C package is an *ab initio* electronic structure program capable of simulating and modeling electrical properties of nanostructured systems coupled to semi-infinite electrodes. The VNL software package (*www.virtualnanolab.com*) is based on ATK and gives access to atomic-scale modeling techniques with a graphical interface for simulation and analysis of the atomic scale properties of nanoscale devices.

5.3. Electronic and Structural Properties of Diamondoids

5.3.1. *Electronic structure*

The first step in the investigation of electronic properties of molecular clusters is to determinate (theoretically or experimentally) the electronic density of states (EDOS), identifying mainly the energies of the highest occupied molecular orbital (HOMO), the lowest unoccupied molecular orbital (LUMO) and the band gap which is the difference between the HOMO and LUMO energies,

$$\text{Band Gap} \equiv \text{ELUMO} - \text{EHOMO}. \tag{5.4}$$

In diamondoids and other organic compounds, the HOMO level plays the role of the valence band, the LUMO level plays the role of the conduction band, and the band gap is inversely proportional to the excitability. The analysis of changes on band gap associated with the size of the molecules provides us important information concerning electrical properties of the matter. To illustrate, Fig. 5.3 shows EDOS spectra for four diamondoid molecules, adamantane, diamantane, tetramantane, and decamantane.

The EDOS spectra results reported in Fig. 5.3 were obtained by using DFT within the local density approximation (LDA). This approximation is essentially a method of calculation of the electronic band structure of solids and it expresses the potential of an electron at a given location as a function of the electron density at the same site. The spectrum consists of discrete eigenvalues associated with the finite size of the molecules. Moreover, some considerations were taken into account as the use of the Troullier–Martins

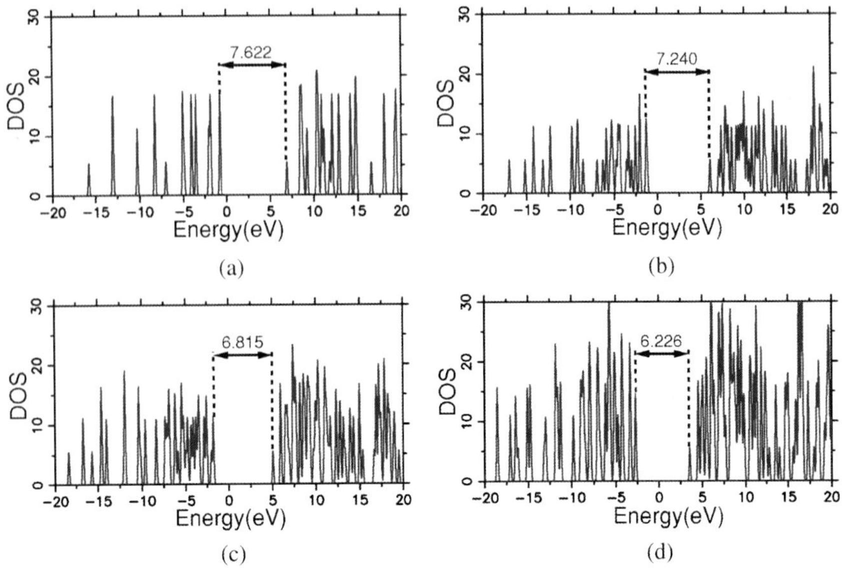

Fig. 5.3. Electronic structure of an (a) adamantane, (b) diamantane, (c) tetramantane, and (d) decamantane. Energy = 0 corresponds to the Fermi level. The fundamental HUMO-LUMO band gaps are indicated in eV (McIntosh *et al.*, 2004).

pseudopotentials to describe the effect of atomic nuclei plus core electrons on the valence electrons and the Perdew–Zunger-parametrized exchange-correlation potential (Ordejón *et al.*, 1997; Ordejón *et al.*, 2000) as implemented in the SIESTA code (www.icmab.es/siesta/).

However, the EDOS spectra of McIntosh *et al.* (2004) model for diamondoids (Fig. 3), and all currently theoretical approaches, differs from the EDOS spectra experimentally obtained by using Soft X-ray emission (SXE) and X-ray absorption (XAS) spectroscopy for HOMO and LUMO mapping, respectively.

5.3.2. *Quantum confinement effect*

Quantum confinement effect (QCE) is the phenomenon of nonzero lowest energy and quantization of the allowed energy levels in low-dimensional structures, arising from the confinement of electrons within a limited space (Borisenko and Ossicini, 2004). This effect predicts that with decreasing particle sizes the band gap increases due to shifting of the band edges (Willey *et al.*, 2005).

Diamondoid molecules, being low-dimensional structures, allow the study of the QCE in the molecular size limit. These molecules are perfectly size-selected, neutral; fully sp^3 hybridized and exhibit a complete hydrogen surface termination which generally is used by most theoretical models describing semiconductor nanoclusters (Willey *et al.*, 2005).

To observe the existence of QCE in diamondoid molecules, it is necessary to assess the dependent-size HUMO and LUMO energy shifts. Experimental band gaps are estimated by comparing the diamondoid inflection points of the both SXE and XAS. Diamondoids' band gaps increase with decreasing size from hexamantane through adamantane, as it is demonstrated by the experimental (SXE and XAS) data in Fig. 5.4 and the theoretical (DFT and QMC) methods in Table 5.1.

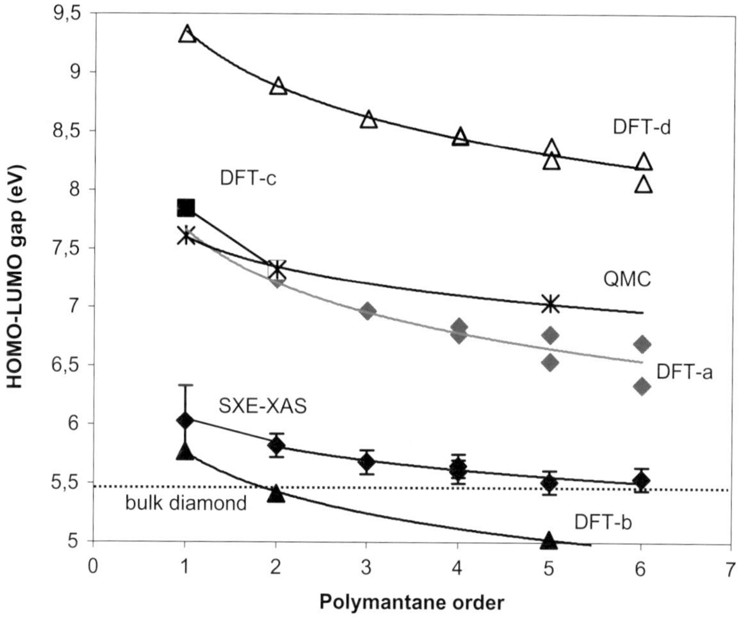

Fig. 5.4. Quantum Confinement effect in diamondoids. Experimental data from SEX–XAS spectroscopy (Willey *et al.*, 2006) and quantum calculations from DFT and QMC models. DFT-a (McIntosh *et al.*, 2004); QMC and DFT-b (Drummond *et al.*, 2005); DFT-c (Xue *et al.*, 2008); DFT-d (Lu *et al.*, 2005).

Table 5.1. HUMO-LUMO gaps [eV] calculated (DFT and QMC methods) and estimated from the SXE and XAS data.

Compound	DFT[a]	DFT[b]	G98-DFT[c]	DFT[d]	QMC[d]	SXE-XAS[e]
adamantane	7.62	7.84	9.33	5.77	7.61	6.03
diamantane	7.24	7.32	8.89	5.41	7.32	5.82
triamantane	6.97		8.61			5.68
tetramantanes	6.77–6.84		8.46–8.47			5.60–5.65
pentamantanes	6.54–6.77		8.26–8.38	5.03	7.04	5.51
hexamantanes	6.34–6.70		8.07–8.26			5.54
bulk diamond	5.4			4.23	5.6	5.47

[a]McIntosh *et al.* (2004); [b]Xue and Mansoori (2008); [c]Lu *et al.* (2005); [d]Drummond *et al.* (2005); [e]Willey *et al.* (2006).

Thus, the experimental and theoretical results confirm the existence of QCE in diamondoid molecules. However, most of the theoretical results of HUMO-LUMO energy gap calculated by models overestimate the experimental data for diamondoids (compare Table 5.1 data with those in Fig. 5.4). One likely reason for this discrepancy is due to the basis set used in the theoretical models which strongly influences the results. Another likely reason is the fact that quantum calculations are done on noninteracting clusters whereas experimental measurements use condensed-phase diamondoids (see Fig. 3.4 in Chapter 3), where particle–particle interaction may play an important role (Bostedt *et al.*, 2004) on the results of measurements.

QCE is expected to push diamondoid band gaps into the UV range, enabling a unique set of sensing applications (Drummond *et al.*, 2005).

5.3.3. *Ionization potential and electron affinity*

In this section, some fundamental properties of diamondoids, such as ionization potentials (IPs) and electron affinity (EA) will be presented.

The IPs are fundamental values for all materials. In fact, they are of central importance for all potentially technological applications as well as in theoretical predictions. The adiabatic (vertical) IP corresponds to the energy difference between the ground state of a neutral molecule and an

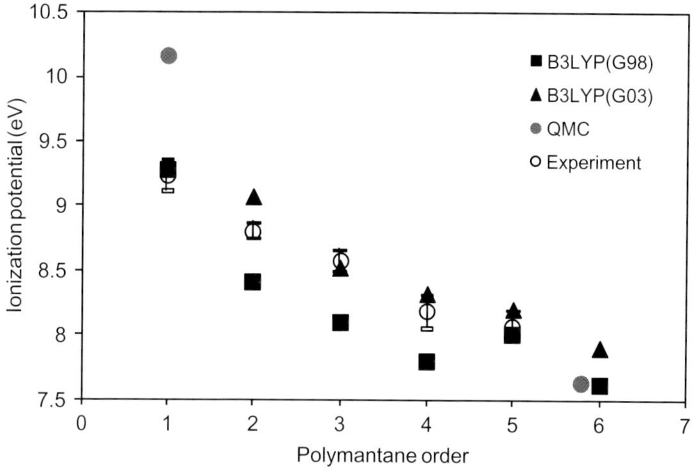

Fig. 5.5. Comparison of the experimental adiabatic (vertical) ionization potential data of diamondoids and DFT [B3LYP/6-31G(d)] calculations with GAUSSIAN03 (Fokin *et al.*, 2005) and GAUSSIAN98 (Lu *et al.*, 2005), as well as QMC (Drummond *et al.*, 2005) predictions (Lenzke *et al.*, 2007).

ionized molecule in the same geometry. Thus, the IPs are experimentally determined from photoion yield measurements of diamondoids in the gas phase (Lenzke *et al.*, 2007). Experiments carried out in the gas phase prevent alterations in the electronic structure of a molecule due to particle–particle and particle–substrate interactions which exist in the condensed phases (Bostedt *et al.*, 2004a).

Figure 5.5 shows the adiabatic (vertical) experimental and theoretical IPs of diamondoids of various polymantane orders.

Here, polymantane order represents the number of adamantane cages that form a given diamondoid molecule. Thus, adamantane, diamantane, and triamantane have polymantane order of 1, 2, and 3, respectively. Fokin *et al.* (2005) and Lu *et al.* (2005), in parallel and independently, performed DFT calculations at the B3LYP/6-31G(d) level of theory using the GAUSSIAN03 and GAUSSIAN98 computer codes, respectively, to calculate the vertical IPs. On the other hand, Drummond *et al.* (2005) used quantum Monte Carlo (QMC) methods in theoretical predictions of diamondoids IPs. According to Fig. 5.5 only the IP of adamantane (polymantane of order one), calculated with QMC, is quite different (more than 1 eV) from the measured and other theoretical values. Additionally, a larger H-terminated diamond molecule, with the chemical formula $C_{29}H_{36}$ (close to polymantane order 6 in Fig. 5.5), is shown for comparison as a higher order diamondoid. This molecule was empirically constructed to be approximately spherical and belonging to the T_d group symmetry. Indeed, its size lies between pentamantane and hexamantane, however, it is not a true diamondoid due to its different round shape and the lack of cage-only structure (Drummond *et al.*, 2005; Lenzke *et al.*, 2007).

Sharp decrease in IPs with increasing polymantane order for, both theoretical predictions and experimental data, can be seen as a consequence of the QCE. The DFT calculations of Fokin *et al.* (2005) are in good agreement with the experimental data, whereas the DFT calculations performed by Lu *et al.* (2005) underestimate the experimental IPs with a maximum absolute error of 0.4 eV.

In contrast to diamondoids which exhibit monotonic decrease of the IPs with increasing molecular size, the IPs of fullerenes exhibit no monotonous trend regarding size dependence (Sánchez *et al.*, 2005).

A number of researchers (Himpsel *et al.*, 1979; Rutter *et al.*, 1998; Prins, 2003; Roth *et al.*, 2010) have pointed out that H-terminated nanodiamond surfaces exhibit negative electron affinities (NEAs), which may suggest that diamondoid molecules would also have NEAs. The electron affinity of an atom or molecule is defined as the change in its energy when an electron is added to it, producing a negative ion. NEA materials are useful for the development of electron emitters like the design of efficient cathodes that can supply electrons to the vacuum with little energy loss (Marsusi *et al.*, 2009a). The NEA feature of diamondoids could open up possibilities in optoelectronic applications such as coating surfaces with diamondoids, to produce new electron-emission devices. QMC (Drummond *et al.*, 2005) and DFT (Drummond *et al.*, 2005; Wang *et al.*, 2008) calculations have demonstrated the existence of NEAs in isolated diamondoid molecules, as can be seen in Fig. 5.6.

For adamantane, Drummond *et al.* (2005) have calculated Electron Affinities (EAs) with similar results using QMC and DFT theoretical methods. Taking a closer look in Fig. 5.6, it can be seen that NEA values of ~0.3 eV are achieved for higher diamondoids, i.e., [121]tetramantane and $C_{29}H_{36}$ (a spherical diamond cluster).

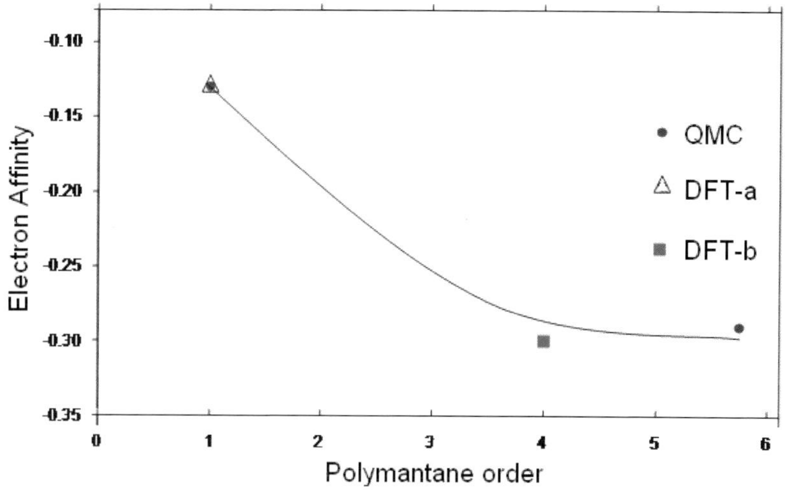

Fig. 5.6. Electron Affinities (in eV) calculations of adamantane, [121]tetramantane and $C_{29}H_{36}$ (spherical particle). QMC and DFT-a calculations by Drummond *et al.* (2005) and DFT-b calculations by Wang *et al.* (2008).

Marsusi and co-workers (2009a,b) have calculated the optoelectronic properties of adamantane ($C_{10}H_{16}$) compared to hydrogen-terminated sila- and germa-adamantane ($Si_{10}H_{16}$ and $Ge_{10}H_{16}$) by the DFT[28]. They have shown that the electronic properties of adamantane in comparison to sila- and germa-adamantane are more affected by hydrogen atoms. Their calculations also show that the electron affinity of $C_{10}H_{16}$ is negative, while those of $Si_{10}H_{16}$ and $Ge_{10}H_{16}$ are positive (see Fig. 5.7).

This also indicates that, adamantane and other diamondoid molecules have an intrinsic negative affinity, which makes their self-assembled structures promising candidates as electron-emission devices.

5.3.4. *Electronic properties*

Quantum mechanical principles (*ab initio* packages) can be used to predict the electronic properties of diamondoids and derivatives. As an example, Fig. 5.8 shows the electronic properties of an elongated hexamantane isomer (isolated) and within an infinite diamondoid chain (as an infinite quasi-one-dimensional crystalline counterpart).

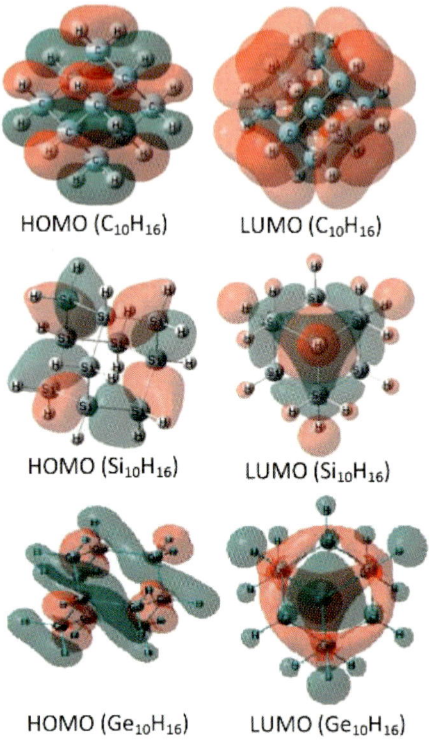

Fig. 5.7. Calculations of the isosurfaces plot of HOMO and LUMO states for $C_{10}H_{16}$, $Si_{10}H_{16}$ and $Ge_{10}H_{16}$ with isodensity value = 0.02 a.u. While LUMO isosurfaces of sila- and germa-adamantane resemble each other, the LUMO isosurface of $C_{10}H_{16}$ has a completely distinct configuration. It has been diffused over the hydrogen atoms, and looks like as if it is dominated by the H-terminated surface. This is a result of the almost uniform charge transfer from hydrogen to carbon in C–H and H–C–H segments. This brings about a potential step that causes a conduction band that lies below the vacuum level and therefore a negative electron affinity. The electron accumulation is indicated with red and the charged depletion with green colors (Marsusi *et al.*, 2009a).

Note that the electronic spectra of hexamantane (Fig. 5.8(b)) and its infinite crystalline counterpart (Fig. 5.8(e)) are similar and characterized by a series of van Hove singularities (Kittel, 2005) in the valence and conduction regions. In fact, the dense sequence of these singularities reflects the dense spectrum of relatively flat bands, as seen in Fig. 5.8(f), showing a wide-gap that features this system as an electrical insulator. However, probably due to the NEA of diamondoids (Himpsel *et al.*, 1979; Rutter and Robertson, 1998; Prins, 2003a, 2003b), associated with a large

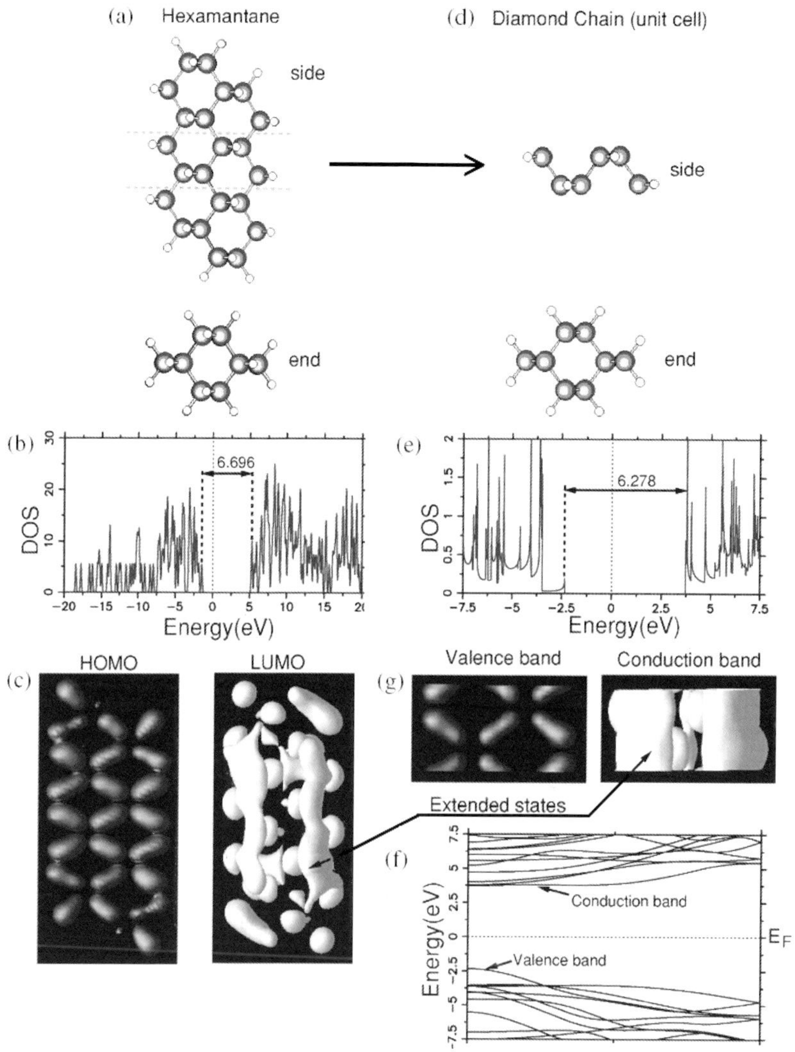

Fig. 5.8. Structural and electronic properties of an elongated haxamantane isomer (left panels) and an infinite diamondoids chain (right panels). The equilibrium structure is presented in (a) and (d), with the large spheres denoting carbon and the small spheres hydrogen atoms. The electronic density of states is shown in (b) and (e), with the Fermi level at $E_F = 0$. The one-dimensional HOMO and LUMO band structures of the infinite diamondoid chain are shown in (f). The charge-distribution of electrons in the highest occupied molecular orbital (HOMO) and the lowest unoccupied molecular orbital (LUMO) of hexamantane is shown in (c). The corresponding charge distribution in the top valence and bottom conduction band of the diamondoid chain is shown in (g) (McIntosh *et al.*, 2004).

length-to-diameter aspect ratio, diamondoid chains may surpass carbon nanotubes in applications such as cold cathodes or low-voltage electron emitters in flat panel displays (McIntosh *et al.*, 2004). Additionally, an infinite diamondoid chain could possibly acquire conductivity by mechanisms of *n*-type (electron or hole) doping.

A simple analysis of the charge delocalization as a prerequisite for electrical conduction could be done, comparing the charge distribution of the top-valence and bottom-conduction bands of a diamondoid chain, Fig. 5.8(g), with those of the HOMO and LUMO of the hexamantane molecule, Fig. 5.8(c). The results for the infinite diamondoid chain suggest that the charges associated with the valence band is localized in pockets near interatomic bonds, hindering "hole transport". Observe that this charge distribution is very similar to that of the HOMO of hexamantane. On the other hand, the conduction band consists of four strongly delocalized states located on the outer perimeter of the structure, which could be used to conduct electrons. Thus, these extended conduction bands find their counterparts in the LUMO of hexamantane, which is also similar in delocalization and across the middle section of this molecule. In addition, the bottom-conduction band is quite flat, suggesting that the electrons in a lightly electron-doped system should have a low mobility.

Diamondoids have wide band gaps, thus their chemical doping is unlikely to provide free carriers, since it would require the introduction of impurities with a very low ionization potential, likely to introduce trap sites. McIntosh *et al.* (2004) suggested that a more likely scenario of *n*-type (electrons or hole) doping would involve enclosing the diamondoid chain inside a carbon nanotube and *n*-type doping the surrounding nanotube. Alternatively, De Azevedo *et al.* (1999) were able to promote physical doping in polyaniline (PANI), by gamma irradiation using ^{60}Co at low absorbed dose (< 1 kGy). PANI is an intrinsically a conducting polymer and its conductivity is associated to its different oxidation states that change with the irradiation.

5.3.5. *Functionalized diamondoid molecules*

An interesting controllable method to connect diamondoids with strong bonds lies in the construction of functional elements of NEMS devices

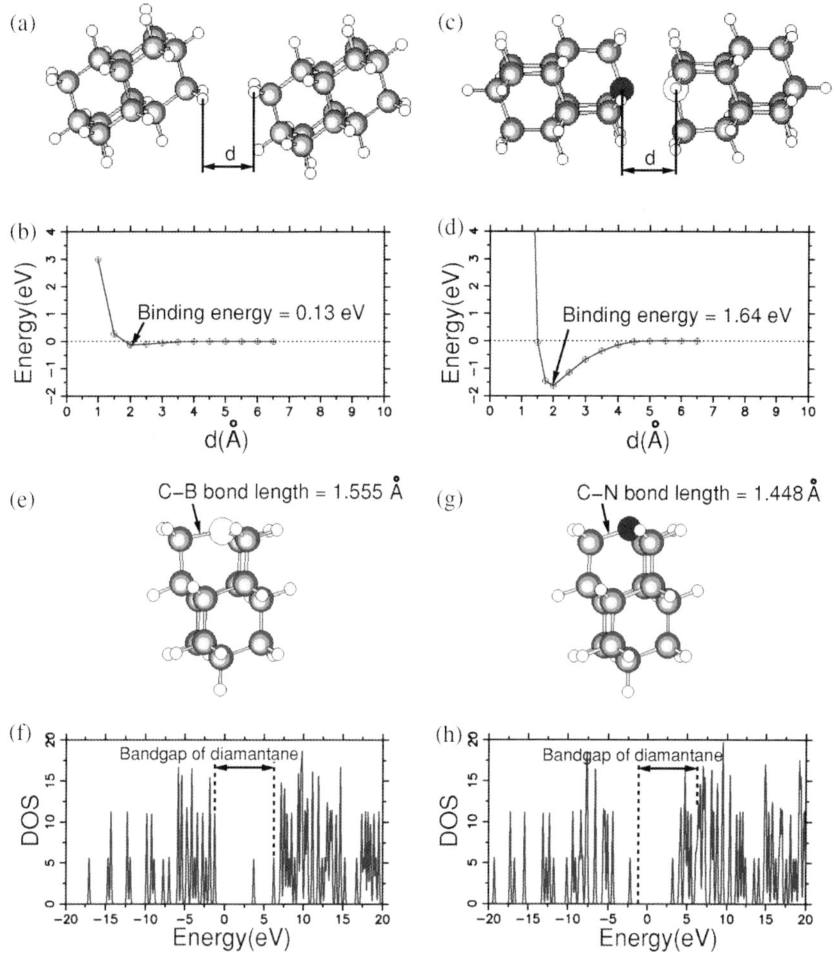

Fig. 5.9. Structure, interaction energy, and electronic properties of unmodified and functionalized diamantanes: (a) Structural arrangement and (b) interaction energy of two unmodified diamantanes $C_{14}H_{20}$, separated by distance d. (c) Structural arrangement and (d) interaction energy of two chemically $C_{13}BH_{19}$ and $C_{13}NH_{19}$, with the B and N sites in the neighboring molecules facing each other. (e) Equilibrium structure and (f) density of states of $C_{13}BH_{19}$. (g) Equilibrium structure and (h) density of states of $C_{13}NH_{19}$. The dashed lines indicate the position of the HOMO and LUMO in the unmodified diamantanes, with bandgap = 7.240 eV. Boron (light shading) and nitrogen (dark shading) sites are emphasized in (c), (e) and (g) (McIntosh *et al.*, 2004).

based on diamondoid molecules. Originally Drexler, in 1992, postulated a new concept in nanotechnology named "mechanosynthesis" that drives to the creation of nanoscale electromechanical systems (NEMS) (Drexler, 1992). He also suggested that diamondoid molecules could be useful for construction of these nanosystems. However, chemically unmodified diamondoid molecules interact rather weakly, leading to the formation of molecular crystals (Dahl *et al.*, 2003).

To illustrate this weakness, Figs. 5.9(a) and (b) depict the arrangement and binding of two interacting diamantine cages.

As can be seen, the binding energy of two diamantanes is only 0.13 eV. On the other hand, an alternative way to improve bonding involves chemical functionalization of diamondoids in order to create more reactive sites. Of course the elements for substitution must be close to carbon in the periodic table, such as boron (B) and nitrogen (N), which have similar atomic sizes as carbon. Moreover, carbon is already successfully substituted by boron and nitrogen in carbon nanotubes and fullerenes (Golberg *et al.*, 1999). Figure 5.9(c) shows that a similar substitution of terminal carbon atoms by boron and nitrogen in adjacent diamondoids, would result in a much stronger polar bond between the modified molecules, as confirmed by the increase in biding energy shown in Fig. 5.9(d). The binding energy of 1.64 eV between this functionalized diamondoid would prevent spontaneous dissociation at room temperature.

In Figs. 9(e)–9(h) results of substitution of the CH group in diamantane ($C_{14}H_{20}$) by a boron and a nitrogen atom are shown, leading to stable structures with geometries closely related to that of unmodified diamantane.

The average calculated C–C bond lengths for both diamondoid molecules and infinite diamondoid chain is 1.530 Å (Drummond *et al.*, 2004). The equilibrium C–C bond length calculated for optimized bulk diamond structure is of ~1.540 Å, which demonstrates good agreement with the experimental value of 1.544 Å (McSkimin and Andreatch, 1972; Kittel, 2005).

The C–B bond length, in $C_{13}BH_{19}$ structure (Fig. 5.10(e)) is somewhat larger than the C–C bond length, suggesting a weaker bond. On the other hand, in $C_{13}NH_{19}$ (Fig. 5.10(g)), the C–N bond is shorter than the C–C bond, suggesting a stronger bond. McIntosh *et al.* (2004) suggested that the substitution of C atom by ether B or N atoms leads, in both cases, to

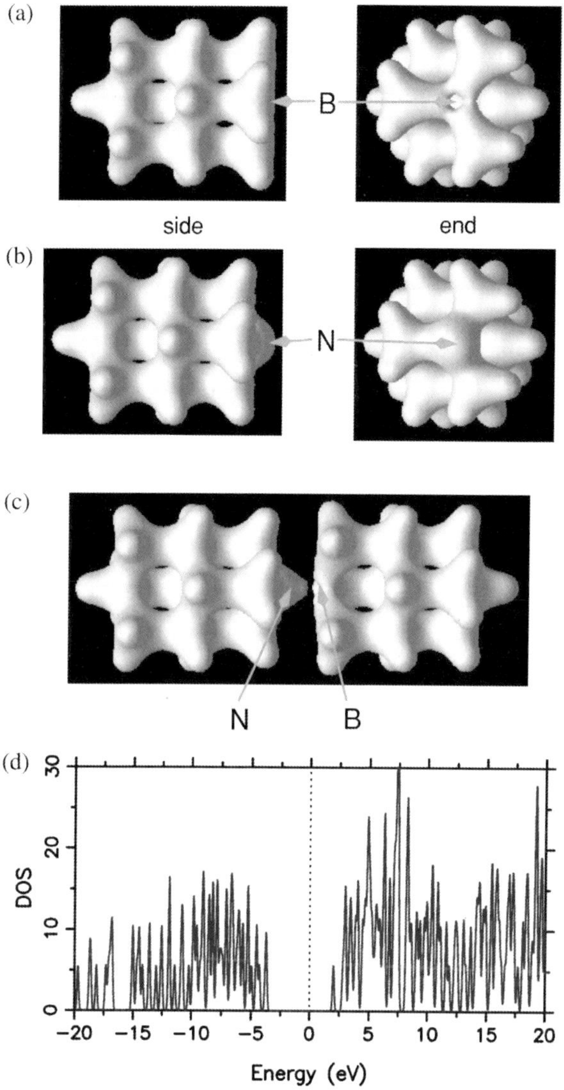

Fig. 5.10. Total charge distribution in (a) $C_{13}BH_{19}$ and (b) $C_{13}NH_{19}$ in their equilibrium structure, shown in Figs. 7(e) and 7(g). (c) Total charge distribution and (d) electronic density of states of $C_{13}BH_{19}$ interacting with $C_{13}NH_{19}$, in the structural arrangement displayed in Fig. 7(c) (McIntosh *et al.*, 2004).

impurity states within the HOMO-LUMO optical gap of their density of states (DOS), as can be seen in Figs. 9(f) and 9(h).

Figure 5.10 depicts the effect of atomic substitution, carbon by either boron or nitrogen, on the overall charge distribution within the functionalized diamondoid molecules. The excess charge at the nitrogen site in $C_{13}NH_{19}$, sharply visible by charge density protrusion in Fig. 5.10(b), when combined with a charge deficit at the boron site in $C_{13}BH_{19}$, visible in Fig. 5.10(a), leads to the formation of a polar bond, since the two sites face each other. According to Fig. 5.10(c), the charge cloud at the nitrogen site extended toward the positively charged boron site in the stable complex consisting of the two functionalized diamondoid molecules. A close look on the charge density distribution in the bond region also reveals the fundamental difference between the polar B–N bond, predominantly weaker, with very little charge in the bond region, and covalent C–C bonds with a strong charge accumulation in the bond region. Figure 5.10(d) depicts the density of states of the bonded complex with a large bandgap of few eV, similar to the individual functionalized diamondoids. Note that the total density of states of the bonded complex is closely related to that of its components, due to the polar nature of the bond, as seen in Figs. 8(f) and 8(h).

Garcia *et al.* (2009) using *ab initio* calculations found that boron or nitrogen when incorporated into adamantane molecules form thermodynamically very stable functional groups as tetra-bora-adamantane. A hypothetical molecular crystal, proposed by these researchers, formed by molecules with structure like tetra-bora-adamantane plus tetra-aza-adamantane, in a zincblende structure, showed large cohesive energy of 1.81 eV/primitive cell as well as bulk modulus of 20 GPa. Note that these values are considerably larger than those in typical molecular crystals and, hence, may provide stability and stiffness at room temperature. Additionally, this hypothetical crystal presented a band gap of 3.9 eV, which suggests potential applications in the optoelectronic field.

5.3.6. *Quantum conductance*

Unmodified diamondoid molecules demonstrate very poor electrical transport features, due to the wide fundamental gap. This limits their utility in applications as optoeletronic devices. Chemical functionalization approximation seems to be a good strategy for building NEMS based on diamondoids, as mentioned above. In a recent investigation, Xue and Mansoori (2008) report the *ab initio* studies of the quantum conductance of lower diamondoid molecules and some of their important derivatives. Specifically seven molecules, two of which are lower diamondoids and the rest are important diamondoid derivatives are chosen. They are classified into three groups as shown in Table 5.2.

Group 1: Adamantane (*ADM*), Diamantane (*DIM*) the lowest two diamondoids. Group 2: Memantine, Rimantadine, and Amantadine, the three derivatives of adamantane which have found tremendous amount of medical applications as antiviral agents. Group 3: ADM•Na, DIM•Na, the two artificial molecules, substituting one hydrogen ion in adamantane and diamantane with a sodium ion. The latter group has potential applications in NEMS. Table 5.2 shows the lower diamondoids and five derivates classified into three groups.

One important goal of molecular electronics is to find functional molecular structures that can be used as the logic units with which one can build nanoscale integral circuits. There have been many studies in the use of single molecules as "functional electronic devices" and finding the transportation properties of different molecules and molecular building blocks (Schön *et al.*, 2001; Reed *et al.*, 2001; Emberly and Kirczenow, 1998). In their quantum conductance calculations, Xue and Mansoori (2008) used the ATK and the VNL software packages. In most cases of calculations reported by Xue and Mansoori (2008), they used LDA-PZ, which stands for the local density approximation (LDA) with the Perdew–Zunger (PZ) parametrization (Perdew and Zunger, 1981) of the correlation energy of a homogeneous electron gas calculated by Ceperly and Alder (1980). In these *ab initio* electronic structure computations, they used the Double Zeta Polarization (DZP) basis set. It is important to choose a basis set large enough to give a good description of the molecular wave function of diamondoids. Double Zeta Polarization (DZP) basis set is one such basis set. A double-zeta basis set for hydrogen has two functions, and a true double-zeta basis set for carbon would have ten functions. Additional flexibility is built in by adding higher-angular momentum basis functions. Since the highest angular momentum orbital for carbon is a p orbital, the

Table 5.2. Molecular formulas and structures of Adamantane, Diamantane, Memantine, Rimantadine, Amantadine, Optimized ADM•Na and Optimized DM•Na molecules. Blacks, whites, and purples balls represent C, H, N, and Na atoms, respectively (Xue and Mansoori, 2008).

Group 1		Group 2			Group 3	
Adamantane	**Diamantane**	**Amantadine**	**Rimantadine**	**Memantine**	**Optimized ADM•Na**	**Optimized DIM•Na**
$C_{10}H_{16}$	$C_{14}H_{20}$	$C_{10}H_{17}N$	$C_{11}H_{20}N$	$C_{12}H_{21}N$	$C_{10}H_{15}Na$	$C_{14}H_{19}Na$
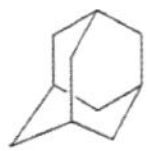	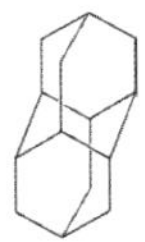	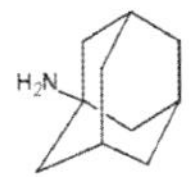	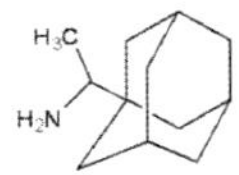	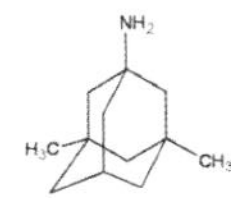	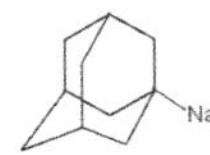	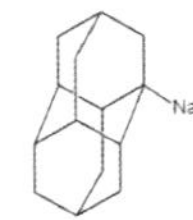
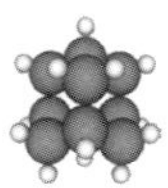	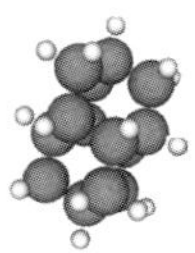	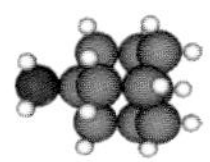	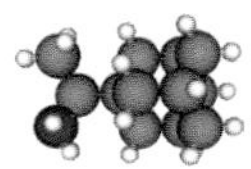	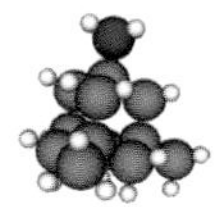	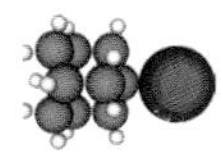	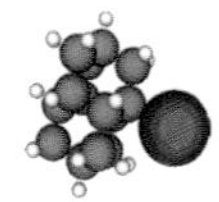

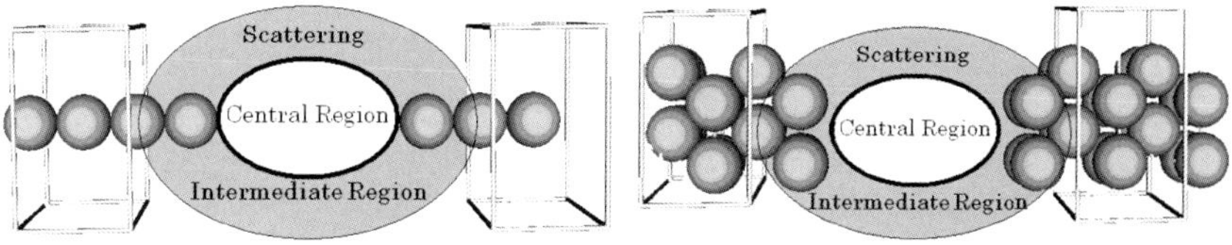

Fig. 5.11. Two semi-infinite linear (left plane) and 2 × 2 (right plane) gold (Au) chain electrodes used by Xue and Mansoori (2008) in their quantum conductance studies of seven lower diamondoids and derivatives.

"polarization" of the atom can be described by adding a set of *d* functions. In their *Ab initio* calculations, the optimal value of the mesh cut-off was found to be 100 Ry, and the optimum *k*-point grid mesh number in the Z direction was found to be 100.

In order to calculate the electronic properties of nanoscale systems, a two-probe simulation system was created. This system consists of two semi-infinite electrodes and a scattering intermediate region which includes several layers of molecules from each electrode. In the practical calculations, only the central region (region between the two electrodes) and the scattering regions are considered, and the bigger the scattering region, the more accurate the calculation will be but also more computation efforts are needed.

Xue and Mansoori (2008) constructed two types of electrodes: First type is semi-infinite linear gold (Au) chains and second one is comprised of Au (100) surface in a 2 × 2 unit cell, as shown in Fig. 5.11. The diamondoid molecules lie into the central region between the two electrodes.

Gold was chosen as electrodes since it is more practical and promising as monatomic nanowire. The constant bond length of Au atom was optimized by ATK. Then by performing separate calculations for the scattering region as well as the central region, where the diamondoid molecule is located, followed by an intelligent recombination of the two subsystems, the quantum transmission spectrums and conductance of diamondoids were calculated.

The HOMO, the LUMO and the band gap of all the seven molecules are reported in Table 5.3.

The smaller the band gap (the difference between the energies of the HOMO and LUMO) is indicative of the fact that the more easily a molecule can be excited. According to Table 5.3, the three members of group 1 are generally in agreement with previous calculations reported by Reed *et al.*

Table 5.3. HOMO and LUMO energies of the three groups. (Fermi energy = 0 eV). (Xue and Mansoori, 2008).

	Group 1		**Group 2**			**Group 3**	
	Adamantane	**Diamantane**	**Memantine**	**Amantadine**	**Rimantadine**	**ADM•Na**	**DIM•Na**
HOMO [eV]	−6.554	−6.226	−5.023	−4.956	−4.975	−3.078	−3.435
LUMO [eV]	1.288	1.097	1.228	1.043	0.974	−1.699	−1.674
Band Gap[eV]	7.842 (7.622*)	7.323 (7.240*)	6.251	5.999	5.949	1.379	1.761

(*) *Calculation by* McIntosh *et al.* (2004).

(2001). Note that group 1 members have the largest band gap, which proves that diamondoids are electrical insulators. Group 2 have smaller band gaps, which indicate that those three molecules could be electrical semiconductors. The results also indicate that the $-NH_2$ is electron donating. This is consistent with the fact that $-NH_2$ has a pair of nonbonded electrons. Group 3 shows not only smaller band gaps than groups 1 and 2 which indicate high conductance, but also interesting electronic behaviors, such as both the HOMO and LUMO are below Fermi energy which is an indicator of metallic properties of those organometallic molecules.

The quantum conductance, G, is calculated by the following equation (Emberly and Kirczenow, 1998),

$$G = G_{0.}T(E, V_b), \tag{5.5}$$

where $G_0 = 2e^2/h = 77.5\mu S$. In this equation, $T(E, V_b)$ is the transmission probability for electrons incident at an energy E through a device under a potential bias $V_{b.}$

The current through the scattering intermediate region is determined by the quantum-mechanical probability for electrons to tunnel through the diamondoid molecule from one electrode to the other. This current is calculated using the Landauer formula which expresses the conductance of a system at $T = 0$ in terms of the quantum mechanical transmission coefficients (Emberly and Kirczenow, 1998),

$$I = \int_{\mu_L}^{\mu_R} T(E, V_b)dE, \tag{5.6}$$

where μ_L and μ_R are the left- and right-side metallic reservoirs electrochemical potentials ($\mu_L > \mu_R$), $\mu_{L/R} = \pm eV_b$ and $T(E, V_b)$ is the transmission probability for electrons incident at an energy E through a device under a potential bias V_b. When $V_b > 0$, it means that positive charge transport from the right electrode. The Landauer equation based on the Green's function method relates the elastic conductance of a junction to the probability that an electron with energy E injected in one electrode will be transmitted to another electrode through a scattering region which in our case is the diamondoid molecule.

Using the above-mentioned procedure, Xue and Mansoori (2008) calculated the energy levels, transmission spectrums and conductance of the seven chosen molecules. They also calculated current-voltage (*I-V*) and conductivity-voltage (*G-V*) characteristics of some of those molecules.

In this study, the focus was on the exhibition of the results obtained and the prospective applications in the field of nanotechnology. As a simulation of real experiments, geometrical optimizations using DFT method is necessary. However, since diamondoids are very stable and stiff molecules, the geometrical structures of these molecules would not, and thus the conductance would not, be affected. For these reasons, one does not need to perform geometrical optimizations.

The quantum conductance and transmission spectra of diamondoid molecules and their derivatives have been calculated by using DFT method, with geometrical optimization of two-probe systems (Xue and Mansoori, 2008). For adamantane and diamantane molecules, the distances between linear Au electrodes (central region width in Fig. 5.11) present optimal values of 0.906 nm and 1.10 nm, respectively. Nevertheless, geometrical optimizations of the derivatives of diamondoids with electrodes were not performed, because structure optimization process of the whole systems, including Au electrodes, would promote undesirable change on molecular orientations, due to their asymmetrical geometry.

Table 5.4 shows the data of quantum conductance of three diamondoid groups at zero bias ($V_b = 0$).

Transmission spectra depicted in Figs. 11(a) and 12(a) for adamantane and diamantane confirm their electrical insulator features.

On the other hand, group 2 diamondoids, memantine, rimantadine and amantadine, show conductances rather higher than their parents (adamantane and diamantane) due to the presence of NH_2 group in their structures, except for rimantadine (orientation 1). Figures 11(b) and 12(b) display the transmission spectra of group 2 with well-defined resonant peaks.

Group 3: ADM•Na, DIM•Na, the two organometallic diamondoids, substituting one hydrogen ion in adamantane and diamantane with a sodium ion, generally exhibit high conductance at both Au linear and Au 2 × 2 electrode systems, as shown in Table 5.4. Fig. 5.12(c) and 13(c) show transmission data for these organometallic molecules.

The HOMO-LUMO band gaps values of 1.379 eV for ADM•Na and 1.761 eV for DIM•Na, as reported in Table 5.3, suggest the utility of these molecules to build molecular electronic devices.

Table 5.4. Quantum conductance (G), in μS, and Central Region Width, in Å, of the three damandoids groups at zero bias and the corresponding orientations (Xue and Mansoori, 2008).

		Quantum Conductance (*Central Region Width*)	
		Au linear Electrodes	**Au 2 × 2 Electrodes**
Molecule	**Orientation**	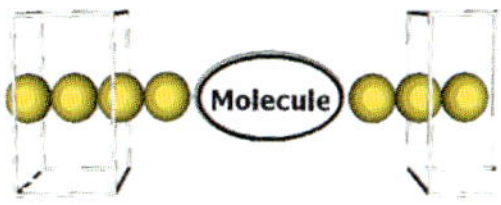	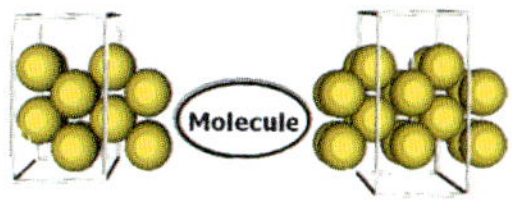
Adamantane		**0.042** (*9.06*)	**2.393** (*8.15*)
Diamantane		**0.137** (*11.00*)	**10.785** (*9.76*)
Memantine (Orientation 1)	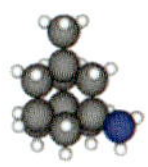	**0.850** (9.08)	**28.668** (*8.46*)
Memantine (Orientation 2)		**1.161** (8.69)	**31.308** (8.65)
Memantine (Orientation 3)		**0.842** (8.53)	**3.629** (9.64)
Amantadine (Orientation 1)		**0.767** (8.84)	**14.206** (8.68)
Amantadine (Orientation 2)		**8.072** (7.83)	**15.039** (8.10)
Rimantadine (Orientation 1)		**0.020** (9.90)	**2.837** (9.85)
Rimantadine (Orientation 2)		**11.818** (7.43)	**40.817** (8.32)
Na		**26.107** (9.06)	**77.494** (8.17)

(*Continued*)

Table 5.4. (Continued)

		Quantum Conductance (*Central Region Width*)	
		Au linear Electrodes 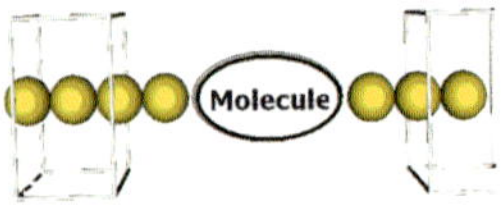	**Au 2 × 2 Electrodes** 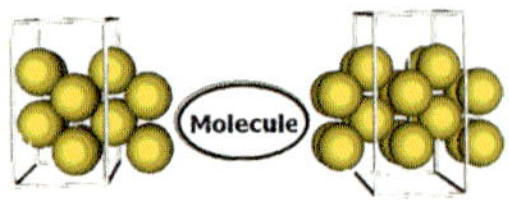
Molecule	**Orientation**		
ADM·Na (Orientation 1)		**18.078** (9.06)	**27.790** (8.36)
ADM·Na (Orientation 2)	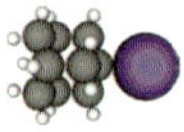	**44.54** (9.90)	**53.60** (9.98)
DIM·Na (Orientation 1)		**1.37** (7.67)	**103.41** (8.48)
DIM·Na (Orientation 2)	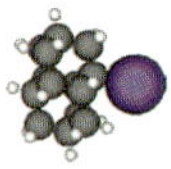	**44.32** (11.06)	**54.24** (10.95)
DIM·Na (Orientation 3)	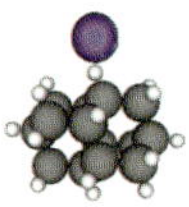	**0.578** (10.63)	**23.654** (9.47)

The high conductance values of 44.54 μS for ADM•Na (orientation 2) and 44.32 μS for DIM•Na (orientation 2), at Au linear electrodes, which are much higher than the metallic single sodium atom conductance, as reported in Table 5.4, also confirm the utility of these molecules for building molecular electronic devices. Also this may open a route for the development of high temperature superconducting materials which would have many useful applications in energy industry (Panek *et al.*, 1988).

Similar effects are observed in the Au 2 × 2 electrode cases as reported in Table 5.4. In fact, Ramezani and Mansoori (2007) and Mansoori *et al.* (2009) have suggested diamondoid molecules and derivatives as molecular building blocks for MEMS and NMES applications. A close inspection in Table 5.4 data reveals that predicted conductance for all diamondoids

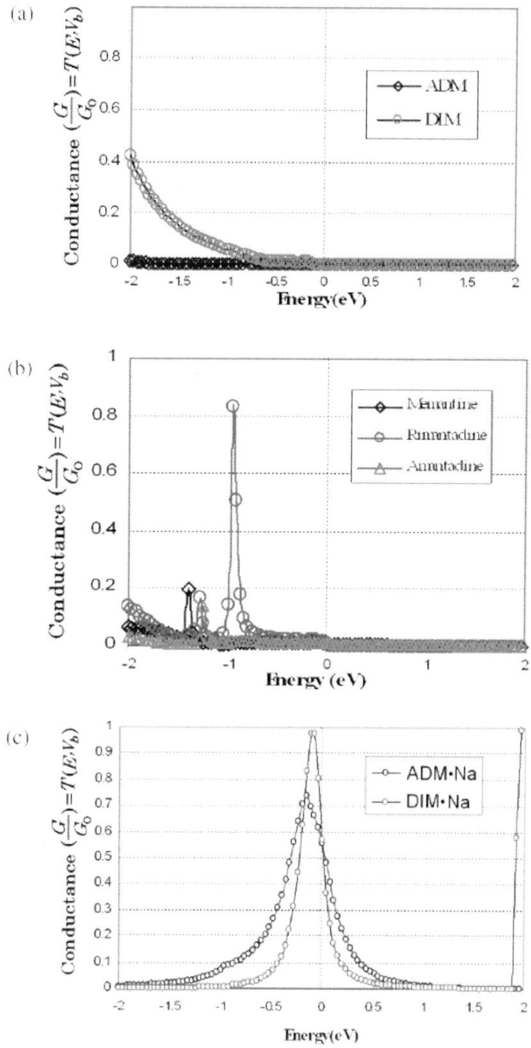

Fig. 5.12. Transmission spectra in Au Linear electrodes cases. (a) Transmission spectra of Group 1. (b) Transmission spectra of Group 2 corresponding to "Orientation 1" in Table 4. (c) Transmission spectra of Group 3 corresponding to "Orientation 1" in Table 4 (Xue and Mansoori, 2008).

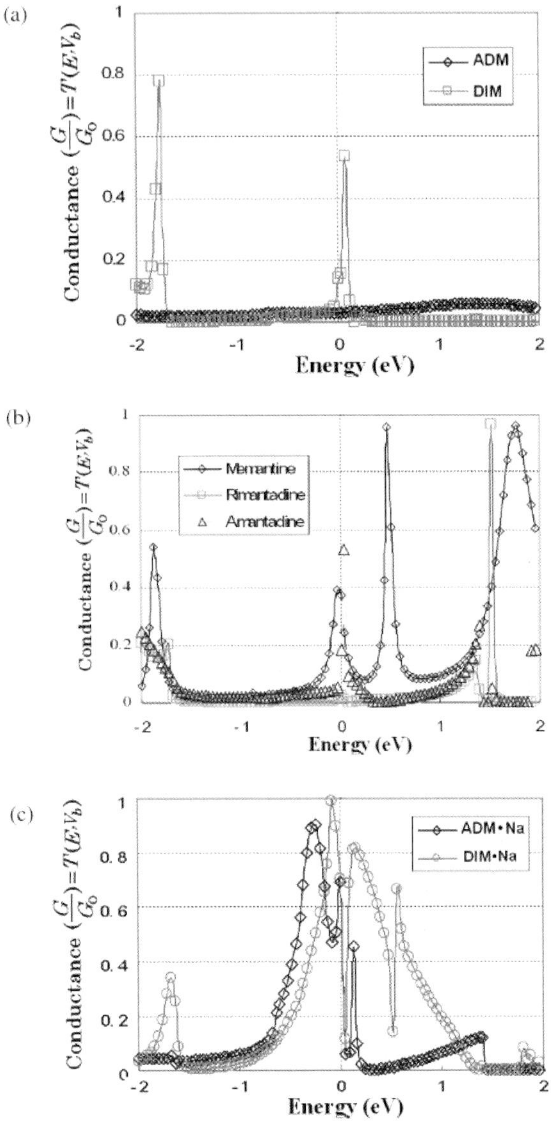

Fig. 5.13. Transmission spectra in Au 2×2 unit cell cases. (a) Molecules from Group 1. (b) Molecules from Group 2 corresponding to "Orientation 1" in Table 4. (c) Molecules from Group 3 corresponding to "Orientation 1" in Table 4 (Xue and Mansoori, 2008).

groups are strongly affected by molecular orientation geometry into central region between the Au electrodes.

The molecules confined in Au 2 × 2 electrodes have higher conductance than those confined in Au linear chains. Even with the same central region widths as the linear Au electrodes cases (0.906 nm and 1.100 nm) the conductance of adamantane and diamantane in Au 2 × 2 electrodes case (1.9375 μS and 0.418465 μS) are still greater than those in linear case (0.042 μS and 0.137 μS). Since diamondoids are 3-dimensional structures, when electrodes also have 3-dimensional structures, the entire systems will have more conductive channels and thus have higher conductance than when electrodes are linear.

The transport of electrons through a molecular system can be investigated as a one-electron elastic scattering problem. Here, a molecule acts as a defect between two metallic reservoirs of electrons. Thus, the current through the scattering intermediate region (Fig. 5.11) can be determined, using Eq. (5.6), by the quantum mechanical transmission probability $T(E)$ for electrons to tunnel from one electrode to the other through the diamondoid molecule. Note that the differential conductance is then given by derivate of the current with respect to voltage.

Fig. 5.13 depicts *I-V* and *G-V* characteristics for ADM•Na and DIM•Na molecules. It is interesting to note the presence of plateaus between 0.5 *V* and 0.9 *V* in the *I-V* graph for both molecular systems.

Currents reach the maxima at 11.90 μA and 11.35 μA. For $V_b > 0$, bias V_b values increase while transmission $T(E,V_b)$ values decrease, leading to maximum current values with *I-V* plateau features, according to Eq. (5.6). This *I-V* characteristic presented by ADM•Na and DIM•Na can be used to build molecular rectifiers or surge protectors.

Recently, structural and electronic properties of solid adamantane in a FCC lattice form were investigated using DFT (Hamadanian *et al.*, 2010). One, two, three, and four tertiary hydrogen atoms were substituted with Na to form $C_{10}H_{16-n}Na_n$. According to this investigation, although, pristine adamantane shows a wide band gap and can be categorized as an insulator, it is shown that under Na doping process, the band gap reduces and finally it becomes semi metallic or metallic.

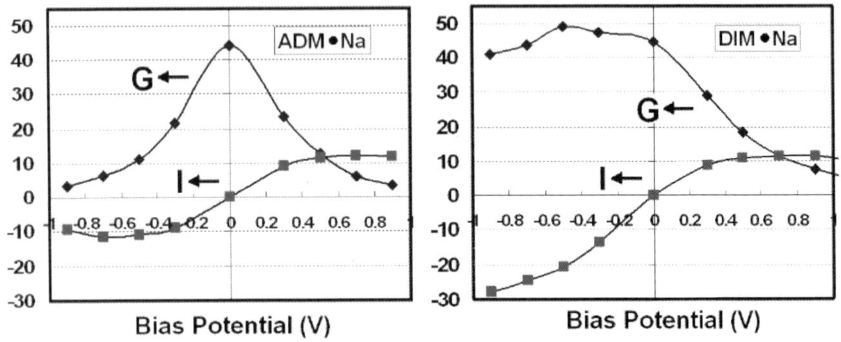

Fig. 5.14. G [μS] versus V(V) and I [μA] versus V [Volt] characteristics of ADM•Na and DIM•Na in linear electrodes corresponding to their horizontal orientations (Orientations 2 in Table 3) (Xue and Mansoori, 2008).

5.4. Intermolecular Interactions

Knowledge about intermolecular interactions is very important in macroscopic statistical mechanical calculation of the behavior of matter and phase transitions (Mansoori, 1980; Benmekki and Mansoori, 1988; Hamad and Mansoori, 1990; Pourgheysar *et al.*, 1996; Firouzi *et al.*, 1998; Modarress *et al.*, 1999). Such knowledge is very critical to designing and fabricating nanostructure-based devices such as thin film transistors, light-emitting nanodevices, semiconductor nanorods, nanocomposites, etc. However, the nature and role of intermolecular interactions in such nanostructures as diamondoids is very challenging and not well understood (Rafii-Tabar and Mansoori, 2004). For a good number of atoms and molecules, quantum mechanical *ab initio* calculation methods (Rafii-Tabar and Mansoori, 2004; Mansoori, 2005) have been successful in producing accurate intermolecular potential functions (Mansoori *et al.*, 2009).

Zhang *et al.* (2007) have recently performed a first-principles simulation of the interaction between adamantane and an atomic force microscope (AFM) tip made up of gold.

The probed atoms are the outside hydrogen atoms. They first hold the AFM tip above adamantane and move along the x-direction. At each of the point, they do three radial scans and then use the finite difference

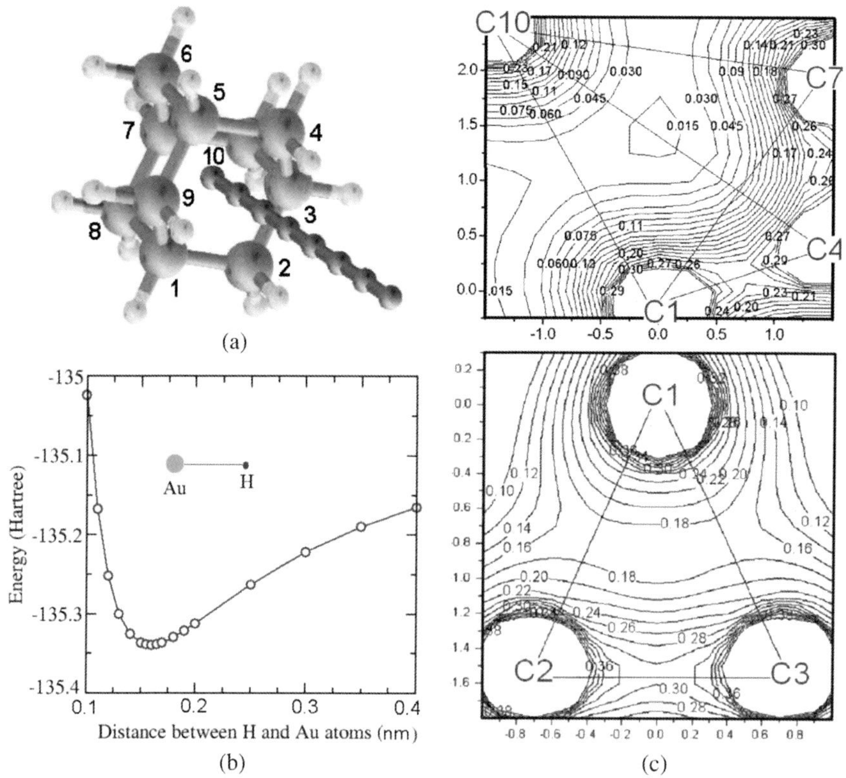

Fig. 5.15. (a) Demonstration of various locations of a gold molecule with respect to an adamantane molecule. (b) Pair interatomic potential energy function between gold and a hydrogen atom. (c). Charts of the distributions of the electron density in adamantane across the C1–C7–C10 and across the C1–C2–C3 planes and along the C1–C4 line (Figures courtesy of Prof. G. Zhang).

method to compute the forces at each point. The number of calculations is huge.

By holding the tip at different distances from adamantane, three scans across two surfaces, one with a carbon atom at the center and the four other equivalent atoms at the corners, and the other with three equivalent carbon atoms in the front and three other atoms in the back forming a hexagon shape, have revealed the detailed morphology of adamantane (see Fig. 5.15).

For the first scan surface, a huge potential energy change is observed when the tip is close to adamantane, which results from the strong

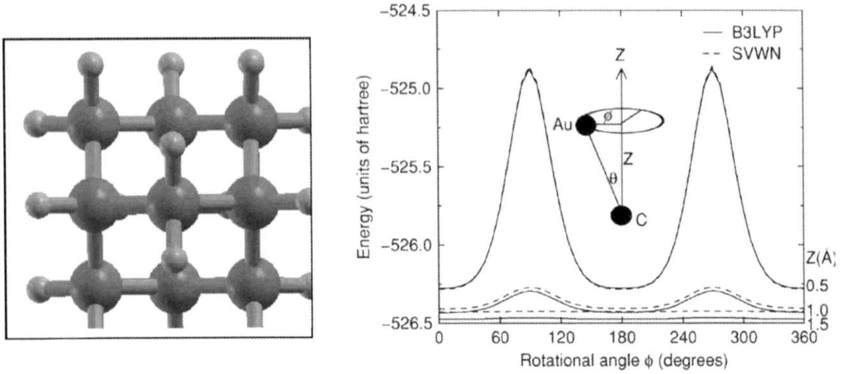

Fig. 5.16. Top: Adamantane oriented along the (100) direction. The z axis points in the (100) direction and through the center carbon atom. Bottom: Energy change as a function of the rotational angle ϕ at $z = 0.5$, 1.0, and 1.5 Å. The solid line is from the B3LYP calculation. The dashed lines, vertically shifted, refer to the SVWN (Slater exchange and Vosko, Wilk, and Nussair correlation functional (Vosko *et al.*, 1980) results. Inset: Scanning geometry of the gold atom. The origin is at the center carbon atom. θ is the polar angle and ϕ is the azimuthal angle (Zhang *et al.*, 2007).

interaction from two hydrogen atoms attached to the center carbon atom.

The *ab initio* result by Zhang *et al.* (2007) demonstrates quantitatively how an AFM tip interacts with adamantane. It is found that the AFM tip is able to detect the sharp potential change in adamantane. For instance, AFM tip along different directions, makes it possible to probe directly the detailed morphology of the molecular structure. To be more specific Zhang *et al.* (2007) directly have computed the xy-scan image by scanning across two different surfaces, where one is hexagonal and the other is square surface.

If we rotate adamantane along the (100) direction so that one face-centered atom is at the center and the other four are at the corners, then we end up with a square-like structure (see the top structure in Fig. 5.16).

The scan in this orientation will be called the (100) scan. On the second scan surface, a radial scan shows the maximum force constant of 2 Hartrees/Å^2 as can be seen from the inset in Fig. 5.17.

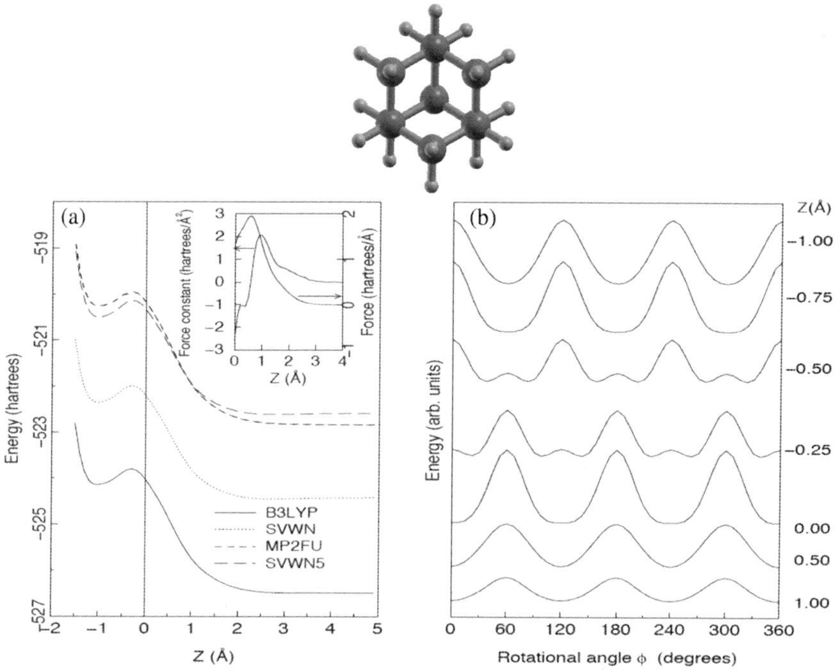

Fig. 5.17. Top: Adamantane oriented along the (111) direction. The z axis is also along this direction. Bottom: (a) Total energy change as a function of the distance z between the tip and adamantane. The gold atom scans radially away from adamantane. Four different methods reveal a similar potential energy change. Inset: Force constant and force versus z. (b) Potential energy change obtained at the B3LYP level as a function of ϕ at several distances z, similar to Fig. 15. A clear peak change is noticed as the tip moves away from adamantane (Zhang *et al.*, 2007).

If the tip scans across the [111] direction (Fig. 5.17), the potential change directly follows the sharp structural difference in two kinds of carbon atoms. One can see a large force change in the vicinity of those atoms. In addition, if the scan is close to the tetragonal direction, a four-fold symmetry emerges. For the scan along the [111] direction, we estimate the force constant to be about 1.5 times that of C_{60}, which is an indication of the excellent mechanical properties of adamantane. This is proof of the hardness of adamantane.

The rotational scan along the second surface reveals a systematic change in the potential energy as the tip is moved away from adamantane.

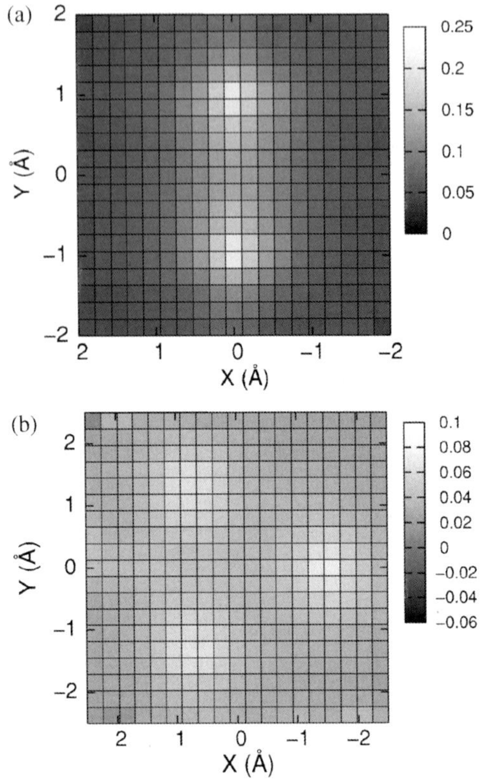

Fig. 5.18. x-y scan image for (a) (100) and (b) (111) surface. The bright spots represent where the hydrogen atoms are located, and the gray scale on the right shows the magnitude of the force. For (a), the tip is 1.2 Å away from the center carbon atom; for (b), the tip is 1.8 Å away from the center of the triangle formed by the three outermost carbon atoms (Zhang *et al.*, 2007).

Due to the existence of two types of carbon atoms in adamantane, the original potential maxima are shifted 60° to new maxima. Between these maxima, there is a flat region.

Figure 5.18 shows the force along the z-direction, where Zhang *et al.* (2007) use the intensity to show the magnitude of the force.

One sees that there are two bright spots, which according to the geometry of adamantane, to correspond to the two outermost hydrogen atoms. Figure 5.18(b) shows a scan over the hexagonal surface, and unlike the square scan, there are three bright spots in this picture, corresponding to the three hydrogen atoms. These distinctive features are expected to show up in real experiments.

5.5. Concluding Remarks

Diamondoid molecules demonstrate great potential for derivatization with interesting properties and as molecular building blocks in various nanotechnology applications. *Ab initio* calculations are quite useful for molecular property prediction of diamondoids and their derivatives as reported in this chapter.

Quantum confinement effects (QCE) (see Sec. 5.3.1.1) and NEA (see Sec. 5.3.1.2) are some examples of the unique characteristics of diamondoids, which have attracted the attention of the research and industrial communities. Several theoretical methods have been developed in order to investigate phenomena and physical properties involving diamondoid molecules. Predictions of vibrational spectra and frequencies (Lu *et al.*, 2005; Richardson *et al.*, 2005; Richardson *et al.*, 2006; Filik *et al.*, 2006; Ramachandran *et al.*, 2007; Jensen, 2004; Oomens *et al.*, 2006) are also useful in understating of vibronic properties of diamondoids. The first-principles simulation of interaction of diamondoids with other systems as carbon nanotubes (McIntosh *et al.*, 2004), metallic surface (Wang *et al.*, 2008), and metallic AFM catilever tips (Zhang *et al.*, 2007; Tyler *et al.*, 2003) is a promising way to design new materials. Additionally, functionalized diamondoids (McIntosh *et al.*, 2004; Xue and Mansoori, 2008; Barnard *et al.*, 2003; Galasso, 2000; Pichierri, 2006; Garcia *et al.*, 2010; Xue and Mansoori, 2010) may lead to the creation of nanosystems with new chemical and physical properties and with suitable applications in pharmacology and electronic devices.

Diamondoids are generally optically transparent in visible light and have high electrical insulating properties as diamond does. The large gaps between the highest occupied molecular orbital and lowest unoccupied molecular orbital (HOMO-LUMO gaps) in diamondoids are the molecular counterparts of a large fundamental gap in diamond, which is responsible for its optical transparency in the visible range and its insulating properties.

The abundant derivatives of diamondoids are expected to have many interesting opto-electronic properties. Despite the numerous extensive investigations into properties of diamondoids and derivatives, their opto-electronic properties are still to be further studied.

Electronic properties of diamondoid-based nanomaterials and MEMS are determined by the behavior of the electrons that bind their carbon and hydrogen atoms together. To make accurate quantitative predictions about such behavior, we need to perform electron distribution calculations in diamondoid molecules and their derivatives. For this purpose, we need to perform highly accurate calculations of the electronic structures and energies of diamondoids-based nanomaterials and MEMS.

Bibliography

Araujo, E. S. D., Mansoori, G. A., Xue. Y. and Lopes Barros de Araujo, P. (2011). *Physics Express*, **1**(2): 67–88.

Areshkin, D. A., Shenderova, O. A., Adiga, S. P. and Brenner, D. W. (2004). *Diam. Relat. Mater.*, **13**: 1826.

Barnard, A. S., Russo, S. P. and Snook, I. K., (2003). *J. Chem. Phys.*, **118**: 10725–10728.

Becke, A. D. (1988). *Phys. Rev. A.*, **38**: 3098.

Becker, U. and Shirley, D. A. (1996). VUV and Soft X-Ray Photoionization, Physics of Atoms and Molecules. Plenum, New York.

Benmekki, E. H. and Mansoori, G. A. (1988). *Fluid Phase Equil.*, **41**: 43–57.

Bostedt, C., van Buuren, T., Willey, T. M., Franco, N., Terminello, L. J. (2004a) *Appl. Phys. Lett.*, **85**: 5334–5336.

Bostedt, C., van Buuren, T., Willey, T. M., Franco, N., Terminello, L. J., Heske, C. and Möller, T. (2004b). *Appl. Phys. Rev. Lett.*, **84**: 4056–4058.

Borisenko, V. E. and Ossicini, S. (2004). What is what in the nanoworld: A handbook on nanoscience and nanotechnology, Wiley-VCH, Weinheim.

Ceperley, D. M and Alder, B. J. (1980). *Phys. Rev. Lett.*, **45**: 566–569.

Dahl, J. E., Liu, S. G., Carlson, R. M. K. (2003). *Science*, **299**: 96–99.

Drummond, N. D., Williamson, A. J., Needs, R. J. and Galli, G. (2005). *Phys. Rev. Lett.*, **95**: 096801.

Drexler, K. E. (1992). *Nanosystems: Molecular Machinery, Manufacturing and Computation.* John Wiley & Sons, Inc. New York, NY.

Emberly, E. G. and Kirczenow, G. (1998). *Phys. Rev. B.*, **58**: 10911.

Endo, K., Koizumi, S., Otsuka, T., Ida, T., Morohashi, T., Onoe, J., Nakao, A., Kurmaev, Z., Moewes, A. and Chong, D. P. (2003). *J. Phys. Chem. A.*, **107**: 9403–9408.

deHeer, W. A. (1993). *Rev. Mod. Phys.,* **65**: 611–676.

De Azevedo, W. M, Lima, A. P. D., De Araujo, E. S. (1999). *Rad. Protec. Dosim.*, **84**: 77–81.

Fermi, E. (1927). *Red. Accad. Naz. Lincei*, **6**: 602–607.

Filik, J., Harvey, J. N., Allan, N. L., May, P. W., Dahl, J. E. P., Liu, S. and Carlson, R. M. K., (2006). *Spectrochim. Acta. A.*, **64**: 681–692.

Firouzi, F., Modarress, H. and Mansoori, G. A. (1998). *Europ. Polymer J.*, **34**: 1489–1498.

Fock, V. (1930). *Z. Phys.*, **61**: 126–148.

Fort Jr., R. C. (1976). *Adamantane: The Chemistry of Diamondoid Molecules*, Drekker, New York.

Foulkes, W. M. C. *et al.*, (2001). *Rev. Mod. Phys.*, **73**: 33–83.

Frisch, M. (2003). Computer Code GAUSSIAN 03, Revision A.1., Gaussian, Inc. Pittsburgh, PA.

Galasso, V. (2000). *THEOCHEM*, **528**: 171–176.

Garcia, J. C., Justo, J. F., Machado, W. V. M. and Assali, L. V. C. (2009). *Phys.l Rev. B.*, **80**: 125421.

Golberg, D., Bando, Y., Han, W., Kurashima, K. and Sato, T. (1999). *Chem. Phys. Lett.*, **308**: 337–342.

Graupner, R., Ristein, J., Ley, L. and Jung, C. (1999). *Phys. Rev. B.*, **60**: 17023.

Hamad, E. Z. and Mansoori, G. A. (1990). *J. Phys. Chem.*, **94**: 3148–3152.

Hamadanian, M., Khoshnevisan, B. and Fotooh, F. K. (2010). *J. Molecular Structure: THEOCHEM.*, **961**: 48–54.

Hertel, I. V., Steger, H., de Vries, J., Weisser, B., Menzel, C. and Kamke, W. (1992). *Phys. Rev. Lett.,* **68**: 784.

Himpsel, F. J., Knapp, J. A., VanVechten, J. A. and Eastman, D. E. (1979). *Phys. Rev. B.*, **20**: 624.

Hitchcock, A. P., Newbury, D. C., Ishii, I., Horsley, J. A, Redwing, R. D., Johnson, A. L. and Sette, F. (1986). *J. Chem. Phys.*, **85**: 4849–4862.

Hoffman, A., Comtet, G., Hellner, L., Dujardin, G. and Petravic, M. (1998). *Appl Phys. Lett.,* **73**: 1152–1154.

Horsfield, A. P. and Bratkovsky, A. M. (2000). *J. Phys.: Condens. Matter,* **12**: R1–R24.

Janssens, E., Neukerman, S., Vanhoutte, F., Silverans, R. E., Lievens, P., Navarro-Vazquez, A. and Schleyer, P. v. R. (2003). *J. Chem. Phys.*, **118**: 5862–5871.

Jensen, J. O. (2004). *Spectrochim. Acta A.*, **60**: 1895–1905.

Kittel, C. (2005). *Introduction to Solid State Physics*, Wiley, New Caledonia, USA.

Lee, C. T., Yang, W. T., Parr, R. G. (1988). *Phys. Rev. B.*, **37**: 785.

Lenzke, K., Landt, L., Hoener, M., Thomas, H., Dahl, J. E., Liu, S. G., Carlson, M. K., Möller, T. and Bosedt, C. (2007). *J. Chem. Phys.*, **127**: 084320–084325.

Lie, E. H. (1981). *Rev. Mod. Phys.*, **53**: 603–641.

Lu, A. J., Pan, B. C. and Han, J. G. (2005). *Phys. Rev. B.*, **72**: 035447.

Mansoori, G. A. (1980). *Fluid-Phase Equil.*, **4**: 61–69.

Mansoori, G. A. (2005). *Principles of Nanotechnology: Molecular Based Study of Condensed Matter in Small Systems.* World Scientific, Pub Co., Hackensak, NJ.

Mansoori, G. A., George, T. F., Zhang, G. and Assoufid, L. (2009). Structure and Opto-Electronic Behavior of Diamondoids, with Applications as MEMS and at the Nanoscale Level. *Encyclopedia of Nanotechnology*, E. D. Carlson (Ed.), Nova Sci Publ's, Hauppauge, NY, **1**: 527–545.

Marsusi, F., Mirabbaszadeh, K. and Mansoori, G. A. (2009a). *Physica E: Low-dimensional Systems and Nanostructures,* **41**: 1151–1156.

Marsusi, F. and Mirabbaszadeh, K. (2009b). *J. Phys.: Condens. Matter*, **21**: 5303–5310.

Modarress, H., Ahmadnia, E. and Mansoori, G. A. (1999). *J. Chem. Phys.,* **111**: 10236–10241.

Morar, J. F., Himpsel, F. J., Hollinger, G., Hughes, G. and Jordan, J. L. (1985). *Phys. Rev. Lett.*, **54**: 1960.

McIntosh, G. C., Yoon, M., Berber, S., Tománek, D. (2004). *Phys. Rev. B.*, **70**: 045401.

McSkimin, M. J. and Andreatch, P. (1972). *J. Appl. Phys.*, **43**: 2944–2948.

Needs, R. J. *et al.*, (2002). CASINO version 1.6 User Manual (University of Cambridge), Cambridge.

Oomens, J., Polfer, N., Pirali, O., Ueno, Y., Maboudian, R., May, P. W., Filik, J., Dhal, J. E., Liu, S. and Carlson, R. M. K. (2006). *J. Molec. Spectroc.*, **238**: 158–167.

Ordejón, P., Artacho, E., Soler, J. M. (1997). *Int. J. Quantum Chem.,* **65**: 453–461.

Ordejón, P., Artacho, E., Soler, J. M. (2000). *Phys. Rev. B.*, **53**: R10441.

Panek, D., Prando, G. and Mansoori, G. A. (1988). *Int'l J. Energy Sources*, **10**: 159–172.

Perdew, J. P. and Zunger, A. (1981). *Phys. Rev. B.*, **23**: 5048.

Pichierri, F. (2006). *Chem. Phys. Lett.*, **421**: 319–323.

Pourgheysar, P., Mansoori, G. A. and Modarress, H. (1996). *J. Chem. Phys.*, **105**: 9580–9587.

Prins, J. F. (2003). *Semicond. Sci. Technol.*, **18**: S125–S130.

Rafii-Tabar, H. and Mansoori, G. A. (2004). Interatomic Potential Models for Nanostructures. *Encyclopedia of Nanoscience & Nanotechnology*. E. D. Carlson (Ed.), Nova Sci Publ's, Hauppauge, NY, **4**: 231–248.

Ramachandran, G., Manogaran, S. (2007). *Theochem*, **816**: 31–41.

Ramezani, H. and Mansoori, G.A. (2007). *Diamondoids as Molecular Building Blocks for Nanotechnology*. In Molecular Building Blocks for Nanotechnology: From Diamondoids to Nanoscale Materials and Applications, (Springer, New York), Topics in Applied Physics, **109**, 44–71b.

Reed, M. A., Chen, J., Rawlett, A. M., Price, D. W. and Tour, J. M. (2001). *Appl. Phys, Lett.*, **78**: 3735–3737.

Richardson, S. L., Baruah, T., Mehl, M. J., Pederson, M. R. (2005). *Chem. Phys. Lett.*, **403**: 83–88.

Richardson, S. L., Baruah, T., Mehl, M. J., Pederson, M. R. (2006). *Diamond and Related Materials*, **15**: 707–710.

Roth, S., Leuenberger, D., Osterwalder, J., *et al.* (2010). *Chemical Physics Letters*, **495**: 102–108.

Rutter, M. J. and Robertson, J. (1998). *Phys. Rev. B.*, **57**: 9241.

Sánchez, G., Díaz-Tendero, S., Alcamí, M. and Martín, F. (2005). *Chem. Phys. Lett.*, **416**: 14–17.

Schön J. H., Meng, H. and Bao, Z. (2001). Nature (London), **413**: 713–716.

Slater, J. C. (1929). *Phys. Rev.*, **34**: 1293.

Stöhr, J *NEXAFS Spectroscopy*, (1996). Springer Series in Surface Sciences, Springer-Verlag, Berlin.

van Buuren, T., Dinh, L. N., Chase, L. L., Sickhaus, W. J. and Terminello, L. J. (1998). *Phys. Rev. Lett.*, **80**: 3803–3806.

Thomas, L. H. (1927). *Phil. Soc.*, **23**: 542–548.

Tyler, T., Zhirnov, V. V., Kvit, A. V., Kang, D. and Hren, J. J. (2003). *App. Phys. Lett.*, **82**: 2904–2906.

Vosko, S. H. Wilk, L. and Nusair, M. (1980). *Can. J. Phys.*, **58**: 1200–1211.

Xue, Y. and Manssori, G. A. (2008). *International J. Nanoscience*, **7**: 63–72.

Xue, Y. and Mansoori, G. A. (2010). *Int. J. Mol. Sci.*, **11**: 288–303.

Zhang, G. P., George, T. F., Assoufid, L., Mansoori, G. A. (2007). *Phys. Rev. B.*, **75**: 035413.

Willey, T. M., Bostedt, C., van Buuren, T., Dahl, D. E., Liu, S. G., Carlson, R. M. K., Meulenberg, R. W., Nelson, E. J., Terminello, L. J. (2006). *Phys. Rev. B.*, **74**: 205432.

6

Biomedical Applications of Diamondoids

6.1. Introduction

Biomedical uses of adamantane and its derivatives represent an important fraction of all the related patents, ranging from 11 to 23% of the total (see Fig. 1.20 black line in Chapter 1).

Starting from the late 50's and early 60's, the hydrophobic nature associated with the compact and highly symmetrical structure of the adamantane molecule have attracted researchers interested in the discovery of structurally unprecedented drugs, such as polyhaloadamantane pesticides (Webber and Harthoorn, 1959); and new or improved biological activities conferred by a caged moiety attached to known bioactive substances, such as orally active hypoglycemic sulfonylureas (Gerzon *et al.*, 1963). Biomedical uses of adamantane and its derivatives represent around 12% of all the adamantane-related patents, ranging from 6 to 25% of the total (Fig. 1.20 empty circles/black line). Nowadays, cage chemical groups, especially adamantyl moieties, keep attracting a crescent number of researches interested in improving biological activity of drugs. This attraction is not without a reason, as it is shown in this chapter. It is also known that adamantly group changes the properties of known drugs or provides an important pharmacophore for the design of new drugs (Lamoureux and Artavia, 2010).

In this chapter, we report on how diamondoid molecules interact with living systems and how these interactions can be useful in their biomedical applications. In addition, we discuss about characteristics of diamondoid molecules as tools for diagnosis and cell metabolism screening.

6.2. Fighting Infectious Diseases with Diamondoids Derivatives

In this section, we present the role of diamondoids and derivatives in the design and action of antiviral and antimicrobial agents for treatment of infectious diseases threatening human health. Infectious diseases are caused by exogenous pathogens, such as prions, viruses, bacteria, fungi, and parasites. Rapid changes in the spectra of infectious diseases occur in a complex globalized scenario and involve many factors.

6.2.1. *Diamondoids-based drugs against influenza viruses*

6.2.1.1. *Amantadine and Rimantadine*

Figure 6.1 shows the chemical structures of amantadine (1-aminoadamantane hydrochloride) and rimantadine (α-methyl-1-adamantane methylamine hydrochloride). These two adamantane amino-derivatives are widely used as antiviral agents. They inhibit influenza propagation by blocking the viral M2 protein ion channel, which prevents fusion of the virus and host-cell membranes and release (uncoating) of viral RNA into the cytoplasm of infected cells. As a consequence of the interaction of these adamantane derivatives with the viral protein, a lowering in pH in the Golgi compartment (an organelle in the eukaryotic cell) also alters hemagglutinin conformation (Hay *et al.*, 1985; Ciampor *et al.*, 1992; Wang *et al.*, 1993).

Amantadine antiviral activity was first reported by Davies and coworkers in 1964 (Davies *et al.*, 1964). This compound, in its soluble hydrochloric salt form, upon *in vitro* and *in vivo* animal testing showed a selective inhibition of influenza A infections in tissue culture, chick embryos, and mice. The initial tissue culture studies indicated that the compound was not virucidal and appeared to act by interfering with the penetration of the host cell by the virus. Amantadine was originally

Fig. 6.1. Amantadine (left) and rimantadine (right), the two diamondoid drugs available for Influenza A prophylaxis and therapy.

approved by the United States Food and Drug Administration (FDA) in 1966 under the trade name Symmetrel®. Now, it is also available as the generic amantadine hydrochloride.

Rimantadine had its first clinical trial in 1969, at the State Penitentiary of Iowa, USA. A group composed of nine patients treated with Rimantadine had a shorter course of influenza A2 disease than a placebo group of nine other patients when judged by an overall "clinical impression" and by analysis of the duration of fever and of the duration and severity of the signs and symptoms characteristic of the disease (Rabinovich *et al.*, 1969). Rimantadine was approved in 1993 with the trade name Flumadine®. Now, it is also available as generic rimantadine hydrochloride.

Common side effects of amantadine and rimantadine observed in controlled studies in young heathy subjects are nervousness, lightheadedness, difficulty concentrating, sleeping disorders (insomnia, fatigue), and gastrointestinal complaints.

Nevertheless, rimantadine recipients reported lower frequency and severity of these symptoms. Amantadine recipients also performed less well on an objective test measuring sustained attention and problem-solving ability (Hayden *et al.*, 1981). Adverse effects related to the central nervous system (CNS) toxicity have also been reported among elderly persons taking amantadine, especially in the presence of renal impairment, and include confusion, personality changes, agitation, aggressiveness, hostility, delusions, hallucinations, dizziness, ataxia (gross muscle movements discoordination), loss of balance, falling, and seizures (Degelau *et al.*, 1990; Somani *et al.*, 1991; Stange *et al.*, 1991).

The emphasis of amantadine and rimantadine antiviral activity has been their selectivity towards influenza A virus. Nevertheless, their effectiveness against a number of other infectious agents has been also reported. These other infectious agents include influenza B (Kolocouris *et al.*, 1996), human immunodeficiency virus (HIV) (Koloucoris *et al.*, 1996; El-Sherbeny, 2000; El-Emam *et al.*, 2004; Balzarini *et al.*, 2007), hepatitis virus (Wagner *et al.*, 2003), herpes simplex virus (Rosenthal *et al.*, 1982), bacteria (see Sec. 6.2.1.2) (El-Emam *et al.*, 2004; Wang and Chern, 1996; Chen *et al.*, 2006; Jia *et al.*, 2006; Kadi *et al.*, 2007; Nayyar *et al.*, 2007; Zhang, 2007;), parasites (see Sec. 6.2.3) (Koloucoris *et al.*, 2007; Zoidis *et al.*, 2008; Kelly *et al.*, 2001; Creek *et al.*, 2007; Keiser *et al.*, 2007; Singh *et al.*, 2007; Tripathi *et al.*, 2007; Xiao *et al.*, 2007), and fungus (El-Emam *et al.*, 2004; Kadi *et al.*, 2007).

Although amantadine and rimantadine are FDA-approved for treatment or prevention of influenza, they were not recommended by the CDC for use in the United States during the 2008–2009 season (CDC, 2008b), due to the increasing number of reports on viruses resistance to these drugs worldwide (Bright *et al.*, 2005, 2006; Deyde *et al.*, 2007; Barr *et al.*, 2007; Tang *et al.*, 2008). In the USA, 92% of H3N2 (a mutated strain of H2N2) influenza A viruses presented gene mutation associated with amantadine and rimantadine resistance (Bright *et al.*, 2006). Resistance to these drugs develops rapidly as about 30% of treated patients excrete resistant mutant virus with preserved virulence and human-to-human transmission capability (Hayden *et al.*, 1989; Englund *et al.*, 1998). A single amino acid substitution at position 26, 27, 30, 31, or 34 within the transmembrane domain of the M2 protein confers adamantane resistance to mutants (Hay *et al.*, 1985; Belshe *et al.*, 1988) that is passed on to multiple virus generation (Hayden *et al.*, 1989; Abed *et al.*, 2005). In spite of biochemical reasons for the prevalence of amantadine and rimantadine resistance in influenza A virus, inappropriate use of drugs containing these molecules in many countries, may have contributed to the rampant number of reported resistances (Bright *et al.*, 2005; Weinstock and Zuccotti, 2006). A more astonishing example of misusage is suspected to have occurred in China during the late 90's when chicken farmers may have administrated amantadine to fight influenza A H5N1 (bird flu) infections in poultry (Cyranoski, 2005), resulting in the development of resistance in this virus strain isolated in the northern China, conferred by a mutation at the position 31 of the M2 protein, where serine is replaced by asparagine (S31N mutation) (He *et al.*, 2008). H5N1 virus isolated from other Asian countries between 1996 and 2005, in both human and birds, showed high incidence of the same mutation (Cheung *et al.*, 2006). In addition, in 2000–2004, amantadine resistance-associated mutations in the M2 gene were also detected in 16.4% of H7HA subtype Influenza A isolated in New York (USA) and 10.6% of H9 HA subtype isolated in Southern Asia. Both virus subtypes are known to infect only birds and were collected from chickens. Among these, 100% of the H9 and 67% of H7HA subtypes exhibited a S31N mutation (Ilyushina *et al.*, 2005). Based on extrapolation from trials in seasonal influenza, amantadine and rimantadine might have net clinical benefit as a first-line agent for chemoprophylaxis of H5N1 infection when neuraminidase inhibitors (see Glossary) are not available and the virus is known or likely to be susceptible. Even though the incidence of resistant in H5N1 is increasing worldwide, amantadine and rimantadine are recommended by the World Health Organizatio (WHO) for treatment and prophylaxis of H5N1 infections in humans when the first-line antiviral agents are not available or they are in short supply (Schunemann *et al.*, 2007).

Moreover, in regions where the H5N1 virus is known or likely to be susceptible, adamantane derivatives might be useful as a low-cost, long shelf life stockpile antiviral drug for pandemic control (Schunemann *et al.*, 2007; Hurt *et al.*, 2007; Wong and Yuen, 2006). Combination chemotherapy, i.e. amantadine plus other drugs, have been used in mice with good results against lethal H5N1 infection (Ilyushina *et al.*, 2007).

Initial tests evidenced that Influenza A H1N1 (S-OIV 2009) viruses are also resistant to amantadine and rimantadine (CDC, 2009b). As of August 2009, the pandemic Swine flu virus remained resistant to amantadine and rimantadine (WHO, 2009a).

Amantadine and rimantadine are the only diamondoid derivatives commercially available for influenza prophylaxis and treatment. But the high incidence of resistant Influenza A viruses to these molecules worldwide practically turned these agents into ineffective drugs for human use. Nevertheless, they still might play an important role in prophylaxis during an epidemic, because most of the present resistant H3N2 virus strains have monotonous S31N mutation in the M2 gene (Bright *et al.*, 2005, 2006). A small change in this genetic profile may reestablish susceptibility among circulating virus, making amantadine and rimantadine useful as antiviral drugs again (Weinstock and Zucotti, 2006).

6.2.1.2. *Other diamondoid-based drugs against influenza viruses*

A number of other nitrogen-containing adamantane derivatives also present antiviral activities and may become alternatives to overcome virus resistance (Stamatiou *et al.*, 2001, 2003; Kolocouris *et al.*, 1994, 1996, 2007; Zoidis *et al.*, 2006). In Fig. 6.2, we show the structures of some of these derivatives and they include the following compounds:

Two-spiro-cyclopropaneadamantylamines: *N*-methyl[spiro[cycloprop ane-1,2' -adamantan]-2-amine] (Compound 1), and *N,N*-dimethyl[spiro[cy clopropane-1,2' -adamantan]-2-amine (Compound 2); one-methanamine: spiro [cyclopropane-l,2-adamantan]-2-methanamine (Compound 3); four-spiro-pyrrolidine adamantanes: spiro[pyrrolidine-2, 2'-adamantane] (Compound 4a), its *N*-methyl (Compound 4b) and *N*-ethyl derivatives (Compound 4c), and l-[[(4-chlorophenyl)amino]carbonyl]spiro [pyrrolidine-2,2′-adamantane] (Compound 5) (Kolocouris *et al.* 1994). Please note that the numbering of compounds are ours and may not correspond to compound numbers in the literature to which they are referred to.

In addition, spiro[azetidine-2,2′-adamantane] (Compound 6a) and its *N*-methyl derivative (Compound 6b), spiro[azetidine-3,2′-adamantane] (Compound 7); four-spiropiperidine derivatives: spiro[piperidine-4,2′-adamantane] (Compound 8) (Kocolouris *et al.*, 2007), spiro[piperidine-2,2′-adamantane] (Compound 9a) and its *N*-methyl (Compound 9b) and *N*-ethyl (Compound 9c) derivatives; three-spiromorpholines: spiro[morpholine-3, 2′-adamantane] (Compound 10a) and its *N*-methyl (Compound 10b) and *N*-ethyl (Compound 10c) derivatives; *N,N*-dimethyl-1-(1-adamantan)-2-pyrrolidinemethanamine (Compound 11); and 1-(1-adamantyl)-1-aminocyclopentane (Compound 12).

Heterocyclic rimantadine analogs, including seven pyrrolidines: 2-(1-adamantyl)pyrrolidine (Compound 13a), and its *N*-*n*-butyl derivative (Compound 13b) (Kocolouris *et al.*, 1996), 2-(1-adamantyl)-2-methylpyrrolidine (Compound 13c) (Zoidis *et al.*, 2006), 3-(2-adamantyl) pyrrolidine (Compound 16a), and its *N*-methyl derivative (Compound

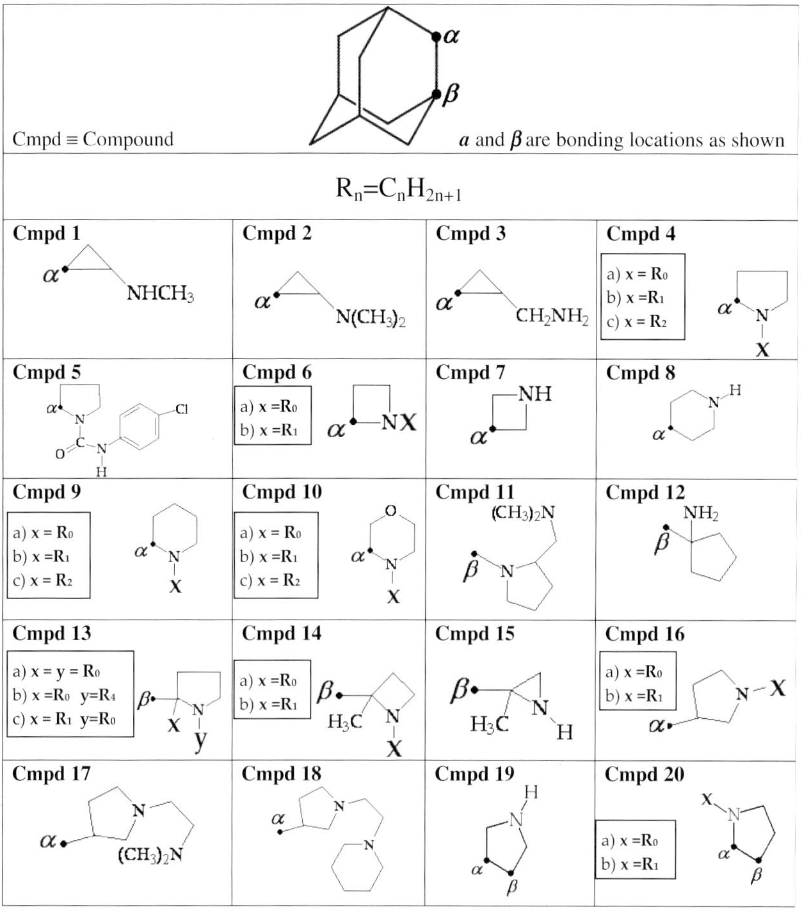

Fig. 6.2. Adamantane derivatives with antiviral activity. **Cmpds** (Compounds) 1–5 (Kolocouris *et al.*, 1994), **6–8** (Kolocouris *et al.*, 2007), 9, 10a, 12, 13a (Kolocouris *et al.*, 1996), 13c, 14, 15 (Zoidis *et al.*, 2006), 16–18 (Stamatiou *et al.*, 2001, 2003), 19 and 20 (Zoidis *et al.*, 2008 present significant anti-influenza A activity (Stamatiou *et al.*, 2001, 2003, 1996, 2007; Zoidis *et al.*, 2006, 2008). **Cmpd 8** also showed antiparasitic action against bloodstream forms of the African trypanosome, *Trypanosoma brucei* (Kolocouris *et al.*, 2007). In addition, **Cmpds 10a**, **12,** and **13a** present activity against influenza B virus (Kolocouris *et al.*, 1996). Borderline anti-HIV-1 action was detected in **Cmpds 9b–c**, **10-a-c** and **11** (Kolocouris *et al.*, 1996). Madin-Darby canine kidney (MDCK) cells were used as influenza A and B host cells and Human T-cells (CEM) were used as HIV hosts.

16b); *N*-[1-ethyl-(2-*N′*,*N′*-dimethylamino)]-3-(2-adamantyl)pyrrolidine (Compound 17) and *N*-[1-ethyl-[2-(*N′*-piperidine)]]-3-(2-adamantyl) pyrrolidine (Compound 18) (Stamatiou *et al.*, 2001, 2003), two azetidines: 2-(1-adamantyl)-2-methylazetidine (Compound 14a), and its (E)-N-methyl derivative (Compound 14b); and 2-(1-adamantyl)-2-methylaziridine (Compound 15) (Zoidis *et al.*, 2006); and three adamantanopyrrolidines: 4-azatetracyclo[6.3.1.$1^{6,10}$.$0^{2,6}$] tridecane (Compound 19), 3-azatetracyclo[6.3.1.$1^{6,10}$.$0^{2,6}$] tridecane (Compound 20a), 3-Methyl-3-azatetracyclo[6.3.1.$1^{6,10}$.$0^{2,6}$] tridecane (Compound 20b and 4-azatetracyclo[6.3.1.$1^{6,10}$.$0^{2,6}$] tridecane (Compound 20) (Zoidis *et al.*, 2008) are also depicted as examples of the potential of diamondoid molecules as antiviral drugs. The von Bayer system of nomenclature (Moss, 1999) is used for reporting the above chemical formulas and in the rest of this book.

The adamantane derivatives with antiviral activity presented in Fig. 6.2 and Table 6.1 were only tested for anti-Influenza A virus activity. Excluding Compound 15, all of them exhibited high potency when compared to amantadine or rimantadine. Nevertheless, Compounds 16a and 17 showed relative minimal cytotoxic concentration (MCC) close or ever lower than relative EC_{50} (Concentration producing 50% inhibition of the virus-induced cytopathic effect) when compared to amantadine, the same happened with Compounds 6a, 6b, 7, 8, 13c, 14a, and 14b when compared to both amantadine and rimantadine. This is an undesirable feature for medicinal drugs as it may elicit toxicity in concentrations close to those used in therapeutic doses.

Apparently, the best combination of *in vitro* high potency and low toxicity arises from *N*-alkyl substituted spiropyrrolidineadamantanes, especially when a methyl group is present (Compound 4b). The nonsubstituted analog Compound 4a is 30 times less potent while N-ethyl derivative Compound 4c is two times less potent than Compound 4b. Compound 5 presents potency equivalent to Compound 4b but its toxicity is five times higher than amantadine (Table 6.1).

Effects of decrease in 2-spirocycle ring size from five to four members were different in nonsubstituted and *N-alkyl* substituted 2-spiropyrrolidines. While *N*-methyl derivative Compound 4b is 30 times more potent than Compound 6b, non-substituted pyrrolidine Compound 4a presents

Table 6.1 Anti-influenza activity of some adamantane derivatives (Cmpd ≡ Compound).

	Relative EC_{50}								
	H2N2			H3N2			Relative MCC		
Virus subtype	I[a]	II[b]	III[c]	I[a]	II[b]	III[c]	I	II	III
Cmpd #									
1	—	—	—	0.01	—	1.00	<1.0	—	<0.5
2	—	—	—	0.06	—	5.00	<1.0	—	<0.5
3	—	—	—	0.01	—	0.66	<0.2	—	<0.1
4a	—	—	—	0.30	—	>25	NT	—	NT
4b	—	—	—	0.01	—	0.47	NT	—	NT
4c	—	—	—	0.02	—	2.00	NT	—	NT
5	—	—	—	0.02	—	2.00	<0.2	—	<0.1
6a	—	—	—	0.18	0.99	0.04	<2.5	<2.5	12.5
6b	—	—	—	0.30	1.63	0.07	<0.5	<0.5	2.5
7a	—	—	—	0.16	0.91	0.04	<1.1	<1.1	5.7
8	—	—	—	0.08	0.44	0.02	<0.2	<0.2	0.9
9a	0.30	≥2	4.9E-3	—	3.57	7.69	NT	<1.0	NT
9b	0.73	≥5	0.01	—	3.57	8.33	NT	<1.0	NT
9c	50.00	≥221	0.63	—	5.26	11.11	NT	<1.0	NT
10a	2.13	≥12	0.04	—	1.09	2.33	NT	<1.0	NT

(Continued)

Table 6.1. *(Continued)*

	Relative EC_{50}								
	H2N2			H3N2			Relative MCC		
Virus subtype	**I[a]**	**II[b]**	**III[c]**	**I[a]**	**II[b]**	**III[c]**	**I**	**II**	**III**
Cmpd #									
10b	5.26	≥33	0.09	—	0.23	9.09	NT	<1.0	NT
10c	12.5	≥100	0.21	—	4.35	100.0	NT	<1.0	NT
11	50.00	≥250	0.67	—	6.25	2.22	NT	<1.0	NT
12	3.13	≥16	0.05	—	0.50	20.0	NT	<1.0	NT
13a	0.75	≥4	0.01	—	1.03	0.93	<1.0	1.0	<1.0
13b	50.00	≥3	0.91	—	8.33	1.08	<1.0	1.0	<1.0
13c	0.04	0.11	0.05	0.30	4.55	0.30	0.16	0.18	<0.19
14a	0.05	0.14	0.06	0.38	5.88	0.38	0.04	0.04	<0.05
14b	0.05	0.29	0.12	0.55	25.0	1.82	0.19	0.21	<0.22
15	1.32	4.00	1.61	0.69	50.0	0.69	0.04	0.05	<0.05
16a	0.23	—	9.75E-4	0.39	—	0.04	<0.23	—	≤0.12
16b	7.23	—	0.03	1.47	—	0.15	<0.29	—	≤0.15
17	0.15	—	6.18E-4	0.73	—	0.08	<0.14	—	≤0.08

(Continued)

Table 6.1. (*Continued*)

	Relative EC_{50}								
	H2N2			H3N2			Relative MCC		
Virus subtype	**I[a]**	**II[b]**	**III[c]**	**I[a]**	**II[b]**	**III[c]**	**I**	**II**	**III**
Cmpd #									
18	0.65	—	2.76E-3	0.03	—	2.8 E-3	<0.03	—	≤0.01
19a	—	—	—	1.10	6.25	0.25	<4.68	<4.68	23.40
19b	—	—	—	1.70	9.09	0.39	<4.39	<4.39	21.95
20	—	—	—	0.24	1.33	0.06	<0.94	<0.94	4.7

EC_{50} = Concentration producing 50% inhibition of the virus-induced cytopathic effect, as determined by measuring the cell viability. H2N2 = A subtype of Influenza A viruses. H3N2 = A mutated strain of H2N2. MCC = Minimal cytotoxic concentration, or concentration that causes microscopically detectable changes in cell morphology, relative to I, Amantadine, II, Rimantadine, III, Ribavirin. EC_{50} (μg/mL) for commercial antivirals varied with the influenza A virus strain used in each particular study; N. T = nontoxic, as neither the new nor the reference compound showed cytotoxicity in any tested concentration (Data reported in this table are from Stamatiou *et al.*, 2001, 2003; Kolocouris *et al.*, 1994, 1996, 2007; and Zoidis *et al.*, 2006).

roughly 66% of azetidine Compound 6a potency. Increasing cytotoxicity in the studied azetidines could also be noticed (Table 6.1).

Increase in the 2-spirocycle size from five to six members, to give Compound series 9a–c, did not result in improvements of EC_{50} relative to amantadine, when compared with Compounds 4a–c. However, piperidine Compounds 9a and 9b are less toxic than their azetidine counterpart Compounds 6a and 6b. Similarly to what was verified for azetidines, N-alkyl substitutions decrease potency against influenza virus, with relative reduction being more severe in H2N2 (a subtype of Influenza A viruses) antiviral activity, as *N*-ethyl azetidine Compound 9c is more than 100 times less potent than non-substituted azetidine Compound 9a, when compared to any of the reference commercial drugs presented. Decrease in potency caused by *N*-alkyl substituents is also present in the Compounds 10a–c spiromorpholine series.

An intriguing increase in cytotoxicity appears to happen when the nitrogen atom of the spiro heterocycle is moved away from the C-2 atom of the adamantyl group. Thus, 2-azetidine Compound 7 is two times more cytotoxic than 3-azetidine 6a. Likewise, elevated toxicity was detected in 4-piperidine Compound 8 when compared with 2-piperidine Compound 9a. Conversely, little, or virtually no, difference in antiviral activity was detected between the two previously cited azetidines, while dislocation of nitrogen atom in piperidines from position 2 to position 4 made Compound 8 eight times more potent than Compound 9a.

These findings reveal a complex relation between antiviral activity and structure of adamantane derivatives, and yet unveil the great potential of these molecules as new drugs against influenza virus.

Provided we know the detailed molecular basis of virius replication in the body, we will be able to design more potent antivirus drugs. For example, according to Wei *et al.* (2009), the replication of influenza A viruses needs a pH-gated proton (or M2) channel. Therefore, to develop effective drugs against influenza A, the most logical strategy is to inhibit the M2 channel. Recently, the atomic structure of the M2 channel was determined by NMR spectroscopy (Schnell and Chou, 2008). Then Wei *et al.* (2009) proposed a benchmark dataset that contains 34 newly-developed adamantane-based M2 inhibitors and covers considerable structural diversities and wide range of bioactivities. As a result, a molecular design algorithm was developed that provides useful insights for dealing with the drug-resistant problem and designing effective adamantane-based antiflu drugs.

6.2.2 *Other antiviral activities of diamondoids derivatives*

As discussed in the Sec. 6.2.1, many nitrogen-containing adamantane derivatives present antiviral activity against influenza A virus. However, other viral infections might also be fought against with diamondoid derivatives.

In addition to anti-influenza A activity (Kolocouris *et al.*, 1996), in adamantane derivative Compounds 9b-c, 10a-c and 11 (Fig. 6.3) a borderline anti-HIV-1 action but no anti HIV-2 activity was detected.

Other promising diamondoid anti-HIV-1 agents are shown in Fig. 6.3. Values of EC_{50} [μM] for Compounds 21a-c (2-Adamantan-1-yl-3-(4-methyl-pyridin-2-yl)-thiazolidin-4-one; 2-Adamantan-1-yl-3-(6-methyl-pyridin-2-yl)-thiazolidin-4-one and (R)-(+)-2-Adamantan-1-yl-3-(4,6-dimethyl-pyridin-2-yl)-thiazolidin-4-one, respectively), Compound 22 (2-Adamantan-1-yl-3-(2-diethylamino-ethyl)-thiazolidin-4-one) (Balzarini *et al.*, 2007), and Compounds 23a,b (2-(1-Adamantyl)-4H-3,1–6-bromobenzoxazin-4-one and 2-(1-Adamantyl)-4H-3,1–6-chlorobenzoxazin-4-one, respectively) (El-Sherbeny, 2000) are presented in Table 6.2. Similarly to what is observed for Compounds 9b–c, 10a–c and Compound 11, Compounds 21a–c, 22, and 23a,b are inactive against HIV-2 virus.

Small differences of methyl-substitution patterns in pyridinyl group resulted in an enhancement of approximately four-fold in *in vitro* anti-HIV-1 activity in Compound 21a when compared to Compound 21b. Both compounds were found to be of low toxicity, as they are unable to inhibit 50% of T-cell proliferation up the concentration of 304 μM.

Compound 21c is the most active adamantane derivative against HIV-1 virus reported. This substance is the only adamantane derivative to present significant inhibition at sub-micromolar concentrations. Moreover, its selectivity index (SI), defined as $SI = IC_{50}/EC_{50}$, is 198. Such high selectivity is a good indication of the great potential of this compound as an antiviral drug. Interestingly, the replacement of adamantyl in Compound 21c by methyl group inactivates the resulting thiazolidinone against the

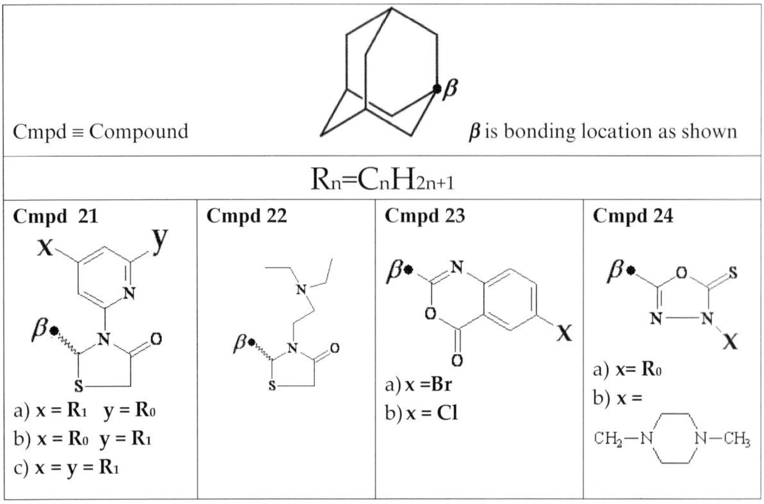

Fig. 6.3. Adamantane derivatives with Anti HIV-1 activity. **Cmpds** (Compounds) **21a,b** and **22** were tested as racemates. (R)-(+) dextrorotary isomer **Cmpd 21c** is 67 times more efficient than its [(S)-(–)] levorotary enantiomer in protecting host T-cell against HIV-1 cytopathicity, while replacement of adamantyl in **Cmpd 21c** by methyl group inactivates the resulting thiazolidinone against the virus (Balzarini *et al.*, 2007). **Cmpds 24a,b** also presented antibacterial and antifungal activity.

virus. This reveals the necessity for adamantane moiety in antiviral activity (Balzarini *et al.*, 2007).

Compounds 21a–c and 22 were also evaluated for their inhibitory activity of recombinant HIV-1 reverse transcriptase (RT-IC_{50}) (Table 6.2). Compound 21c exhibited the highest inhibitory activity to reverse transcriptase (RT) compared to Compounds 21a, 21b, and 22. The less active Compounds 21b and 22 required concentrations more than two and two hundred times higher, respectively, to achieve similar level of inhibition on RT. Nevertheless, a direct comparison between RT-IC_{50}/IC_{50} for Compound 21a (~7) and Compound 22 (~62) suggests that these substances may act through different inhibition mechanism. A considerable part of Compound 22 antiviral activity can hardly be associated with RT inhibition. Data collected on kinetic interaction of Compound 21c with HIV-1 RT (Balzarini *et al.*, 2007), associated with its inactivity against HIV-2 and its stereoselectivity {active (+) *versus* inactive (–) derivative}, place this compound and probably similar adamantyl thiazolidinone(s) in

Table 6.2. Anti HIV-1 activity of some selected adamantane derivatives and two licensed drugs, AZT and Nevirapine. Cmpds (Compounds) 21a,b and 22 were tested as racemates, Cmpd 21c was tested.

Compound	EC_{50} [μM]	IC_{50} [μM]	SI	$RT\text{-}IC_{50}$[a] [μM]	Ref.
Cmpd 21a	5.6	>304	>54	40	Balzarini *et al.* (2007)
Cmpd 21b	21	>304	>14	40	Balzarini *et al.* (2007)
Cmpd 21c	0.178	35	198	19	Balzarini *et al.* (2007)
Cmpd 22	7.7	56	7	475	Balzarini *et al.* (2007)
Cmpd 23a	50.3	>200	>4	—	El-Sherbeny (2000)
Cmpd 23b	69.7	>200	>3	—	El-Sherbeny (2000)
Cmpd 24a[b]	—	>288	—	—	El-Emam *et al.* (2004)
Cmpd 24b[c]	—	>287	—	—	El-Emam *et al.* (2004)
AZT	0.073	35.6	488	—	El-Sherbeny (2000)
Nevirapine	0.12	>250	>417	2.5	Balzarini *et al.* (2007)

[a] Polycytidylate-oligodeoxiguanylate (Poly cR. dG) was used as homopolymeric template.
[b] EC_{100} = 210 μM and EC_{43} = 42 μM.
[C] EC_{53} = 143.5 μM and EC_{19} = 28.7 μM
The notations used in this table have the following meanings:
EC_{50} = Effective concentration required to protect T4 Lymphocytes host cell against the cytopathicity of HIV by 50%,
IC_{50} = Inhibitory concentration required to reduce host cell proliferation by 50%,
SI = Selectivity index; ratio $= IC_{50}/EC_{50}$,
$RT\text{-}IC_{50}$ = Inhibitory concentration required to inhibit the HIV-1 reverse transcriptase-catalyzed reaction by 50%.

a novel subclass of NNRTIs (nonnucleoside reverse transcriptase inhibitors). The presence of a bulky lipophilic adamantyl moiety is unprecedented among the known NNRTIs, and is required for antiviral and anti-RT activity of Compound 21c.

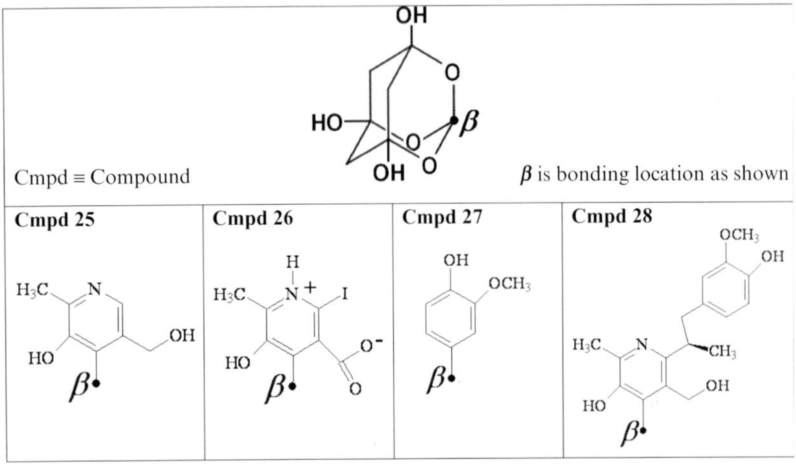

Fig. 6.4. Trioxaadamantane triols (TATs), also called bananins, present potential use in antiviral chemotherapy of severe acute respiratory syndrome (SARS) coronavirus (SARS-CoV)-induced human disease conditions. **Cmpd (compound) 25** = bananin, **Cmpd 26** = iodobananin, **Cmpd 27** = vanillinbananin, and **Cmpd 28** = eubananin. (Tanner *et al.*, 2005)

Novel classes of anti-viral agents may also emerge from heteroadamantane compounds. For example, Compounds 25–28 (Fig. 6.4) are trioxa-adamantane triols (TATs) conjugated with pyridoxal (Kesel, 2003; Kesel, 2005).

These compounds are commonly named as bananins, because of the striking color change during acid-catalyzed rearrangement from precursor pyridoxylidenephloroglucinol (yellow) to prototype bananin Compound 25 (brown-black) resembling the color change observed in ripening banana peels. (Kesel *et al.*, 2003). Bananin (Compound 25), iodobananin (Compound 26), vanillinbananin (Compound 27), and eubananin (Compound 28) present inhibitory action on severe acute respiratory syndrome (SARS) corona virus (SARS-CoV) helicase, a key enzyme in SARS virus viability. In addition, Compound 25 inhibits SARS-CoV *in vitro* replication in a much lower concentration than the cytotoxicity level (Tanner *et al.*, 2005). Therefore, bananins, especially Compounds 25, are good drug candidates for SARS-CoV infection treatment in humans.

6.2.3. *Diamondoids-based drugs for bacterial infection treatment*

Bacterial infections are responsible for a vast number of human diseases. Bacteria alone are associated with eighteen out of the fifty more concerning emerging/re-emerging infectious diseases as listed by the U.S. Centers for Disease Control and Prevention (CDC) (CDC 2008a). In this section we present the diamondoids derivatives which are found to be effective against bacterial infections.

6.2.3.1. *Adamantaplatensymicin*

The natural new antibacterial drug platensimycin (Compound 29 in Fig. 6.5) is a very effective antibiotic which blocks enzymes involved in

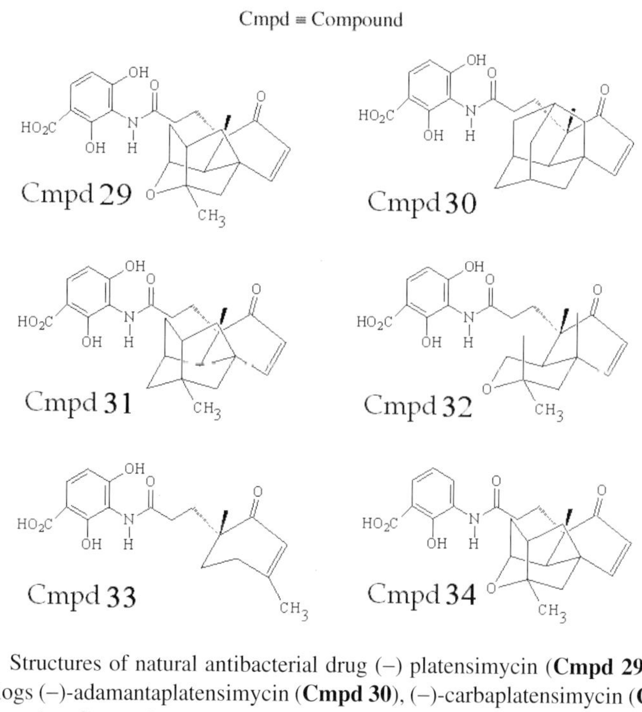

Fig. 6.5. Structures of natural antibacterial drug (−) platensimycin (**Cmpd 29**) and synthetic analogs (−)-adamantaplatensimycin (**Cmpd 30**), (−)-carbaplatensimycin (**Cmpd 31**), (−)-pyran analog **Cmpd 32**, cyclohexenone analog (**Cmpd 33**), and (−)-4-deoxyplatensimycin (**Cmpd 34**) (Nicolaou *et al.*, 2008).

the condensation steps in fatty acid biosynthesis, which Gram-positive bacteria need to biosynthesise cell membranes (Wang *et al.*, 2006).

However, due to limited availability of natural platensimycin and the fact that large amounts of it would be needed as antibiotic in its medical application has been delayed. Therefore emphasis has been on the development of adamantane-based resembling chemical forms with similar properties. Adamantaplatensymicin (Compound 30 in Fig. 6.5) is a synthetic antibacterial drug with a similar chemical form as the natural antibiotic platensymicin (Nicolaou *et al.*, 2007). Adamantaplatensymicin and other representative analogs of natural platensimycin are showed in Fig. 6.5. As the synthetic design of the cage-like domain of natural platensimycin poses a challenge, the more synthetically accessible adamantyl group was used to mimetize the pentacyclic ketolide moiety architecture, resulting in adamantaplatensimycin (Compound 30).

The antibiotic importance of oxygen heteroatom and CH_2 groups in the cage-like domain of natural platensimycin was the reason to design platensimycin isoster carbaplatensimycin (Compound 31) and cageless mimetic pyran (Compound 32) and cylohexenone (Compound 33) derivatives. Compounds 32 and 33 served as probes to evaluate the degree of tolerance of platensimycin lead-structure antibacterial activity to alterations in its lipophilic cagelike domain. Alternatively, functional group deletion in the hydrophilic benzoic acid domain, with preservation of the pentacyclic ketolide motif, as is exhibited by 4-deoxyplatensimycin (Compound 34) structure, were performed in order to verify detrimental effects on antibacterial action caused by possible loss of ability in binding with the FabF enzyme.

Results of minimum inhibitory concentration values (MIC) for Compounds 29 to 34 against Gram-positive bacteria MRSA, VREF (vancomycin-resistant *enterococcus faecium*), and Gram-negative bacterium *E. coli* (Table 6.3) revealed that adamantyl analog Compound 30 and homogenous tetracyclic analog Compound 31 exhibit antibacterial potency similar to "natural" platensimycin on MRSA and VREF, although they are inactive against *E.coli*. Compound 32, in which a bicyclic structure replaces the cage moiety, also shows significant antibacterial activity. Conversely, cyclohexenone Compound 34 is totally depleted in action. These observations indicate that platensimycin lead structure is able to preserve its remarkable *in vitro* biological activity when submitted to quite

Table 6.3. Minimum Inhibitory Concentration Values (IC_{50} = Inhibitory concentration required to reduce host cell proliferation by 50%) of cage natural product (–)-platensimycin (Cmpd 29) and synthetic analogs (–)-adamantaplatensimycin (Cmpd 30), (–)-carbaplatensimycin (Cmpd 31), (–)-pyran 32, cyclohexenone (Cmpd33),and(–)-4-deoxyplatensimycin(Cmpd34)againstMethicillin-Resistant *Staphylococcus aureus* (MRSA), Vancomycin-Resistant *Enterococcus faecium* (VREF), and *E. coli* (Nicolaou *et al.*, 2008).

Cmpd ≡ Compound

	IC_{50} (μg/mL)		
Cmpd #	**MRSA**	**VREF**	***E. coli***
29	0.2–0.4	0.4–0.8	>88
30	1.3–1.8	1.3–1.8	>88
31	1.1–2.2	1.1–2.2	>88
32	3.5–4.3	3.5–8.6	>86
33	>69	>69	>69
34	>85	>85	>85

extensive alterations in its lipophilic cagelike domain. This feature is advantageous when it comes to improving pharmacokinetics of synthetic derivatives. By allowing more room to different molecular designs, specific characteristics such as drug absorption and distribution might be tuned to turn platensimycin derivatives into medicines suitable for oral administration in bacterial infection treatment. Therefore it is possible to design simplified synthetic pathways to create derivatives inspired by natural products scaffold.

For the synthesis of adamantaplatensimycin (Compound 30) sequence researchers took advantage of the commercial availability of 1-bromoadamantane as its starting cage material. This was associated with the high symmetry and hydrophobicity of the adamantyl group in order to create an appending cage that mimics platensimycin complex heterogeneous tetracyclic cagelike moiety, without interfering with antibiotic potency of the final product (Nicolaou *et al.*, 2007). Future *in vivo* investigations may reveal newer and possibly more intriguing details on pharmacokinetics of adamantaplatensimycin and other platensimycin cage derivatives.

Similar probing approach strategies to examine the importance of specific functional groups on antibacterial action may lead to new synthetic drugs that combine the uniqueness of platensimycin mode of action and optimized pharmacological performance imparted by diamondoid moieties to provide treatment for present and future multidrug-resistant bacterial infections.

6.2.3.2. *Diamondoid derivatives action against MTB (Mycobacterium tuberculosis)*

Tuberculosis is considered the most threatening re-emerging disease and is a major public health risk worldwide. MTB *(Mycobacterium tuberculosis)* is the main pathogen responsible for TB (tuberculosis) infection (Youmans, 1963). MTB presents an extremely slow division rate and very particular wall cell structure which allows the baccilum to remain dormant in the host organism for several years.

Renewed interest in TB chemotherapy has led, since the late 90's, to many new promising candidate-drugs and a number of ongoing clinical trials (Rivers and Mancera, 2008; Yew *et al.*, 2008; USA, 2011). One promising adamantane-derivative TB drug candidate for which Clinical Phase I is completed and Phase II protocol is submitted is N-Geranyl-N′-(2-adamantyl)ethane-1,2-diamine reported in Fig. 6.6 as Compound 35a (Lee *et al.*, 2003). Its commercial name is SQ-109, manufactured by Sequella Inc. (Rockville, MD).

To the best of our knowledge, Compound 35a (Fig. 6.6) is the only diamondoid-derivative drug included in clinical trials as antitubercular agent and is presently entering the Phase II evaluation (Sequella, 2010). In October 2010, Sequella has filed an application for an US Investigational New Drug (IND) process, in order to include Compound 35a as a therapeutic drug against *Heliobacter pylori*, which is a bacteria related to duodenal ulcera and gastric carcinomas. Moreover, Compound 35a is under pre-clinical trials as an antifungal agent to treat *Candida glabrata* infections (Sequella, 2010). N-(2-(*o*-anisyl)ethyl))-N′-(2-adamantyl)ethane-1,2-diamine, which is Compound 35c with commercial name SQ73, also developed by Sequella, Inc., is presently under pre-clinical evaluation as an anti-tuberculosis drug (Protopopova *et al.*, 2005; WGND, 2008). These two Compounds (35a and c) were discovered through a large study that

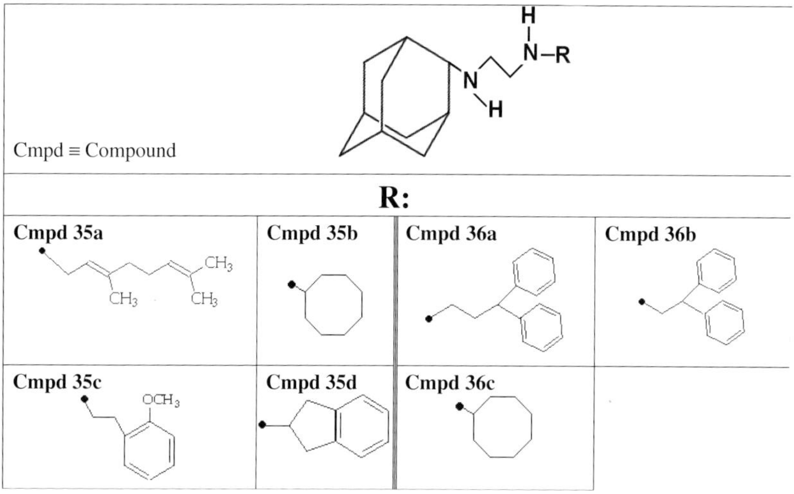

Fig. 6.6. Selected adamantyl-1,2-ethylenediamines having antimicobacterial activity against *Mycobacterium tuberculosis*. **Cmpd (Compound) 35a** (SQ-109) is currently under Phase I Clinical trials, while **Cmpd 35c** (SQ-73) is in preclinical phase (Sequella, 2010).

started in 1999 when a group of researches from the the US National Institute of Allergy and Infectious Diseases (NIAID) screened more than 100,000 compounds for anti-tuberculosis activity. In the following years, a cooperative research and development agreement was established with Sequella Inc. to continue work on the most promising compounds. In 2001, supported by a NIAID/NCI (the US National Cancer Institute) AIDS-related therapies development program, Sequella Inc. selected Compound 35a within a subset of over 63,000 substances from a combinatorial library based on the 1,2-ethylenediamine pharmacophore of ethambutol. A total of 26 compounds from this library showed *in vitro* anti-MTB activity equal to or higher than ethambutol. Interestingly, six were adamantyl derivatives (Compounds 35a,b, 35d and 36a–c). Compound 35a was the most potent of all, exhibiting a 14–35-fold improvement in *in vitro* activity against MTB, when compared to ethambutol (Lee *et al.*, 2003). Further investigations, performed by Protopopova and co-workers (2005) using a different protocol, revealed variations in the minimal inhibitory concentration (MCI) values but identified the same

Table 6.4. Results for *in vitro* studies on adamantyl derivatives of NIAID/Sequella, Inc. ethambutol-based combinatorial library. Activity against *Mycobacterium tuberculosis* (H37Rv strain) expressed as MIC = Minimun inhibitory concentration, the lowest concentration that prevents visible cellular growth. Cytotoxicity of the compounds was determined with monkey kidney Vero cell line ATCC CCL-81, IC_{50} = Inhibitory concentration required to reduce cell proliferation by 50%. SI = Selectivity index; ratio IC_{50}/MIC. Intracellular antimicrobial activity was determined by bioluminescent assay, in MIC concentration, using murine macrophage cell infected with *Mycobacterium tuberculosis* containing an integrated luciferase gene (Protopopova *et al.*, 2005).

Compound	**MIC [μM]**	**IC_{50}[μM]**	**SI**	**Intracellular Antimycobacterial Activity [%]**
Cmpd 35a	1.56	26	16.7	99.2
Cmpd 35b	7.8	80	10.3	82.7
Cmpd 35c	12.5	80	6.4	95.6
Cmpd 35d	15.6	130	8.3	88.3
Cmpd 36a	7.8	13	1.7	ND
Cmpd 36b	7.8	32	4.1	98.8
Cmpd 36c	15.6	34	2.2	ND
Isoniazid[c]	0.91	—	—	96.7
Ethambutol	7.25	—	—	83.8

Cmpd ≡ Compound.

active compounds from the ethambutol derivatives library. Compound 35c was then added to the list of drug candidates.

In vitro activity against MTB, cytotoxicity (IC_{50}) and intracellular activity against MTB of Compounds 35a-d and 36a-c, as determined by Protopopova and co-workers (2005), are shown in Table 6.4. The three 2-adamantyl derivatives, Compounds 35b-d, presented the lowest cytotoxicity among the top 27 hits of the ethambutol library (IC_{50} of 80, 80 and 130 μM, respectively), while Compound 35a was in an intermediate category, with IC_{50} = 26 μM.

Conversely, Compound 36a was the most toxic of all drug candidates, with IC_{50} = 13 μM, probably, in part, due to the presence of the *N*-diphenylalkyl group, as compounds bearing such groups also presented higher cytotoxicity. The overall selectivity index (SI) values of drug candidates from ethambutol library fell between 16.7 and 1.1. This range placed four adamantyl derivatives (Compounds 35a-d) among the five best candidates for *in vivo* efficacy studies. As MTBs are efficiently internalized into host's macrophage phagosomes, any viable anti-MTB agent must be capable of entering infected phagocytic cells. When compared with ethambutol, Compound 35a was one of the three more active compounds, exhibiting 99.2% of intracellular MTB infection inhibition. In addition, Compound 35a demonstrated significant *in vitro* activity against ethambutol-resistant, isoniazid-resistant or rifampicin-resistant strains of MTB (Protopopova *et al.*, 2005).

The mechanism of action of Compound 35a appears to involve inhibition of cell wall synthesis. Its ability in killing ethambutol-resistant MTB suggests that SQ109 aims different targets in cell methabolism or has mechanisms of action different from those of ethambutol. Microarray studies by Boshoff *et al.* (2004) suggest that Compound 35a is, in fact, a new TB drug, instead of just an ethambutol analog.

When tested *in vitro*, in concentrations equivalent to 0.5 MIC, Compound 35a showed strong synergistic inhibitory action on MTB growth with isoniazid (0.5 MIC) and with rifampicin (0.1 MIC), while exhibited additive effect with streptomycin. Combinations of Compound 35a and ethambutol or pyrazinamide did not result in synergic or additive actions (Chen *et al.*, 2006). TB treatment with multiple drugs is an important strategy to minimize the risk of development of drug-resistance (Caminero, 2008). The possibility of including this new, potent drug in the first-line anti-MTB formulations makes it one of the most promising anti-tuberculosis drugs of the decade. *In vivo* oral administration of Compound 35a (0.1–25 mg/kg/day, 28 days) to the mice showed dose-dependent reductions of mycobacterial load in both spleen and lung. These results are comparable to those obtained by ethambutol administration at 100 mg/kg/day, but they are inferior to isoniazid at 25 mg/kg/day. Compound 35a also shows large volume of distribution into various tissues, with persistently high concentrations in the lungs, spleen and kidneys, at levels at least 10-fold higher than its MIC (Jia *et al.*, 2005). After a single dose, administred intravenously, concentration in these tissues are significantly higher than those in plasma measured at the same time, specially in the lungs, where the concentration of 35a after 1 h is 25 times higher than peak plasma concentration after 2 minutes of administration. This ability of rapid transfer from circulation to lung tissue may be related to the presence of adamantyl group. It should be pointed out that major adamantyl marketed drugs amantadine and rimantadine also share similar specific lung distribution (Hoffman *et al.*, 1988). However, direct participation of adamantyl group in anti-MTB activity may also play a role in 35a overall performance. An indication of the latter is the frequent presence of adamantyl derivatives among both *in vitro* and *in vivo* best hits of the NIH/NIAID ethambutol-based combinatorial library.

In 2006, NIH and Sequella Inc. signed worldwide licensing agreement, giving Sequella Inc. exclusive rights to continue development of Compound 35a. In the same year, Sequella Inc. received Investigational New Drug status for this compound from the US Food and Drug Administration. In 2007, the US Food and Drug Administration and the European Medicines Agency granted the Orphan Drug Status for 35a (USA, 2007a) entitling Sequella Inc. to seven years of market exclusivity for 35a for the treatment of patients with TB (USA, 2007b).

This brief description on the ongoing process of research and development of 35a drug candidate exemplifies the many potentialities of diamondoid groups in exerting new and powerful biomedical properties to molecules to which they are attached. This huge potential is been continuously explored and new adamantyl drug candidates are been discovered.

In 2007, Nayyar and collaborators reported a series of 2-substituted 4-(1-adamantyl)-quinolines displaying promising activities against drug-sensitive and drug-resistant MTB H37Rv strains and were considered important lead molecules (Fig. 6.6). Compound 38a was the most potent drug and is comparable to isoniazid, presenting 99% inhibition growth at 1 μg/mL against isoniazid-susceptible MTB H_{37}Rv strain. In addition, Compound 38a presented MIC = 3.125 μg/mL (99% Inhibition) against isoniazid-resistant MTB. Compounds 37a-d and 38b-e presented 99% inhibition at MIC = 3.125 μg/mL against susceptible strains (Nayyar *et al.*, 2007).

6.2.3.3. *Other diamondoid derivatives with antibacterial action*

A variety of biological activities could be identified in diamondoids derivatives containing heterocyclic rings such as oxadiazoles, thiadiazoles and maleimides. Figure 6.7 shows structures of biologically active

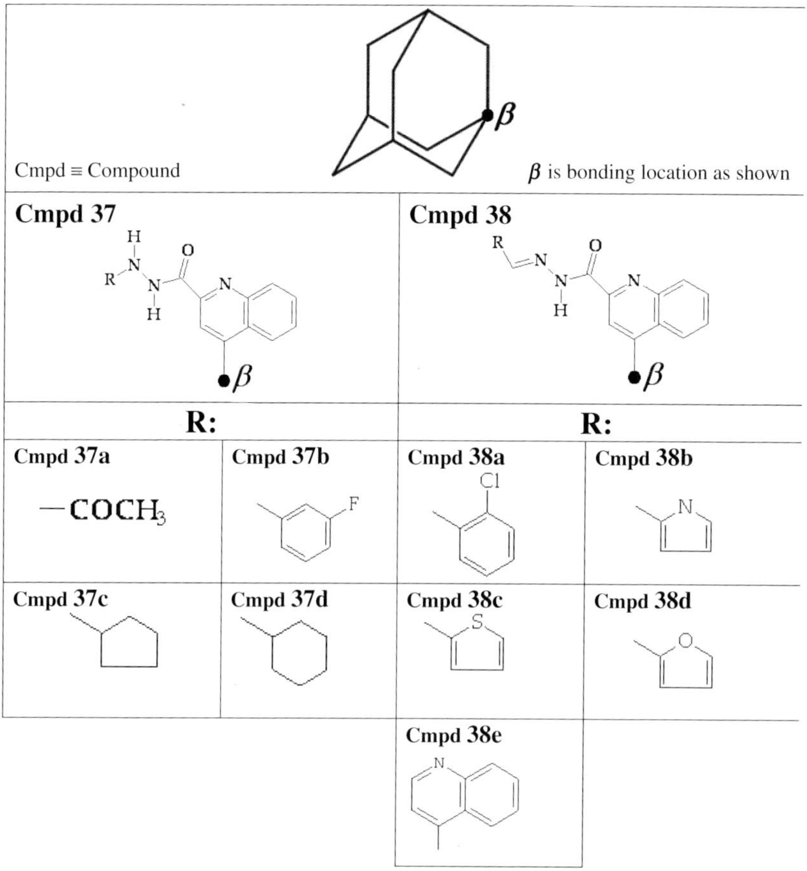

Fig. 6.7. Selected 1-adamantylquinolines possessing antimycobacterial activity against Gram-positive and Gram-negative bacteria (Nayyar *et al.*, 2007).

diamondoid derivatives (1-adamantylquinolines) possessing anti-MTB activity.

Figure 6.8 shows selected adamantane derivative molecules having antibacterial activity. *N*-1-adamantylmaleimide (Compound 39) and *N*-1-diamantylmaleimide (Compound 40) were prepared by reaction of maleic acid with respective primary diamondoid amines, followed by cyclization by dehydration with acetic anhydride and sodium acetate. Compounds 39 and 40 were also used as monomers in radical polymerization performed in benzene with 2,2′-azoisobutyronitrile as initiator. Corresponding

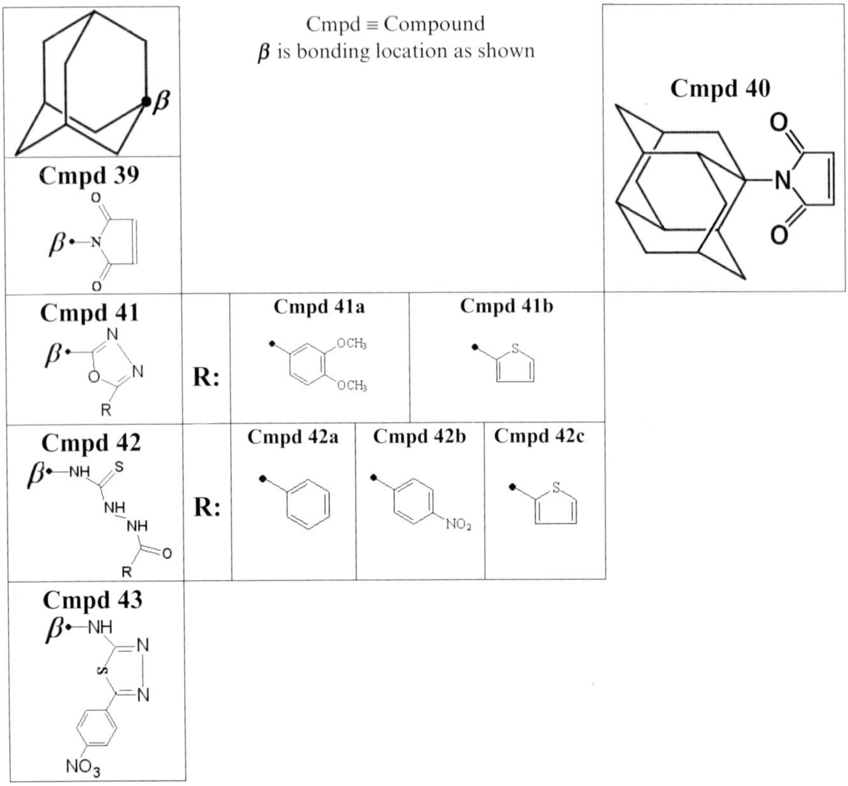

Fig. 6.8. Selected adamantane derivative molecules having antibacterial activity. Maleimides **39** and **40** presented moderated *in vitro* antibacterial activity on *S. aureus* ATCC 25923 but exhibited strong antitumoral activity (See Sec. 6.3 for more diamondoid derivatives presenting anticancer activity) (Wang and Chern, 1996). Additional adamantyl derivatives showed *in vitro* antibacterial activities equivalent or superior to ampicillin against Gram-positive bacteria *Bacillus subtilis* (oxadiazoles **41a,b;** thiosemicarbazides **42b,c** and thiadiazole **43**) and *S. aureus* (**42b**), and Gram-negative bacteria *E. coli* (**42a,b** and **43**) and *Pseudomonas aeuroginosa* (**42b**). **Cmpds (Compounds) 41a,b** produced good dose-dependent anti-inflammatory action. In addition, **Cmpd 41b** also presented good antifungal activity against *Candida albicans*.

poly (*N*-1-adamantylmaleimide) and poly (*N*-1-diamantylmaleimide) possess average molar masses of 9.5 and 1.9 kg/mol, respectively (Wang and Chern, 1996).

Both of these diamodoid maleimides and their respective polymers exhibit *in vitro* activity against Gram-positive bacterium *S aureus* (strain

Table 6.5. Comparison of diameters of inhibitory zones. That is, the clear area around a disc paper saturated with an antimicrobial agent on agar surface (standard, 10 μg per disk). Antibiotic-susceptable test microorganism does not grow where antibiotic is.

Compound	**Diameter of inhibitory zone**
Cmpd 39 (*N*-1-adamantylmaleimide)	21 mm
Cmpd 40 (*N*-1-diamantylmaleimide)	16 mm
Poly (*N*-1- adamantylmaleimide)	11 mm
Poly (*N*-1-diamantylmaleimide)	10 mm
Ampicillin (*A broad spectrum antibacterial drug*)	29 mm

ATCC 25923), as determined by the diameter of the inhibitory zone for this microorganism on a 8 mm disk paper containing 30 μg of the tested diamondoid substance as shown in Table 6.5.

Lower antibacterial activities of diamondoid polymers may be attributed, in some extent, to more difficult migration of these macromolecules through the paper disk, when compared to their respective monomers. All previously-cited compounds were almost inactive against Gram-negative bacteria *Pseudomonas aeruginosa* and *E. coli* (Wang and Chern, 1996). In addition, Compounds 39 and 40 showed *in vitro* antitumoral activity against several tumor cell lines higher than 5-Fluorouracil (an important commercial anticancer drug) activity while poly (*N*-1-adamantylmaleimide) presented antitumoral activity similar to 5-Fluorouracil (Wang and Chern, 1996) (See Sec. 6.3 for diamondoid anticancer drugs).

In order to synthesize Compound 41a [2-(1-Adamantyl)-5-(3,4-dimethoxyphenyl)-1,3,4-oxadiazole] and Compound 41b {2-(1-Adamantyl)-5-(2-thienyl)-1,3,4-oxadiazole} 1-adamantanecarbonyl chloride and an appropriate carboxylic acid hydrazide were brought into reaction. Then the resulting compounds were added to phosphorus oxychloride and refluxed to yield Compound 41a and 41b after cooling and purification (80–88% overall yielding) (Kadi *et al.*, 2007).

Compound 42a [1-Benzoyl-4-(1-adamantyl)-3-thiosemicarbazide], Compound 42b [1-(4-Nitrobenzoyl)-4-(1-adamantyl)-3-thiosemicarbazide], and Compound 42c [1-(2-thienyl-carbonyl)-4-(1-adamantyl)-3-thiosemicarbazide] were obtained from reactions between the corresponding carboxylic acid hydrazide and 1-adamantylisothiocyanate under reflux. On cooling, the precipitated crystalline solid was filtered, washed, dried and crystallized from ethanol to give Compounds 42a–c (75–85% yields) (Kadi *et al.*, 2007).

Compound 43 [2-(1-Adamantylamino)-5-(4-nitrobenzoyl)-1,3,4-thiadiazole] was obtained from Compound 42b by dehydration with concentrated sulfuric acid at room temperature for 24 h. Yields after final purification were between 66 and 82% (Kadi *et al.*, 2007).

These Compounds (41a,b, 42a,b,c and 43) were selected within a pool of 24 derivatives screened using the agar disc-diffusion method. These compounds showed *in vitro* antibacterial activities (200 μg per 8 mm disc) equivalent or superior to ampicillin (100 μg per 8 mm disc) against *Bacillus subtilis*, which is a Gram-positive bacteria.

Moreover, Compounds 42a,b and 43 showed activity against Gram-negative bacteria *E. coli* and Compound 42b showed activity against *P. aeuroginosa*. Compounds 41a,b produced good dose-dependent anti-inflammatory action against carrageenin-induced paw oedema in rats. Compound 41b also presented good antifungal activity against *Candida albicans* (Kadi *et al.*, 2007).

The MICs (in μg/ml) of Compounds 41a,b, 42a–c, and 43 are reported in Table 6.6. According to these data, Compound 42b can be considered as the most potent antibacterial drug found in this study. Its *in vitro* antibacterial activity is comparable to the action of the broad spectrum antibacterial drug ampicillin against *S. aureus*, *B. subtilis*, and *E. Coli*.

The brief description of synthetic protocols and biological activities presented in the above paragraphs highlights the relative simplicity for the preparation of adamantyl oxadiazoles, thiosemicarbazides and thiadiazoles derivatives and their good potential as antibacterial, antifungal, and anti-inflammatory drugs. Thus, many other similar derivatives may be obtained through small variations of these short synthetic routes, leading to a variety of potent new drugs.

Table 6.6. The minimun inhibitory concentration (MIC, μg/mL) of Cmpds (Compounds) 41a,b, 42a–c, and 43 and the broad spectrum antibacterial drug ampicillin which is used extensively to treat bacterial infections since 1961 (Kadi *et al.*, 2007).

	MIC [μg/mL]		
Compound	***Staphylococus aureus***	***Bacillus subtilis***	***Escherichia coli***
Cmpd 41a	2.0	0.5	ND
Cmpd 41b	ND	2.0	ND
Cmpd 42a	ND	1.0	4.0
Cmpd 42b	1.0	0.5	2.0
Cmpd 42c	ND	4	ND
Cmpd 43	2.0	2.0	1.0
Ampicillin	1.0	0.5	2.0

MIC determined as the lowest concentration that completely inhibited visible growth of the microorganism as detected by unaided eye. Microorganism strains are *Staphylococus aureus* IFO 3060, *Bacillus subtilis* IFO 3007, and *Escherichia coli* IFO 0583. ND, not determined. Cmpd ≡ Compound

6.2.4. *Diamondoid drugs for parasitic infection treatment*

6.2.4.1. *Adamantyl antimalarial drugs*

Tropical diseases are all diseases that occur mainly or exclusively in hot, humid conditions. Most common tropical diseases are malaria, leishmaniosis, schistosomiasis, onchocerciasis, lymphatic filariosis, Chagas disease, African trypanosomiasis, and dengue (WHO, 2009b).

In humans, malaria is caused by four types of *Plasmodium* parasites: *P. falciparum*, *P. vivax*, *P. malariae*, *and P. ovale*. *P. falciparum and P. vivax* are the most common infectious agents of the disease. *P. falciparum* is responsible for a larger number of fatal cases than *P. vivax* (Collins and Paskewitz, 1995; WHO, 2009c).

Many new adamantyl-4-amino-7-quinolines which are active against *P. falciparum* strains were recently disclosed by Solaja and co-workers (2008). Compounds 44a–d (Fig. 6.9) presented *in vitro* antimalarial activity higher than chloroquine (an antimalaria drug) against chloroquine-susceptible *P. falciparum* African D6 clone (D6), chloroquine-resistant *P. falciparum* Indochina W2 clone (W2) and multidrug-resistant *P. falciparum* TM91C235 strain (TM).

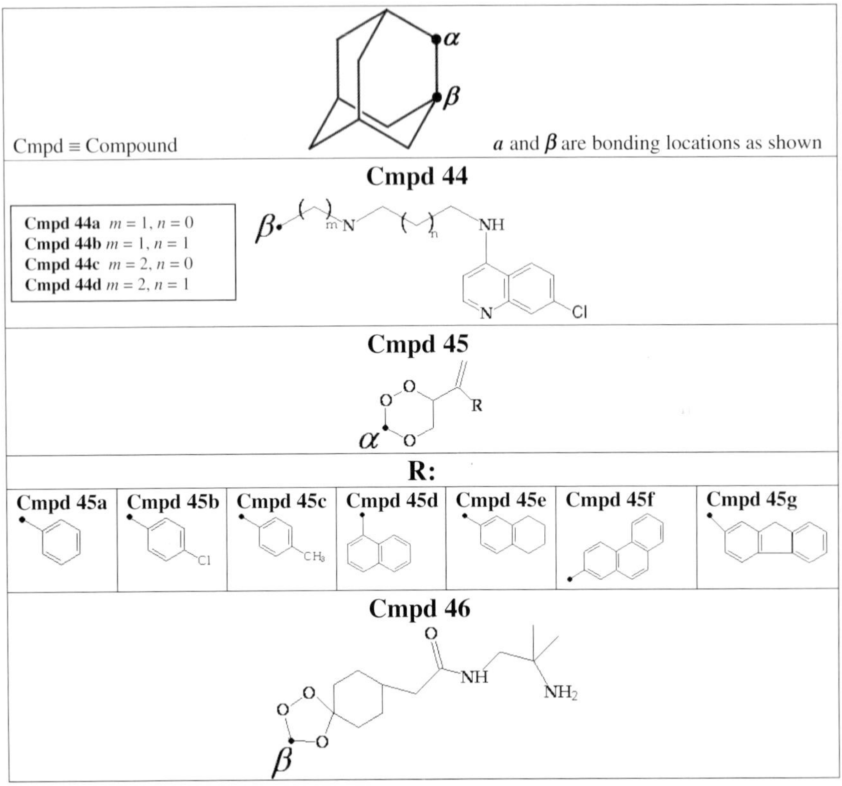

Fig. 6.9. Representative adamantyl antimalarial drugs. **Cmpds** (Compounds) **44a–d** are chloroquine derivatives (Solaja *et al.*, 2008), while **Cmpds 45a–g** (Singh *et al.*, 2007; Tripathi *et al.*, 2007) and **Cmpd 46** (OZ277/RB×11160) (Vennerstrom *et al.*, 2004) are peroxides that mimic artemisinin unique pharmacophoric 1,2,4-trioxane heterocycle.

In addition, Compounds 44a–c were found to be more active than mefloquine (another antimalarial drug) against D6 and TM strains, while Compound 44d presented activity equivalent to mefloquine against D6 strain. Compounds 44a,b were more potent than artemisinin (another anti-malarial drug) against TM and D6 strains, as well. Compound 44a can be considered as having potency equivalent to artemisinin against the highly drug-resistant W2 strain while Compound 44c showed antimalarial activity higher than artemisinin only against drug-susceptible D6 strain.

IC_{50} (nM) for parasite inhibitory activity of Compounds 44a-d are shown in Table 6.7. A general inspection of IC_{50} values of

Table 6.7. *In vitro* activity of adamantyl -4-amino-7-quinoline Cmpds (Compounds) 44a–d and standard drugs chloroquine, mefloquine and artemisinin against *P. falciparum* strains and against botulinum neurotoxin serotype A (Solaja *et al.*, 2008).

	IC_{50} [nM]			IC_{50} [μM]
Compound	**D6**	**TM**	**W2**	**BoNT/ALC**
Cmpd 44a	4.26	8.73	8.40	18.92
Cmpd 44b	3.83	7.39	12.10	31.98
Cmpd 44c	8.52	17.76	15.66	11.81
Cmpd 44d	11.43	42.11	33.66	50.10
Chloroquine	12.44	124.24	392.09	—
Mefloquine	11.25	33.46	4.96	—
Artemisinin	9.00	13.04	6.70	—

Cmpd ≡ Compound; IC_{50} = Inhibitory concentration required to reduce parasite proliferation by 50%; D6 = chloroquine-susceptible *P. falciparum* African D6 clone; TM = multidrug-resistant *P. falciparum* TM91C235 strain (Thailand); W2 = chloroquine-resistant *P. falciparum* Indochina W2 clone (W2); BoNT/A LC = neurotoxin serotype A light chain.

adamantyl-4-amino-7-quinolines reveals that all four compounds are more active against chloroquine-susceptible than chloroquine-resistant strains. Nevertheless, many similarities in antimalarial activity between Compounds 44a and 44b can be noticed, while Compound 44c is 1.5 times more potent against D6 and 2.4 times more potent against multidrug-resistant TM strain than homologue Compound 44d. Compounds 44a and 44b exhibit equivalent antiparasitic activity against each of the studied strains, particularly against *P. falciparum* TM.

The detrimental effect of the number of cabon atoms increment in the diamine spacer (n) on antimalarial activity of these adamantyl-4-amino-7-quinolines (Compounds 44 on Fig. 6.9) becomes clear when "n" increased to $n = 3$. Such alteration led to 3- to 10-fold decrease in antimalarial activity in the 1′-adamantyl-methyl series (Compounds 44a and 44b), exhibiting $m = 1$, and 1.5- to 10-fold decrease in the (1′-adamantyl)-2-ethyl series (Compounds 44c and 44d) which are those derivatives having $m = 2$. Interestingly, the presence of an amide link between the chloroquine-like moiety and adamantyl group, inactivates the compounds synthesized with $m = 1$ or 2 and $n = 0$ to 3 carbon atoms (Solaja *et al.*, 2008).

Compounds 44a–d also presented inhibition of the metalloprotease activity of the botulinum neurotoxin serotype A light chain (BoNT/A LC), a highly toxic by-product of a naturally occurring, spore-forming anaerobic bacterium, *Clostridium botulinum*. BoNT/A LC is a 50 kDa zinc endopeptidase that specifically cleaves neuronal proteins responsible for acetylcholine release. Neurotoxin serotype A, together with serotype B, E, and F provoke botulism-associated neuroparalysis in humans (Humeau *et al.*, 2000). BoNT/A LC inhibition activity is also present in other 4-amino-7-quinoline antimalarial drugs (Burnett *et al.*, 2003; 2007). IC_{50} (μM) for BoNT/A LC inhibition are shown in Table 6.6. Curiously, Compounds 44a and 44c, both with an ethylene spacer between nitrogen atoms ($n = 0$), are 1.7 and 4.2-fold more potent than their respective analogs, Compounds 44b and 44d, both bearing propenyl spacer. This is contrary to the findings discussed in the previous paragraph.

Some interesting results are published in recent years on *in vivo* antimalarial activity of synthetic six-membered 1,2,4-trioxanes spiroanellated at C-3 with adamantane group (Fig. 6.9, Compounds 45a–g) (Tripathi *et al.*, 2007; Singh *et al.*, 2007). Compounds 45a–c were tested in rhesus monkeys by intravenous inoculation with *P. knowlesi* (W1 strain). This parasite strain causes a fatal infection in rhesus monkeys and is used as model for severe malaria in humans (Waters *et al.*, 1993). While W1 strain infection resulted in 100% mortality rate of control-infected rhesus monkeys, treatment with Compound 45a *via* intramuscular injection in oil suspension (80 mg/kg/day during 5 days) led to complete cure rate of 71% (five out of seven individuals) in the testing group, with no signs of recrudescence until 60 days after treatment. The same protocol followed with Compounds 45b or 45c resulted in 100% cure rate (two out of two), with no recurrence within the same period of post-treatment observation.

Standard antimalarial drug α, β-arteether showed results equivalent to Compounds 45b and 45c when administered at 12 mg/kg/day in 5 days. Replacement of *spiro*-adamantyl group for *spiro*-cyclopentane in Compounds 45a and 45b basic structures resulted in inactive drugs against *P. knowlesi* W1 strain (Tripathi *et al.*, 2007). Thus, the presence of the adamantyl group has direct influence on antimalarial activity of 1,2,4-trioxanes at carbon 3 of the trioxane ring.

Table 6.8. *In vivo* antimalarial activity of adamantyl -1,2,4-trioxane Cmpds (Compounds) 45d-g against Plasmodium yoelii nigeriensis in Swiss mice by the oral and intramuscular routes (Singh *et al.*, 2007).

Cmpd #	Log P	Route	Dose [mg/kg/day][a]	Suppression on day 4 [%][b]	Survival on day 28 [%]
45d	5.65	Oral	96	100	100
		Oral	48	100	50
		Im	96	87	0
45e	5.96	Oral	96	100	100
		Oral	48	100	20
		Im	96	100	20
45f	6.65	Oral	96	100	91
		Oral	48	100	45
		Im	96	100	20
45g	6.39	Oral	96	100	100
		Oral	48	100	100
		Oral	24	100	80
		Im	96	100	82
45a	4.65	Oral	96	99	0
		Im	96	97	0
Arteether	3.84	Oral	48	100	100
		Oral	24	100	20
Artesunic acid	3.04	Oral	48	100	100
		Oral	24	100	0

Log P = (octanol-water partition coefficient) as defined by Eq. (6.1). Results obtained on arteether and artesunic acid are shown for comparison.

[a]Total of four days of treatment.

[b]Suppression [%] = [(C − T)/C] × 100; where C = parasitemia in control group and T = parasitemia in treated group.

Previous observations within a given peroxide class revealed that higher potency and better oral antimalarial activity in animal models were associated with higher lipophilicity (Tang *et al.*, 2004; Singh *et al.*, 2006). In addition, good results obtained with Compounds 45a–c led to preparation and testing of Compounds 45d–g (Fig. 6.9) for antimalarial activity against multidrug-resistant *P. yoelii nigeriensis* in mice, either by oral or intramuscular route.

These four new adamantyl-6-arylvinyl-substituted 1,2,4-trioxanes (Compounds 45d–g) are more lipophylic than Compounds 45a–c. Therefore, they are expected to present good bioavailability by the oral route. In fact, results shown in Table 6.8 revealed 100% of parasite elimination on the fourth day of treatment and 100% survival after 28 days of observation, when an orally administered dose of 96 mg/kg/day of Compounds 46d, 46e, or 46g were used.

The same parasitaemia suppression followed by a slight decrease in survival percentage was observed when the same dose of the highly lipophilic Compound 45f was administred under the same conditions. Therefore, an enhancement in antimalarial potency with the increase in lipophilicity is noticed in this 1,2,4-trioxane series (Compounds 45d–g), as treatment with Compound 45a, which is struturally similar to Compounds 45d–g, but presents lower lipophilicity, resulted in no survival of treated specimens after 28 days. At 48 mg/kg/day oral dose, all four Compounds 45d–g showed total clearance of parasitaemia, but only Compound 45g provided the same level of protection after 28 days.

Decrease in survival percentage for Compounds 45d–f suggests that increase in lipophilicity is beneficial to some extent, as seen in data for of octanol-water partition coefficient (P, expressed as log P) for Compounds 45a and 45d–g, when compared to survival rates (Table 6.8). Log P clearely reveals the relation of high lipophilicity (high log P) and increase in antimalarial action from Compound 45a to Compound 45g, with decrease in survival rate for Compound 45f, which has the higher log P, where P is the partition coefficient (ratio of concentrations of a solute (drug) between two phases (organic and aqueous). The logarithm of partition coefficient,

$$\mathrm{LogP} = \mathrm{Log}\left(\frac{[\mathrm{Solute}]_{\mathrm{Organic}}}{[\mathrm{Solvent}]_{\mathrm{Aqueous}}}\right), \qquad (6.1)$$

is a measure of lipophilicity. The octanol–water partition coefficient (P) is defined as the ratio of a substance concentration in the octanol phase to its concentration in the aqueous phase of a two phase octanol/water system. This parameter is used to describe lipophilicity for neutral compounds and is useful in environmental and biological studies, as it helps to predict, among other things, soil absorption (Karickhoff *et al.*, 1979) and bioaccumulation (Neely *et al.*, 1974).

Further increase in lipophilicity in this peroxide family did not result in improved biological activity. Attempts to replace 6-arylvynil group by 1-adamantyl group (R = 1-adamantyl), either by keeping original C-3 spiroannellated adamantyl or by making further alterations using *spiro*-cyclopentyl or *spiro*-cyclohexyl at this position, yielded compounds with Log P between 4.38 and 5.43 and infection suppression on day 4 of oral treatment between 70 and 52%. Nevertheless, no survival was observed on day 28. Thus, the role of the adamantyl group in increasing biological activity in this family of peroxides is closely related to spiroanellation with 1,2,4-trioxane ring. The presence of diamondoid group at the position C-6, in its turn, showed adverse effect on parasitaemia suppression (Singh *et al.*, 2007).

Compound 45 g is the most promising drug candidate of this adamantyl six-membered 1,2,4-trioxanes family. The trioxane ring present in Compound 45 series has six members, 3 carbons and 3 oxygens. Note that Compound 46 has a five-angled trioxane ring. This substance was further evaluated in 24 mg/kg/day oral dose and exhibited superior performance when compared to standard drugs arteether and artesunic acid (Table 6.8). In addition, Compound 45 g was the only substance to show significant action *via* intramuscular administration.

Compound 45 g was also tested by oral route in rhesus monkeys infected with *P. cinomolgi*, providing 100% clearance of parasite within five days of observation. Intramuscular administration of Compound 46 g in oil suspension was ineffective in providing parasite clearance. (Singh *et al.*, 2007).

Compound 46, also known as OZ277/RBx11160 (nicknamed OZ) was first prepared by Vennerstrom and co-workers by a short and economical synthetic method (Vennerstrom *et al.*, 2004). This five-membered 1,2,3-trioxane was initially part of the Medicine for Malaria Venture (MMV) portfolio. MMV is a not-for-profit public–private partnership established in 1999 and dedicated to reducing the burden of malaria in disease-endemic countries by discovering, developing and facilitating delivery of new, effective, and affordable antimalarial drugs (MMV, 2009a). After November 2006, MMV decided to stop funding further development of Compound 46, leaving to its pharmaceutical partner,

Table 6.9. Average and range, in parentesis, of inhibitory concentration (IC in nM) for Cmpd (Compound) 46 and standard drugs artesunic acid, chloroquine and mefloquine. (Kreidenweiss *et al.*, 2006).

Compound	No. isolates[a]	IC_{50}	IC_{90}	IC_{99}
Cmpd 46	38	0.47 (0.13–2.23)	1.29 (0.26–5.00)	3.73 (0.37–6.84)
Artesunic acid (salt)	43	0.96 (0.20–5.95)	2.47 (0.33–29.9)	5.76 (0.57–49.0)
Chloroquine[b]	43	113 (12.4–332)	241 (20.2–737)	544 (40.2–967)
Mefloquine	44	1.94 (0.24–21.2)	4.02 (0.27–48.8)	10.1 (0.30–124)

[a] Samples collected from malaria patients in Lambaréné, Gabon.
[b] All isolates were resistant to chloroquine.

Ranbaxy, the mission of completing the task (MMV, 2009b). In July 2008, Ranbaxy confirmed that Phase II trials were progressing in Thailand and India, and that regulatory approval had been sought for trials involving children in Kenya and Tanzania (Mann, 2008).

Compound 46 is the first fully synthetic antimalarial peroxide to enter Phase II clinical trial (Maerki *et al.*, 2006). This drug candidate present good bioavailability and is highly active against laboratory-adapted *P. falciparum* strains and rodent parasites *in vivo* (Vennerstrom *et al.*, 2004). It also showed *in vitro* activity against parasites isolated from malaria patients in Lambaréné, Gabon (Kreidenweiss *et al.*, 2006). Results are presented in Table 6.9. All parasites isolated from these patients were chloroquine-resistant. Compound 46 showed the highest molar activity, and the narrower range of activity, indicating a lower potential for development of resistance to this peroxide antimalarial in the tested isolates.

6.2.4.2. *Trypanosomiasis, Schistosomiasis and Fascioliasis-New treatments based of adamantyl peroxides*

Figure 6.10 shows chemical structures of Compound 47a (OZ78) and Compound 47b (OZ288), two promising drug candidates presenting

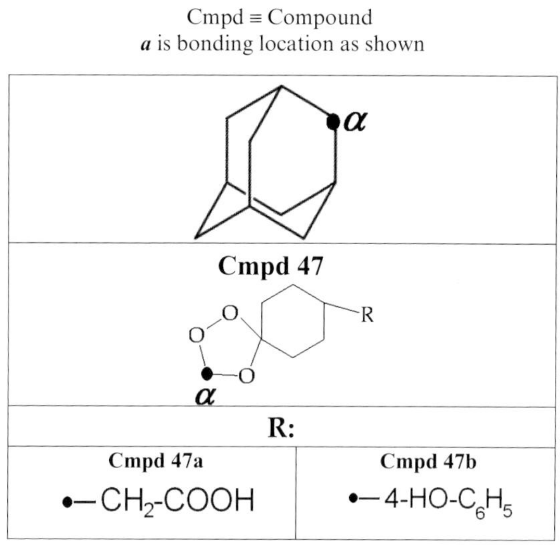

Fig. 6.10. Two members of the OZ family (1,2,4-trioxolanes), Compounds **47a** (OZ78) and **47b** (OZ288), presenting promising antiparasitic action. R = *para*-hydroxyphenyl or 4-hydroxyphenyl group (Keiser *et al.*, 2006a, 2006c, 2007; Xiao *et al.*, 2007).

in vivo activity against parasites that afflict millions of people around the world (Keiser *et al.*, 2006a, 2006c, 2007).

Table 6.9 shows some highlights about selected parasites, areas of incidence, standard medicines for treatment and main results obtained with adamantyl 1,2,4-trioxolanes. Neither of these two compounds have entered clinical trials related to the cited diseases, but such ability in reducing parasite burden *in vivo* is a good indication that adamantyl 1,2,4-trioxanes have much to offer as new drugs against parasitic infections.

6.3. Fighting Cancer with Diamondoids Derivatives

In this section, we present the main aspects of new diamondoid derivatives capable of inducing relevant suppressive action on cancer cells. Comparisons between these new diamondoid-based drug candidates and standard drugs will be also presented.

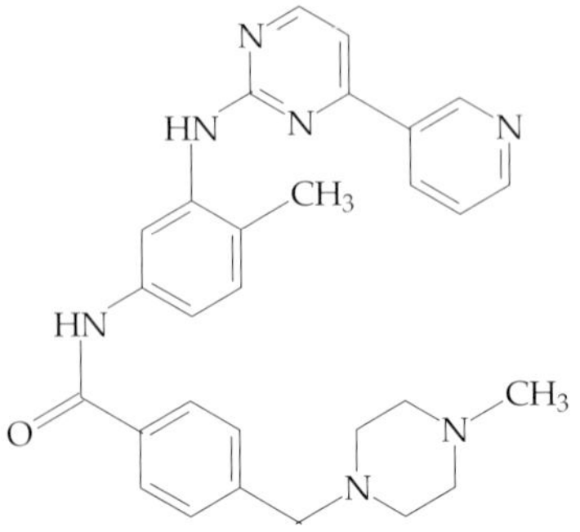

Imatinib

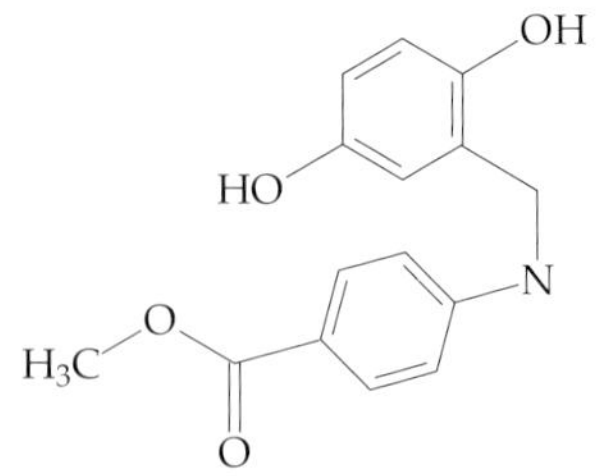

Tyrphostin AG957

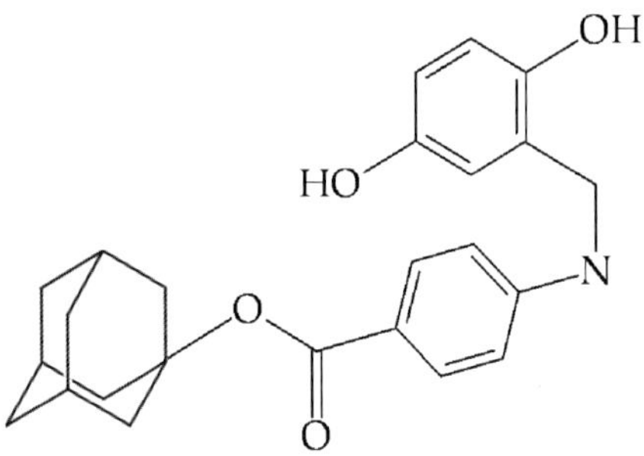

Adaphostin

Fig. 6.11. Anticancer drugs for targeted therapy (Svingen *et al.*, 2000).

6.3.1. *Adaphostin: An adamantane derivative for cancer chemotherapy*

The adamantyl ester of tyrphostin AG957, known as adaphostin (Fig. 6.11) has shown more potency than its parent drug (tyrphostin AG957) in inhibiting myelogenous leukemia K562 cell line proliferation, although it was less potent than AG957 as a kinase inhibitor *in vitro* (Svingen *et al.*, 2000).

Experiments have demonstrated that adaphostin also induces apoptosis in blasts from patients with acute myelogenous leukemia (AML) and lymphocytes from patients with chronic lymphoblastic leukemia (Chandra *et al.*, 2003; Shanafelt *et al.*, 2005). These observations suggested the possibility of a different cytotoxic mechanism that does not depend on the inhibition of BCR-ABL gene, a mutant fusion gene created after chromosomal traslocation by juxtapositioning the *Abl1* gene on chromosome 9 to a part of the *BCR* ("breakpoint cluster region") gene on chromosome 22. This is reciprocal translocation that creates an elongated chromosome 9, and a truncated chromosome 22, which is called the Philadelphia chromosome. This is a mutant fusion gene found in most patients with a chronic and/or acute leukemia and its mutation plays an important role in human leukemia (Witte, 1988).

An alternative mechanism of action for adaphostin involves production of intracellular peroxide. Adaphostin induces a relatively rapid rise in intracellular reactive oxygen species (ROS) in both Bcr-Abl genes, and in normal cells, as well (Chandra *et al.*, 2003). This unique mechanism of action motivated further studies on this adamantyl derivative in several imatinib (medicine for treatment of some leukemias)-resistant cell lines. Adaphostin induced cell death on imatinib-resistant CML (Chronic Myeloid Leukemia) and Ph+ ALL (Philadelphia-Positive Acute Lymphocytic Leukemia) cell lines, suggesting that ROS production is an efficient mechanism to overpower imatinib-resistance in these cell lines (Chandra *et al.*, 2006). Imatinib is an anticancer drug (see Fig. 6.11).

In addition to leukemia treatment, adaphostin has also been tested in solid tumor cells and exhibited anti-proliferative effects in the human prostate cancer cell line PC-3 by inducing a G1 phase cell cycle arrest (Mukhopadhyay *et al.*, 2006). These findings demonstrate that adaphostin may have a broader action against cancer than originally observed.

6.3.2. *Other promising adamantane-based anticancer drugs*

All-*trans*-retinoic acid (ATRA) and retinoid derivatives (Fig. 6.12) exert selective antiproliferative and cytodifferentiating effects that are mediated predominantly by specific nuclear receptors of the retinoic acid receptor (RARα, β, and γ) and retinoid X receptors (RXRα, β, and γ) families (Altucci and Gronenmeyer, 2001). These actions are of special interest in the search for new antineoplasic agents (Garattini and Terao, 2004).

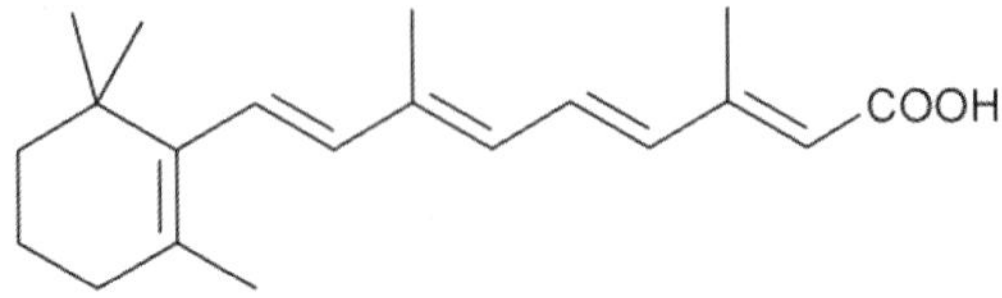

ATRA

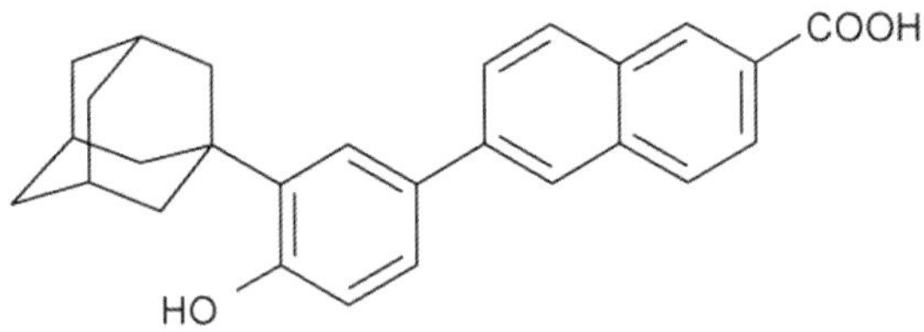

CD437

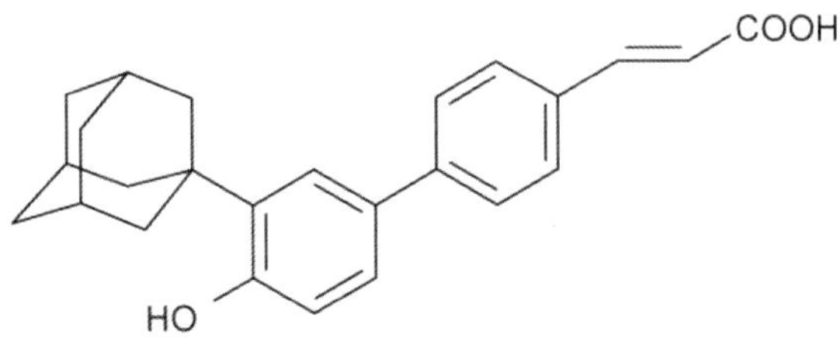

ST1926

Fig. 6.12. All-*trans* retinoic acid (ATRA), adamantyl retinoids CD437 and ST 1926, apoptosis-inductors in malignant cells (Di Francesco *et al.*, 2007).

ST1926 [(E)-3-(4-Hydroxy-3-adamantylbiphenyl-4-yl) acrylic acid] and CD437 [6-[3-(1-Adamantyl)-4-hydroxyphenyl]-2-naphthalene carboxylic acid] and ST1926 [(E)-3-(4-Hydroxy-3-adamantylbiphenyl-4-yl) acrylic acid], as reported in Fig. 6.12, are prototypes atypical synthetic retinoids with better RARγ-agonist characteristics than ATRA (Parrela *et al.*, 2006). ST1926 is orally administered (Di Francesco *et al.*, 2006) and it was undergoing phase I clinical trials in 2008 (Valli *et al.*, 2008). This retinoid derivative appears to have a mode of action different from retinoic acid. *In vitro* tests showed that ST1926 does not induce morphological differentiation in neuroblastoma cells; instead, it produces indirect DNA damage, cell cycle arrest in late S-G2 phases and apoptosis (Di Francesco *et al.*, 2006).

Good *in vivo* results were obtained in experimental combinatorial treatment (with a medicine of two or more active components) of human ovarian carcinoma models with ST1926 and cisplatin (a platinum-based chemotherapy drug and a DNA-damaging agent that is known to induce apoptosis). Tumor cells were inoculated in healthy mice either subcutaneously or by peritoneal passages. ST1926 was administered orally and cisplatin was delivered intravenously. The combination with cisplatin resulted in a substantial improvement of the ST1926 drug, with a more than additive effect in both cisplatin-sensitive and cisplatin-resistant tumors. While treatment with each agent alone produced a marginal activity ($\leq$50% of tumor volume inhibition, TVI, in comparison with control group), the combination resulted in a significant increase in antitumor activity (TVI in the range of 75–90%) (Pisano *et al.*, 2007). Positive results obtained with this model of combinatorial therapy may provide a rational basis for more effective treatments of human ovarian cancer.

6.4. Other Diamondoids-based Drugs

6.4.1. *Memantine as a neuroprotective agent for Alzheimer's disease*

Alzheimer's disease is a progressive brain disease that destroys memory and thinking skills. There is presently no complete cure for Alzheimer's disease, but nanotechnology approaches have a potential for early detection and possibly cure of the disease (Nazem and Mansoori, 2008). A diamondoid-based drug (memantine) slows down the progression of the Alzheimer's disease.

Memantine [1-amino-3, 5-dimethyladamantane] with Namenda as its commercial name (Fig. 6.13) is an FDA-approved drug for treatment of moderate to severe Alzheimer`s disease (USA, 2003). Memantine acts as a neuroprotective agent against excitotoxicity, an excessive exposure to the neurotransmitter glutamate or overstimulation of its membrane receptors, leading to neuronal injury or death. Excitotoxic neuronal cell death is mediated in part by overactivation of N-methyl-d-aspartate (NMDA)-type glutamate receptors. Nevertheless, NMDA receptor activity is also essential for normal neuronal function. This means that potential

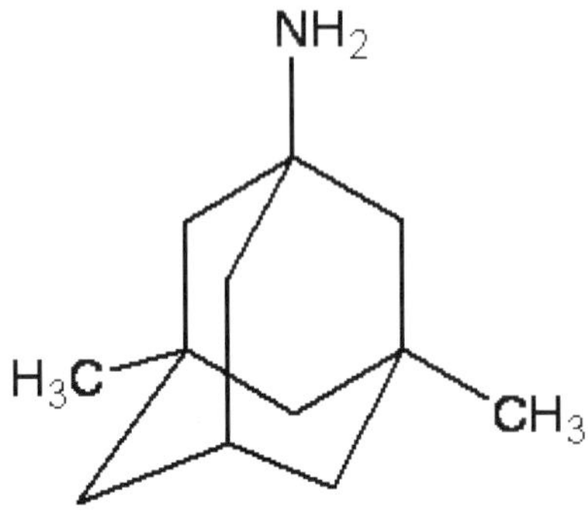

Fig. 6.13. Memantine (1-amino-3, 5- dimethyladamantane), commercially known as Namenda, a synthetic adamantane derivative with biological activity on the central nervous system.

neuroprotective agents that block virtually all NMDA receptor activity will very likely have unacceptable clinical side effects. Memantine preferentially blocks excessive NMDA receptor activity without disrupting normal activity (Lipton, 2004).

Although memantine is only approved for Alzheimer's disease, many studies evaluated different application for the treatment of diseases related to glutamatergic dysfunction. A recent systematic revision published by Zdanys and Tampi (2008) collected results from studies on treatment of major depression disorder; obsessive compulsive disorder; several substance abuse disorders (alcohol dependence, opioid dependence, cocaine abuse); Schizophrenia, bipolar disorder, binge-eating disorder, among others. Although interesting outcomes were achieved, the amount of data available for these off-label uses is still too limited to support routine use of memantine in the treatment of these and other psychiatric disorders.

6.4.2. *Tamorf and its antidotal effects*

An antidote is a substance that counteracts the effect of a poison. New antidotes candidates were reported in recent years which include diamondoid derivatives. One such antidote is tamorf ((1-[2-(2-thienyl)-2-adamantyl] morpholine) (Skare *et al.*, 2002; Lucic-Vrdoljak *et al.*, 2006; Radic *et al.*, 2006; Petek *et al.*, 2008) (Fig. 6.14) which has shown promising results, particularly in Soman poisoning treatment.

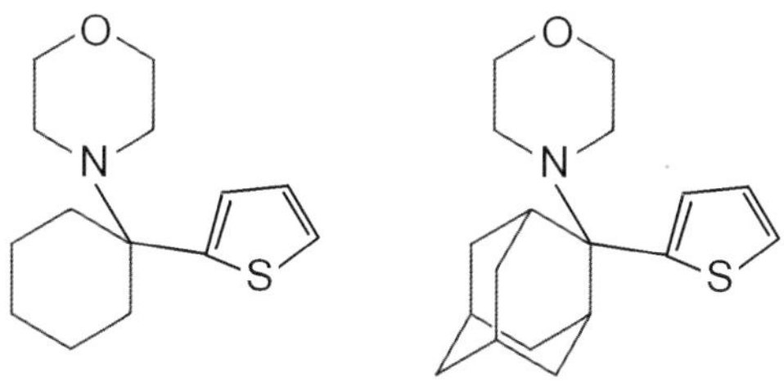

Fig. 6.14. Structure of tenocyclydine (TCP) (left) and its adamantyl-morpholine derivative (Tamorf) (right).

Tamorf was also used alone as Soman poisoning pretreatment, in 2.5 mg.kg^{-1} dose, or in combination with atropine (10.0 mg kg^{-1} dose) as a therapy in rats poisoned with sublethal doses of Soman (in 1/4 and 1/2 of LD_{50}) (Lucic-Vrdoljak *et al.*, 2006). The effect of tamorf was compared with physostigmine (0.1 mg.kg^{-1}), an efficient prophylactic drug in Soman poisoning (Philippens *et al.*, 2000).

Tamorf stopped Soman-induced seizures when administered 5 min's before poisoning and protected Acetylcholinesterase (AChE) activity in the brain. Nonetheless, carbamate physostigmine has a slightly better effectiveness than tamorf in the elimination of Soman-induced poisoning in rats (Lucic-Vrdoljak *et al.*, 2006).

The study also determined the possible genotoxic effects of tamorf and physostigmine, especially primary DNA damage using the alkaline comet assay (Singh *et al.*, 1988) on white blood cells, liver, and brain tissue after treatment with tamorf and carbamate physostigmine. No significant genotoxic effects to rats were detected in both tested doses. Thus, the pretreatment with diamondoid drug tamorf seems to be a good alternative for current pretreatment in Soman poisoning.

6.4.3. *Adamantane derivatives with hypoglycemic action*

6.4.3.1. *Diamondoids-based drugs for diabetes treatments*

Recently three diamondoid-based drugs related to incretin biology have appeared in the market for lowering blood glucose levels in type 2 diabetes. Their chemical structures are reported in Fig. 6.15.

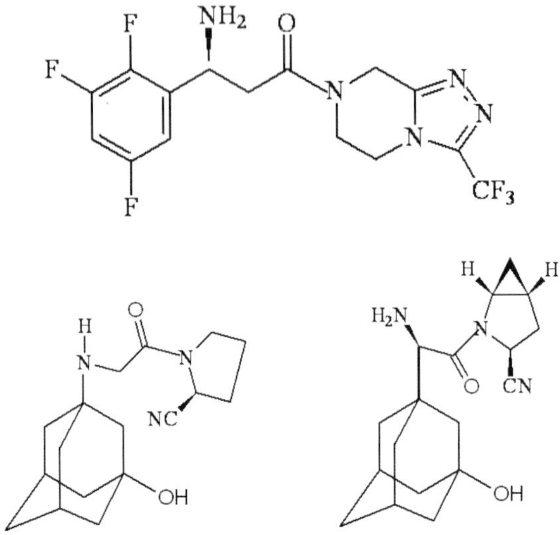

Fig. 6.15. Structure of dipeptidyl peptidase-4 (DPP-4) inhibitors, novel approved drugs for type 2 diabetes treatment. The top compound is Sitagliptin {(R)-4-oxo-4-[3-(trifluoromethyl)-5,6-dihydro[1,2,4]triazolo[4,3-a]pyrazin-7(8H)-yl]-1-(2,4,5-trifluorophenyl)butan-2-amine}. The bottom two compounds are diamondoid derivatives. In bottom left is vildagliptin [(S)-1-[N-(3-hydroxy-1-adamantyl)glycyl]pyrrolidine-2-carbonitrile] and the bottom right is [saxagliptin ((1S,3S,5S)-2-[(2S)-2-amino-2-(3-hydroxy-1-adamantyl)acetyl]-2-azabicyclo[3.1.0]hexane-3-carbonitrile)].

They are: (1) Saxagliptin (Onglyza®), manufactured by Bristol-Myers Squibb Co., Princeton, N.J., and marketed by Bristol-Myers and AstraZeneca Pharmaceuticals LP, Wilmington, Del (USA 2009); (2). Vildagliptin (Galvus®), Novartis Pharma AG, Basel, Switzerland; (3). Eucreas®, a combination of vildagliptin and metformin (a sulphonylurea — an older anti-diabetic drug). The latter two drugs have received approval from the European Medicines Agency (Emea) in 2007 (Emea, 2009a, 2009b).

(1). Saxagliptin and vildagliptin are orally administered diabetic drugs (DPP-4 inhibitors). Sitagliptin was the first diabetic drug to achieve FDA regulatory approval in the US, and paved the way to this new class of anti-diabetic drugs. These diabetic drugs (known as DDP-IV inhibitors) increase endogenous blood levels of active incretins, thus leading to prolonged incretin action. They reduce HbA1c levels (measures of the amount of glycated hemoglobin in blood) with few adverse events and no weight gain (Henness and Keam, 2006). These new anti-diabetic agents, similarly to incretin mimetics also expand β-cell (in the pancreas making and releasing insulin) mass in mice (Xu *et al.*, 1999; Rolin *et al.*, 2002; Mu *et al.*, 2006).

Saxagliptin {(1S,3S,5S)-2-[(2S)-2-amino-2-(3-hydroxy-1-adamantyl) acetyl]-2-azabicyclo[3.1.0]hexane-3-carbonitrile} was disclosed by Augery and co-workers in 2005 as a highly efficacious, stable, and long-acting DPP-IV inhibitor, obtained within a series of *beta*-quaternary cycloalkylglycine amino acid residues incorporated into a 4,5-methanoprolinenitrile scaffold (2-ciano pirrolidinyl group) (Augeri *et al.*, 2005). The 3-hydroxy-adamantyl derivative saxagliptin exhibited *in vivo* potency and duration of action superior to that of any compound from this series and was chosen for the development for the treatment of type 2 diabetes.

(2). Vildagliptin {(S)-1-[N-(3-hydroxy-1-adamantyl)glycyl]pyrrolidine-2-carbonitrile} is a potent, selective DDP-IV (Dipeptidyl peptidase IV) inhibitor (Villhauer *et al.*, 2003). This diamondoid-derivative drug also reversibly inhibits DPP-4 enzyme through covalent bond (Demuth *et al.*, 2005). The majority of adverse reactions associated with its use are mild and transient, nasopharyngitis, dizziness and headaches, and, as a result, do not require treatment discontinuations. No association was found between adverse reactions and age, ethnicity, duration of exposure or daily dose. Vildagliptin improved glycemic control when given as monotherapy or when used in combination with metformin and a thiazolidinedione (thiazolidine-2,4-dione, 2,4-dioxothiazolidine), as measured by clinically relevant reductions in HbA1c levels (nearly equate to glucose levels) (Henness and Keam, 2006). As of February 2011, regulatory approval of vildagliptin in the US awaited the submission of additional data. If approved, vildagliptin, marketed as Galvus, joins Januvia and Onglyza as the third member of the DDP-IV inhibitors for type 2 diabetes treatment in the US.

(3). Eucreas contain two active ingredients, metformin and vildagliptin. Metformin is a type of antidiabetic medicine known as a biguanide. Biguanides have many good uses, but the most important is that it will help the liver stop its production of the glucose in the blood. Eucreas is licensed for use in people with type 2 diabetes whose blood sugar is not controlled by the maximum tolerated dose of metformin alone or who are already taking metformin and vildagliptin as separate tablets.

6.5. Diamodoids Derivatives in Drug Delivery and Drug Targeting

Adamantane has been the subject of many research projects in the field of pharmacophore-based drug design. Although, its application to drug delivery and drug targeting systems is a new matter of considerable importance.

In Chapter 4, we presented cyclodextrin-adamantane inclusion compounds as components of supramolecular gene delivery systems (Sec. 4.4.1). In the present section, adamantyl unique structure, with its highly unusual physical and chemical properties, will be presented as backbone structures or covalently attached groups to drugs. This could have many applications in drug design and drug delivery. For example, attachment of diamondoid groups to drugs with low hydrophobicity could lead to a substantial increase of drug solubility in lipidic membranes due to diamondoid cage's high lipophilicity. Such alteration may increase drug cell uptake (Ramezani and Mansoori, 2007).

When a drug does not accumulate in its exact sight of action, it would not be able to produce the intended therapeutic effects even by using it at a high concentration. The nanometer size of diamondoid molecules makes them possible to enter living cells while carrying along the drug into the cells. Common particles with less than 100 nm size can enter the cells (Geiser *et al.*, 2005), whereas diamondoids are generally smaller than 10 nm in size. Furthermore, due to their high stability and solubility in blood plasma they are expected to play an important part in the future drug delivery systems (Ramezani and Mansoori, 2007). Furthermore, diamondoids have the potential to be utilized in the rational design of multifunctional drug systems and drug carriers.

The site of action for memantine is the CNS and it has higher CNS affinity than amantadine (Hesselink *et al.*, 1999). Nevertheless, both amantandine and rimantadine can also penetrate into the CNS and cause some adverse effects (see Sec. 6.2.1.1). These observations lead one to the conclusion that the adamantane nucleus, which is present in all of the aforesaid drugs, might be responsible for their penetration into the blood-brain barrier (BBB) and their accumulation in the CNS. In other words, the adamantane nucleus is likely to possess a so-called intimate CNS tropism (affinity to and tunneling through CNS), because of its hydrophobicity (Tsuzuki *et al.*, 1994).

Furthermore, because the half-lives of amantadine and rimantadine in the blood stream are long (12–18 hours for amantadine and 24–36 hours for rimantadine in young adults), utilization of adamantane derivative carriers can probably prolong drug presence time in blood circulation.

Extensive investigations for synthesis of new adamantane derivatives with better therapeutic actions and less adverse effects have been performed. It has been proven that adamantylamino-pyrimidines and-pyridines are strong stimulants of tumor necrosis factor-α (TNF-α) (Kazimierczuk *et al.*, 2001). TFN is referred to a substance that can improve the body's natural response to cancer by killing the cancer cells.

Some derivatives of adamantane with antagonist or agonist effects have also been synthesized. For instance, monocationic and dicationic adamantane derivatives block the following important recptors function: AMPA (α -amino-3-hydroxy-5-methylisoxazole-4-propionic acid) recptor, NMDA (N-methyl-D-aspartate) receptor (Samoilova *et al.*, 1999; Bolshakov *et al.*, 2000; Buldakova *et al.*, 1999) and 5-HT3 (5-hydroxytryptamine) and 5-HT4 receptors (Rammes *et al.*, 2001; Flynn *et al.*, 1992). The monocationic and dicationic adamantane derivatives have been employed to investigate the topography of the channel binding sites of AMPA and NMDA receptors (Bolshakov *et al.*, 2000). A dicationic adamantane derivative has been exploited as a selective and specific marker of native AMPA receptor assembly to determine the distribution of AMPA receptor subtypes among populations of the rat brain cells (Samoilova *et al.*, 1999; Buldakova *et al.*, 1999). Other examples include antagonism of 5-HT3 and agonism of 5-HT4 receptors by aza(nor)adamantanes (Flynn *et al.*, 1992), P2X7 receptors antagonism by adamantane amides (Baxter *et al.*, 2003), antagonism of voltage gated calcium channels and probably activation of γ-aminobutyric acid (GABA) receptors by an adamantane amine derivative that results in its anticonvulsive and antinociceptive actions (Zoidis *et al.*, 2005), and inhibition of glycosylceramidase enzyme and glycolipid bio-synthesis by a deoxynojirimycin-bearing adamantane derivative leading to the strong anti-inflammatory and immunosuppressive activities (Shen *et al.*, 2004).

Attaching some short peptidic sequences to adamantane makes it possible to design novel antagonists. The bradykinin antagonist, which is used as an anti-cancer agent is an example. The adamantane-based peptidic bradykinin analog was utilized in structure-activity relationship (SAR) studies on the bradykinin receptors and showed a potent activity in inhibition of bradykinin-induced cytokines release and stimulation of histamine release (Reissmann *et al.*, 2000).

In an attempt to design the β-turn-peptide-mimics, aspartic acid and lysine were attached to each amine group of 1, 3-diaminoadamantane through amide bonds. The term β-turn refers to a peptide chain which forms a tight loop such that the carbonyl oxygen of one residue is hydrogen bonded with the amide nitrogen of a residue located three positions down the chain. This β-turn-peptide-mimic displayed some degree of fibrinogen-GPIIB/IIIA antagonism (Hoekstra *et al.*, 1994).

Furthermore, the six primary and four secondary linking-site possibilities of adamantane (see Fig. 4.1) and larger number of linking sites on higher diamondoids makes it possible to introduce several functional groups consisting of drug, targeting part, linkers, etc., to them without undesirable interactions between them. In fact, adamantane derivatives can act as a central core for such drug systems.

As an example, short peptidic sequences can be bound to adamantane and provide a binding site for connection of macromolecular drugs (like proteins, nucleic acids, lipids, polysaccharides, etc.) as well as small molecules. Hence, short amino acid sequences can have linker roles, which are capable of drug release in the target site.

There are some successful examples of adamantyl (adamantane core) application for delivery of drugs into the brain (Tsuzuki *et al.*, 1994). For this purpose, 1-adamantyl group was attached to several AZT (azidothymidine) drugs via an ester spacer and these prodrugs could pass through the BBB easily to reach the brain. The prodrugs concentration after using such lipophilized prodrugs was measured in the brain tissue and showed an increase of 7–18 fold in comparison with AZT drugs without adamantane vector. The ester spacer link is resistant to the plasma esterases, but it is cleaved after passing through the BBB by brain tissue esterases.

Overall, it is important to note that adamantane is now considered a successful brain-directing drug carrier. Another example of adamantane

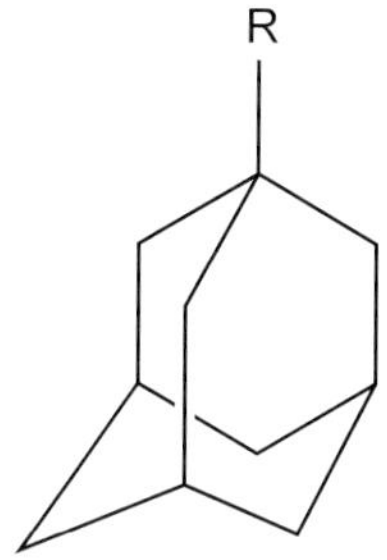

(I) R = NH_2-Tyr-(D-Ala)-Gly-Phe-Leu-CO-O-
(II) R = NH_2-Tyr-(D-Ala)-Gly-Phe-Leu-CO-NH

Fig. 6.16. The adamantane-conjugated [D-Ala2] Leu-enkephalin prodrugs (Tsuzuki *et al.*, 1991).

utilization for poorly-absorbed-drug delivery to the brain is the conjugation of [D-Ala2] Leu-enkephalin derivatives with a 1-adamantyl (Tsuzuki *et al.*, 1991). The anti-nociceptive effect of Leu-enkephalin disappears when it is administered peripherally since proteolytic enzymes would decompose it. As a result it cannot penetrate into the CNS.

It is feasible to conjugate the [DAla2] Leu-enkephalin with a 1-adamantane vector via an ester, amide or a carbamate linkage in order to enhance the drug lipophilicity and thus facilitate its delivery across the BBB to the brain (Tsuzuki *et al.*, 1991). The adamantane-conjugated [D-Ala2]Leu-enkephalin prodrugs (Fig. 6.16) are highly lipophilic and show a significant anti-nociceptive effect because of their ability to cross the BBB (Kazimierczuk *et al.*, 2001). These results suggest that adamantyl is a promising brain-directing drug vector providing a high lipophilicity, low toxicity and high BBB permeability for sensitive and poorly absorbed drugs (Tsuzuki *et al.*, 1991; Kitagawa *et al.*, 1997).

Adamantane has also been used for lipidic nucleic acid synthesis as a hydrophobic group. Two major problems in gene delivery are the low uptake of nucleic acids by cells and their instability in blood medium. An increase in lipophilicity using hydrophobic groups would probably lead to improvement of uptake and an increase in intracellular concentration of nucleic acids. In this case, an amide linker is used to attach the adamantane derivatives to a nucleic acid sequence. Such a nucleic acid derivatization has no significant effect on hybridization with the target RNA. Lipidic nucleic acids possessing adamantane derivative groups can be also exploited for gene delivery (Manoharan *et al.*, 1995).

A polyamine adamantane derivative presenting a special affinity for binding to the major grooves in double-stranded DNA was prepared by Lomadze and co-workers (2002). It should be pointed out that most of the polyamines have affinity for binding to doublestranded RNA, thus making RNA stabilized. DNA selectivity is one of the outstanding features of this ligand. This positive nitrogen-bearing ligand has a tendency to establish hydrophobic interactions with deeper DNA grooves due to its size and steric properties. Such an exclusive behavior occurs because the ligand fits better in the DNA major grooves. This bulky ligand size is the same as zinc-finger protein, which also binds to DNA major grooves (Lomadze and Schneider, 2002).

Higher affinity of adamantane-bearing ligand to DNA, instead of RNA, probably arises from the presence of adamantane and leads to DNA stabilization. This property may be exploited for using such ligands as stabilizing carriers in gene delivery. Adamantyl increases ligand lipophilicity as well as DNA stabilization. Furthermore, ligand/groove size-based targeting may also be possible with less specificity by changing the bulk and conformation of the ligand (Lomadze and Schneider, 2002).

Polymers conjugated with 1-adamantyl moieties as lipophilic pendent groups can be utilized to design nanoparticulate drug delivery systems. Polymer 1 in Fig. 6.17, which is synthesized by homopolymerization of ethyladamantyl malolactonate, can be employed as a highly hydrophobic block to construct polymeric drug carries.

In contrast, polymer 2 (Fig. 6.17), which is synthesized by copolymerization of polymer 1 with benzyl malolactonate, is water-soluble. Its lateral carboxylic acid functions can be used to bind it to biologically active molecules in order to achieve targeting characteristics. The above-mentioned two polymers represent possibilities of producing adamantane-based pH-dependant hydrogels and intelligent polymeric systems (Moine *et al.*, 1997).

6.6. Concluding Remarks

In this chapter, we presented the biomedical applications of diamondoids and derivatives. Among all diamondoids, adamantane and its derivatives

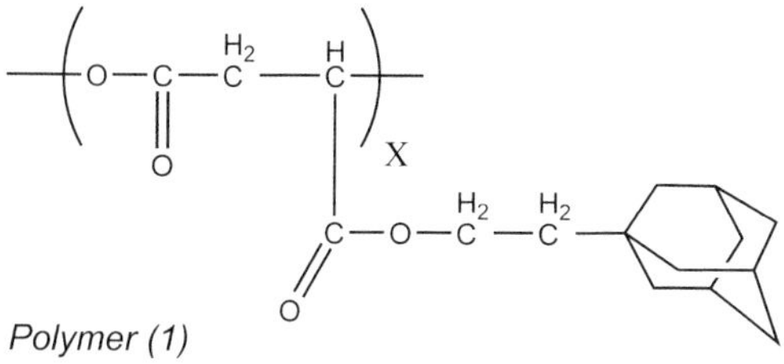

Polymer (1)

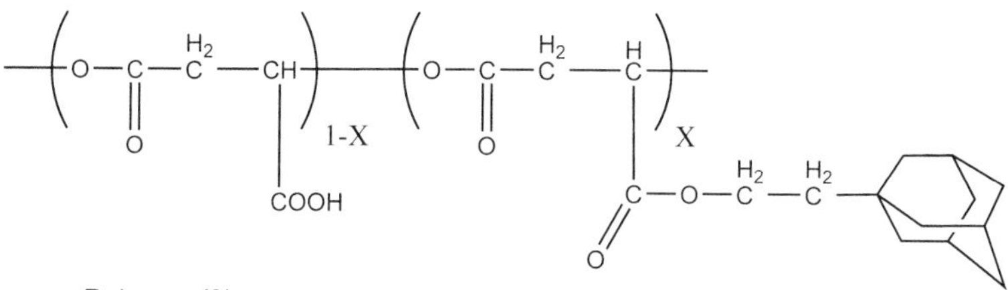

Polymer (2)

Fig. 6.17. Polymer 1 {poly (ethyladamantyl β-malate)} is hydrophobic and polymer 2 {poly (β-malic acid-co-ethyladamantyl β-malate} is hydrophilic. Both of these polymers are used as carriers for different drugs (Moine *et al.*, 1997).

have been most widely used as the molecules for fighting infectious diseases including influenza (by for example amantadine and rimantadine) and other viral diseases, for treatment of bacterial infectious diseases such as tuberculosis (by, for example SQ 109). for parasitic infectious disease treatment such as malaria, trypanosomiasis, schistosomiasis, and fascioliasis (by for example OZ78 and OZ288 compounds).

Fighting cancer with diamondoids derivatives is another important application of these interesting compounds. Among such derivatives adaphostin, a retinoid derivative, is used for chemotherapy and exert selective antiproliferative and cytodifferentiating effects. These actions are of special interest in the search for new antineoplasic agents.

Neuroprotective effect of memantine and some other diamondoids derivatives for Alzheimer's disease is another important set of biomedical applications of these compounds.

Design of new antidotes to counteract the effect of a poison by diamondoid derivatives is another important application of these substances.

Table 6.10. Highlights on other parasitic diseases and antiparasitic action of adamantyl 1,2,4-trioxanes (Cmpd ≡ Compound).

Parasitic Disease (Agent)	Country or Region of higher incidence	People Infected/ at risk	Standard drug for treatment	Adamantyl drug candidate and main antiparasitic action
Clonorchiasis (liver fluke, i.e., *Clonorchis sinensis*)	People's Republic of China, Republic of Korea, Taiwan, and Vietnam.	35 million/601 million (Lun *et al.*, 2005)	Praziquantel (WHO, 1995)	Cmpd 47a — (300 mg/kg) — 98.5% of worm burden reduction in rats infected with adult *Clonorchis sinensis* (Keiser *et al.*, 2007)
Facioliasis (liver flukes, i.e., *Fasciola hepatica* and *Fasciola gigantica.*	South America, Cuba, Western Europe, Egypt and the Islamic Republic of Iran (Keiser and Utzinger, 2005).	17 million infected, 91 million at risk (WHO, 1995).	Triclabendazole (Keiser *et al.*, 2005)	Cmpd 47a — (100 mg/kg) — 100% of worm burden reduction in rats infected with adult *Fasciola hepatica* (Keiser *et al.*, 2006a)
Schistosomiasis (Schistosoma haematobium, Schistosoma japonicum, and Schistosoma mansoni)	Africa, Middle East and East Asia, South America, Philippines	200 million/779 million (Steinmann *et al.*, 2006).	Praziquantel (Cioli and Pica-Mattoccia, 2003).	Cmpd 47a — (200 mg/kg) — 94.2% of worm burden reduction in Hamsters infected with adult *Schistosoma japonicum* (Keiser *et al.*, 2006b) Cmpd 47b — (200 mg/kg) — 95.4% of worm burden reduction in mice infected with juvenile *Schistosoma mansoni* (Keiser *et al.*, 2006b).

A number of adamantane derivatives possess hypoglycemic action. And they are used as drugs for diabetes treatments.

Diamondoids derivatives in drug delivery and drug targeting, especially for the fact that due to their lipophilicity/hydrophobicity they can easily pass through BBB. Adamantane nucleus, which is present in all adamantane derivatives, might be responsible for their ability to penetrate into the BBB and their accumulation in the CNS. Their applications for pharmacophore-based drug design, for drug delivery and drug targeting systems is a new matter of considerable importance.

In conclusion, diamondoid derivatives are important compounds able to render diverse, strong, and sometimes unique biological properties to a multitude of drug classes. As many of these molecules are still in initial development phases, it will not be; surprising if a considerable amount of new cage compounds start to be a good addition to our medical supplies in a near future.

Bibliography

Abed, Y., Goyette, N. and Boivin, G. (2005). *Antimicrob. Agents Chemother.*, **49**: 556–559.

Altucci, L. and Gronemeyer, H. (2001). *Nat. Rev. Cancer*, **1**: 181–193.

Arora, S. K. Sinha, N., Jain, S., Upadhayaya, R. S., Jana, G., Ajay, S., Sinha, R. K. (2004). Preparation of pyrrole derivatives as antimycobacterial compounds. Patent WO2004026828 to Lupin Limited, India.

Augeri, D. J., Robl, J. A., Betebenner, D. A., Magnin, D. R., Khanna, A., Robertson, J. G., Wang, A., Simpkins, L. M., Taunk, P., Huang, Q., Han, S.-P., Abboa-Offei, B., Cap, M., Xin, L., Tao, L., Tozzo, E., Welzel, G. E., Egan, D. M., Marcinkeviciene, J., Shu Y. Chang, S. Y., Biller, S. A., Kirby, M. S., Parker, R. A. and Hamann, L. G. (2005). *J. Med. Chem.*, **48**: 5025–5037.

Balzarini, J., Orzeszko, B., Maurin, J. K., Orzeszko, A. (2007). *Eur. J. Med. Chem.*, **42**: 993–1003.

Barr, I. G., Hurt, A. C., Deed, N., Iannello, P., Tomasov, C., Komadina, N. (2007). *Antivir. Res.*, **75**: 173–176.

Baxter, A., Bent, J., Bowers, K., Braddock, M., Brough, S., Fagura, M., Lawson, M., McInally, T., Mortimore, M. and Robertson, M. (2003). *Bioorg. Med. Chem. Lett.*, **13**: 4047–4050.

Belshe, R. B., Smith, M. H., Hall, C. B., Betts, R. and Hay, A. J. (1988). *J. Virol.*, **62**: 1508–1512.

Bolshakov, K. V., Tikhonov, D. B., Gmiro, V. E. and Magazanik, L. G. (2000). *Neuroscience Lett.*, **291**: 101–104.

Boshoff, H. I. M., Myers, T. G., Copp, B. R., McNeil, M. R., Wilson, M. A. and Barry III, C. E. (2004). *J. Biol. Chem.*, **279**: 40174–40184 (and Supplemental Material).

Bright, R. A., Medina, M.-J., Xu, X., Perez-Oronoz, G., Wallis, T. R., Davis, X. M., Povinelli, L., Cox, N. J. and Klimov, A. I. (2005). *Lancet,* **366**: 1175–1181.

Bright, R. A., Shay, D. K., Shu, B., Cox, N. J., Klimov, A. I. (2006). *JAMA,* **295**: 891–894.

Buldakova, S. L., Vorobjev, V. S., Sharonova, I. N., Samoilova, M. V., Magazanik, L. G. (1999). *Brain Res.,* **846**: 52–58.

Burnett, J. C., Schmidt, J. J., Stafford, R. G., Panchal, R. G., Nguyen,T. L., Hermone, A. R., Vennerstrom, J. L., McGrath, C. F., Lane, D. J., Sausville, E. A., Zaharevitz, D. W., Gussio, R., Bavari, S. (2003). *Biochem. Biophys. Res. Commun.,* **310**: 84–93.

Burnett, J. C., Opsenica, D., Sriraghavan, K., Panchal, R. K., Ruthel, G., Hermone, A. R., Nguyen, T. L., Kenny, T. A., Lane, D. J., McGrath, C. F., Schmidt, J. J., Vennerstrom, J. L., Gussio, R., Bogdan, A., Solaja, B. A., Bavari, S. A. (2007). *J. Med. Chem.,* **50**: 2127–2136.

Caminero, J. A. (2008). *Int. J. Tuberc. Lung. Dis.,* **12**: 869–877.

CDC. (2008a). Infectious Disease Information: Emerging Infectious Diseases. Available at www.cdc.gov/ncidod/diseases/eid/disease_sites.htm. Accessed in 10/02/11.

CDC. (2008b). Antiviral Drugs for Seasonal Flu. Available at www.cdc.gov/flu/about/qa/antiviral.htm. Access in 11/13/2008.

Chandra, J., Hackbarth, J., Le, S., Loegering, D., Bone, N., Bruzek, L. M., Narayanan, V. L., Adjei, A. A. Kay, N. E., Tefferi, A., Karp, J., Sausville, E. A. and Kaufmann, S. H. (2003). *Blood,* **102**: 4512–4519.

Chandra, J., Tracy, J., Loegering, D., Flatten, K., Verstovsek, S., Beran, M., Gorre, M., Estrov, Z., Donato, N., Talpaz, M., Sawyers, C., Bhalla, K., Karp, J., Sausville, E. and Kaufmann, S. H. (2006). *Blood*, **107**: 2501–2506.

Chen, P., Gearhart, J., Protopopova, M., Einck, L. and Nacy, C. A. (2006). *J. Antimicrob. Chemother.,* **58**: 332–337.

Cheung, C.-L., Rayner, J. M., Smith, G. J. D., Wang, P., Naipospos, T. S. P., Zhang, J., Yuen, K.-Y., Webster, R. G., Peiris, J. S. M., Guan, Y. and Chen, H. (2006). *J. Infect Dis.*, **193**: 1626–1629.

Ciampor, F., Bayley, P. M., Nermut, M. V., Hirst, E. M. A., Sugrue, R. J., Hay, A. J. (1992). *Virology,* **188**: 14–24.

Cioli, D. and L. Pica-Mattoccia (2003). *Parasitol. Res.,* 90: S3–S9.

Collins, F. H. and Paskewitz, S. M. (1995). *Annu. Rev. Entomol.,* **40**: 195–219.

Creek, D. J., Charman, W. N., Chiu, F. C. K., Prankerd, R. J., McCullough, K. J., Dong, Y., Vennerstrom, J. L., Charman, S. A. (2007). *J. Pharm. Sci.,* **96**: 2945–2956.

Cyranoski, D. (2005). *Nature,* **435**: 1009–1009.

Davies, W. L., Grunert, R. R., Haff, R. F., McGahen, J. W., Neumayer, E. M., Paulshock, M., Watts, J. C., Wood, T. R., Hermann, E. C. and Hoffmann, C. E. (1964). *Science,* **144**: 862–863.

Degelau, J., Somani, S., Cooper, S. L., Irvine, P. W. (1990). *J. Am. Geriatr. Soc.,* **38**: 428–432.

Demuth, H.-U., McIntosh, C. H. S. and Pederson, R. A. (2005). *Biochim. Biophys. Acta — Prot. Proteom.,* **1751**: 33–44.

Deyde, V. M., Xu, X., Bright, R. A., Shaw, M., Smith, C. B., Zhang, Y., Shu, Y., Gubareva, L. V., Cox, N. J. and Klimov, A. I. (2007). *J. Infect. Dis.,* **196**: 249–257.

Di Francesco, A. M., Meco, D., Torella, A. R., Barone, G., D'Incalci, M., Pisano, C., Carminati, P. and Riccardi, R. (2007). *Biochem. Pharmacol.,* **73**: 643–655.

El-Emam, A. A., Al-Deeb, O. A., Al-Omara, M. and Lehmann, J. (2004). *Bioorg. Med. Chem.,* **12**: 5107–5113.

El-Sherbeny, M. A. (2000). *Arch. Pharm. Pharm. Med. Chem.,* **333**: 323–328.

Emea. (2009a). European Medicines Agency. EPARs for authorised medicinal products for human use. Galvus. Available at www.emea.europa.eu/humandocs/Humans/EPAR/galvus/galvus.htm. Acessed on 02/05/2011.

Emea. (2009b). European Medicines Agency. EPARs for authorised medicinal products for human use. Eucreas. Available at www.emea.europa.eu/humandocs/Humans/EPAR/eucreas/eucreas.htm. Acessed on 03/05/2011.

Englund J. A., Champlin, R. E., Wyde P. R., Kantarjian, H., Atmar, R. L., Tarrand, J., Yousuf, H., Regnery, H., Klimov, A. I., Cox, N. J. and Whimbey, E., (1998). *Clin Infect Dis.*, **26**: 1418–1424.

Flynn, D. L., Becker, D. P., Spangler, D. P., Nosal, R., Gullikson, G. W., Moummi, C. and Yang, D.-C. (1992). *Bioorg. Med. Chem. Lett.,* **2**: 1613–1618.

Garattini, E. and Terao, M. (2004). *J. Chemother.,* **16**: 70–73.

Gerzon, R., Krumkalns, E. V., Brindle, R. L., Marshall, F. J., Root, M. A. (1963). *J. Med. Chem.,* **6:** 760–763.

Hay, A. J., Wolstenholme, A. J., Skehel, J. J. and Smith, M. H. (1985). *EMBO J.,* **4:** 3021–3024.

Hayden, F. G., Belshe, R. B., Clover, R. D., Hay, A. J., Oakes, M. G., Soo, W. (1989). *N. Engl. J. Med.,* **321**: 1696–1702.

Hayden, F. G., Gwaltney Jr., J. M., DeCastle, R. v., Adams, K. F. and Giordani, B. (1981). *Antimicrob. Agents Chemotherap.,* **19**: 226–233.

Hayden, F. G., Treanor, J. J., Fritz, R. S., Lobo, M., Betts, R. F., Miller, M., Kinnersley, N., Mills, R. G., Ward, P., Straus, S. E. (1999). *J. Am. Med. Assoc.,* **282**: 1240–1246.

He, G., Qiao, J., Dong, C., He, C., Zhao, L. and Tian, Y. (2008). *Antivir. Res.,* **77**: 72–76.

Henness, S. and Keam, S. J. (2006). *Drugs,* **66**: 1989–2001.

Hesselink, M. B., De Boer, B. G., Breimer, D. D. and Danysz, W. (1999). *Pharm. Res.,* **16**: 637–642.

Hoekstra, W. J., Press, J. B., Bonner, M. P., Andrade — Gordon, P., Keane, P. M., Durkin, K. A., Liotta, D. C., Mayo, K. H. (1994). *Bioorg. Med. Chem. Lett.,* **4**: 1361–1364.

Hoffman, H. E., Gaylord, J. C., Blasecki, J. W., Shalaby, L. M. and Whitner Jr., C.W. (1988). *Antimicrob. Agents Chemother,* **32**: 1699–1704.

Humeau, Y., Doussau, F., Grant, N. J. and Poulain, B. (2000). *Biochimie,* **82**: 427–446.

Hurt, A. C., Selleck, P., Komadina, N., Shawa, R., Brown, L., Barr, I. G. (2007). *Antivir. Res.,* **73**: 228–231.

Ilyushina, N. A., Govorkova, E. A., Webster, R. G. (2005). *Virology,* **341**: 102–106.

Ilyushina, N. A., Hoffmann, E., Salomon, R., Webster, R. G. and Govorkova, E. A. (2007). *Antivir. Therap.,* **12**: 363–370.

Jia, L., Noker, P. E., Coward, L., Gorman, G. S., Protopopova, M. and Tomaszewski, J. E. (2006). *Brit. J. Pharm.,* **147**: 476–485.

Jia, L., Tomaszewski, J. E., Hanrahan, C., Coward, L., Noker, P., Gorman, G., Nikonenko, B., Protopopova, M. (2005). *British J. Pharm.,* **144**: 80–87.

Kadi, A. A., El-Brollosy, N. R., Al-Deeb, O. A., Habib, E. E. Ibrahim, T. M., El-Emam, A. A. (2007). Synthesis, antimicrobial and anti-inflammatory activities of novel 2-(1-adamantyl)-5-substituted-1,3,4 — oxadiazoles and 2-(1-adamantylamino)-5-substituted-1, 3, 4-thiadiazoles. *Eur. J. Med. Chem.* **42**: 235–242.

Kazimierczuk, Z., Gorska, A., Switaj, T. and Lasek, W. (2001). Adamantylaminopyrimidines and -pyridines are potent inducers of tumor necrosis factor-*alpha*. *Bioorg. Med. Chem. Lett.*, **11**: 1197–1200.

Keiser J., Engels D., Buscher G. and Utzinger, J. (2005). *Expert Opin. Invest. Drugs.*, **14**: 1513–1526.

Keiser, J. and Utzinger, J. (2005). *Emerg. Infect. Dis.*, **11**: 1507–1514.

Keiser, J., Utzinger, J., Tanner, M., Dong, Y. and Vennerstrom, J. L. (2006b). *J. Antimicrob. Chemother.*, **58**: 1193–1197.

Keiser, J., Xiao, S. H. and Tanner, M. and Utzinger, J. (2006a). *J Antimicrob. Chemother.*, **57**: 1139–1145.

Keiser, J., Xiao, S.-H., Dong, Y., Utzinger J. and Vennerstrom, J. L. (2007). *J. Parasitol.*, **93**: 1208–1213.

Kelly, J. M., Quack, G. and Miles, M. M. (2001). *Antimicrob. Agents Chemother.*, **45**: 1360–1366.

Kesel, A. J. (2003). *Bioorg. Med. Chem.*, **11**: 4599–4613.

Kesel. A. J. (2005). *Curr. Med. Chem.*, **12:** 2095–2162.

Kitagawa, K., Mizobuchi, N., Hama, T., Hibi, T., Konishi, R. and Futaki, S. (1997). *Chem. Pharm. Bull.*, (Tokyo), **45**: 1782–1787.

Kolocouris, N., Foscolos, G. B., Kolocouris, A., Marakos, P., Pouli, N., Fytas, G., Ikeda, S., De Clercq, E. (1994). *J. Med. Chem.*, **37**: 2896–2902.

Kolocouris, N., Kolocouris, A., Foscolos, G. B., Fytas, G., Neyts, J., Padalko, E., Balzarini, J., Snoeck, R., Andrei, G. and De Clercq, E. (1996). *J. Med. Chem.*, **39**: 3307–3318.

Kolocouris, N., Zoidis, G., Foscolos, G. B., Fytas, G., Prathalingham, S. R., Kelly, J. M., Naesens L. and De Clercq, E. (2007). *Bioorg. Med. Chem. Let.*, **17**: 4358–4362.

Kreidenweiss, A., Mordmuller, B., Krishna, S. and Kremsner, P. G. (2006). *Antimicrob. Agents Chemother*. **50**: 1535–1537.

Lamoureux, G. and Artavia, G. (2010). *Current Medicinal Chemistr.*, **17**: 2967–2978.

Lee, R. E., Protopopova, M., Crooks, E., Slayden, R. A., Terrot, M. and Barry, C. E. (2003). *J. Comb. Chem.*, **5**: 172–187.

Lipton, S. A. (2004). *J. Alzheimer Dis.*, **6**: S61–S74.

Lomadze, N. and Schneider, H. J. (2002). *Tet. Lett.*, **43**: 4403–4405.

Lucic-Vrdoljak, A., Radic, B., Garaj-Vrhovac, V., Kopjar, N. and Zlender. V. (2006). *J. Appl. Toxicol.*, **26**: 56–63.

Ludwig, R. E. (1960). Chemical Process. US Patent n. 2,937,211.

Lun, Z. R., Gasser, R. B., Lai, D. H., Li, A. X., Zhu, X. Q., Yu X. B., and Fang, Y. Y. (2005). *Lancet Infect. Dis.*, **5**: 31–41.

Maerki, S., Brun, R., Charman, S. A., Dorn, A., Matile, H. and Wittlin. S. (2006). *J. Antimicrob. Chemother.*, **58**: 52–58.

Mann, S. (2008). Will this save millions of lives? *Age* **Jul.**: 1 — 1. Available at www.theage.com.au/national/will-this-save-millions-of-lives-20080711–3dsd.html?page=-1 . Accessed on 02/25/2011.

Manoharan, M., Tivel, K. L. and Cook, P. D. (1995). *Tetrahedron Lett.*, **36**: 3651–3654.

MMV. (2009a). Medicine for Malaria Venture. Available at www.mmv.org/rubrique.php3?id_rubrique=15 . Acessed on 02–24–09.

MMV. (2009b). Medicine for Malaria Venture . Past projects: Tablet, pediatric and intravenous RBx11160. Available at www.mmv.org/article.php3?id_article=46&recherche=OZ277/RBx11160 Acessed on 02–24–09.

Moine, L. Cammas, S. Amiel, C., Guerin, P., Sebille, B. (1997). *Polymer,* **38**: 3121–3127.

Moss, G. P. (1999). www.chem.qmul.ac.uk/iupac/vonBaeyer/.

Mu, J., Woods, J., Zhou, Y. P., Roy, R. S., Li, Z., Zycband, E., Feng, Y., Zhu, L., Li, C., Howard, A. D., Moller, D. E., Thornberry, N. A. and Zhang, B. B. (2006). *Diabetes,* **55**: 1695–1704.

Mukhopadhyay, I., Sausville, E. A., Doroshow, J. H. and Roy, K. K. (2006). *J. Biol. Chem.,* **281**: 37330–37344.

Nayyar, A., Monga, V., Malde, A., Coutinho, E. and Jain, R. (2007). *Bioorg. Med. Chem.,* **15**: 626–640.

Nazem, A., Mansoori, G. A. (2008). *J. Alzheimers Dis.*, **13**: 199–223.

Nicolaou, K. C., Lister, T., Denton, R. M., Montero, A. and Edmonds, D. J. (2007). *Angew. Chem. Int. Ed.*, **46**: 4712–4714.

Nicolaou, K. C., Stepan, A. F., Lister, T., Li, A., Montero, A., Tria, G. S., I., Turner, C. I., Tang, Y., Wang, J., Denton, R. M. and Edmonds, D. J. (2008b). *J. Am. Chem. Soc.*, **130**: 13110–13119.

Petek, M. J., Berend, S., Kopjar, N., Želježić, D., Mladinić, M., Radić, B. and Vrdoljak, A. L. (2008). *Acta Biochim. Polonica,* **55**: 97–105.

Philippens, I. H. C. H. M., Vanwersch, R. A. P., Groen, B., Olivier, B., Bruijnzeel, P. L. B., Melchers, B. P. C. (2000). *Toxicol. Sci.*, **53**: 84–91.

Pisano, C., Vesci, L., Fodera, R., Ferrara, F. F., Rossi, C., De Cesare, M., Zuco, V., Pratesi, G., Supino, R., Zunino, F. (2007). *Annals Oncol.*, **18**: 1500–1505.

Protopopova, M., Hanrahan, C., Nikonenko, B., Samala, R., Chen, P., Gearhart, J., Einck, L. and Nacy, C. A. (2005). *J. Antimicrob. Chemother.*, **56**: 968–974.

Rabinovich, S., Baldini, J. T. and Bannister, R. (1969). *Am. J. Med. Sci.*, **257**: 328–335.

Radic, B., Lucic-Vrdoljak, T., Petek, M. J., Kopjar, N. and Zeljezic, D. (2006). *Toxicology in Vitro,* **20**: 1455–1464.

Ramezani, H. and Mansoori, G. A. (2007). Diamondoids as molecular building blocks for nanotechnology. *In: Molecular Building Blocks for Nanotechnology: From Diamondoids to nanoscale Materials and Applications.* Mansoori, George, Assoufid and Zhang, editors. Springer: New York.

Rammes, G., Rupprecht, R., Ferrari, U., Zieglgansberger, W. and Parsons, C. G. (2001). *Neuroscience Lett.*, **306**: 81–84.

Reiner, E., Radic, Z. and Simeon-Rudolf, V. (2007). *Arh Hig Rada Toksikol.*, **58**: 329–338. Available at hrcak.srce.hr/index.php?show=clanak&id_clanak_jezik=25052. Accessed on 06/01/2011.

Reissmann, S., Pineda, F., Vietinghoff, G., Werner, H., Gera, L., Stewart, J. M. and Paegelow, I. (2000). *Peptides,* **21**: 527–533.

Rivers, E. C. and Mancera, R. L. (2008). *Drug Discovery Today*, **13**: 1090–1098.

Rolin, B., Larsen, M. O., Gotfredsen, C. F., Deacon, C. F., Carr, R. D., Wilken, M. and Knudsen, L. B. (2002). *Am. J. Physiol. Endocrinol. Metab.*, **283**: 745–752.

Rosenthal, K. S., Sokol, M. S., Ingram, R. L., *et al.* (1982). *Antimicrob. Agents Chemother.*, **22**: 1031–1036.

Samoilova, M. V., Buldakova, S. L., Vorobjev, V. S., Sharonova, I. N. and Magazanik, L. G. (1999). *Neuroscience,* **94**: 261–268.

Schleyer P. v. R. (1957). *J. Am. Chem. Soc.,* **79**: 3292–3292.

Schnell, J. R. and Chou, J. J. (2008). *Nature* **451**: 591–595.

Schunemann, H. J., Hill, S. R., Kakad, M., *et al.* (2007). *Lancet Infect. Dis.,* **7**: 21–31.

Shanafelt, T. D., Lee, Y. K., Bone, N. D., *et al.* (2005). *Blood,* **105**: 2099–2106.

Shen, C., Bullens, D., Kasran, A., *et al.* (2004). *Int. Immunopharmacol.*, **4**: 939–951.

Singh, C., Chaudhary, S. and Puri, S. K. (2006). *J. Med. Chem.,* **49**: 7227–7233.

Singh, C., Kanchan, R., Sharma, U. and Puri, S. K. (2007). *J. Med. Chem.*, **50**: 521–527.

Singh, N. P., McCoy, M. T., Tice, R. R. and Schneider, E. L. (1988): *Exp. Cell Res.*, **175**: 184–191.

Singh, S. B., Herath, K. B., Wang, J., Tsou, N and Ball, R. G. (2007). *Tet. Let.*, **48**: 5429–5433.

Skare, D., Radic, B., Lucic, A., Peraica, M., Domijan, A. M., Milkovic-Kraus, S., Bradamante, V. and Jukic, I. (2002). *Arch. Toxicol.*, **76**: 173–177.

Solaja, B. A., Opsenica, D., Smith, K. S., Milhous, W. K., Terzic, N., Opsenica, I., Burnett, J. C., Nuss, J., Gussio, R. and Bavari, S. (2008). *J. Med. Chem.*, **51**: 4388–4391.

Somani, S. K., Degelau, J., Cooper, S. L., Guay, D. R. P., Ehresman, D., Zaske, D. (1991). *Pharmacotherapy,* **11**: 460–466.

Stamatiou, G., Foscolos, G. B., Fytas, G., Kolocouris, A., Kolocouris, N., Pannecouque, C., Witvrouw, M., Padalko, E., Neyts, J. and De Clercq, E. (2003). *Bioorg & Medicinal Chemistry*, **1**: 5485–5492.

Stamatiou, G., Kolocouris, A., Kolocouris, N., Fytas, G., Foscolos, G. B., Neyts, J. and De Clercq, E. (2001). *Bioorg. Med. Chem. Lett.*, **11**: 2137–2142.

Stange, K. C., Little, D. W., Blatnik, B. (1991). *J. Am. Geriatr. Soc.,* **39**: 700–705.

Steinmann, P., Keiser J., Bos, R., Tanner, M. and Utzinger, J. (2006). *Lancet Infect. Dis.,* **6**: 411–425.

Tang, Y., Dong, Y. and Vennerstrom, J. L. (2008). *J. Med. Virol.,* **80**: 895–901.

Tanner, J. A., Zheng, B.-J., Zhou, J., Watt, R. M., Jiang, J.-Q., Wong, K.-L.2, Lin, Y.-P., Lu, L.-Y, He, M.-L., Kung, H.-F., Kesel, A. J. and Huang, J.-D. (2005). *Chem. Biol.,* **12**: 303–311.

Tripathi, R., Mishra, D., Rizvi, A., Singh, C. (2007). *Life Sci.,* **81**: 1544–1548.

Tsuzuki, N., Hama, T., Hibi, T., Konishi, R., Futaki, S., Kitagawa, K. (1991). *Biochem. Pharmacol.*, **41**, R5–R8.

Tsuzuki, N, Hama, T., Kawada, M., Hasui, A., Konishi, R., Shiwa, S., Ochi, Y., Futaki, S., Kitagawa, K. (1994). *J. Pharmaceutical Sci.*, **83**, 481–484.

USA. (2003). Food and Drug Administration. FDA news: FDA Approves Memantine (Namenda) for Alzheimer's Disease. Available at www.fda.gov/bbs/topics/NEWS/2003/NEW00961.html Acessed on 03–01–2009.

USA. (2003). Department of Health and Human Services — FDA. Drug Detail — Cubicin. www.accessdata.fda.gov/scripts/cder/drugsatfda/index.cfm?fuseaction=Search.DrugDetails. Accessed on 04/08/2011.

USA. (2007a). U. S. Food and Drug Administration. List of orphan designations and approvals. Available at www.fda.gov/orphan/designat/list.htm. Accessed on 01/21/2011.

USA. (2007b). NIAID Timeline of Events in Development of SQ109. Available at www3.niaid.nih.gov/topics/tuberculosis/Research/treatment/Sq109Timeline.htm Accessed in 01/12/2011.

USA. (2009). DEPARTMENT OF HEALTH AND HUMAN SERVICES — FDA. Drug Approval Package — Onglyza.

www.accessdata.fda.gov/scripts/cder/drugsatfda/index.cfm?fuseaction=Search.Label_ApprovalHistory#apphist. Accessed on 07/03/2011.

Valli, C., Paroni, G., Di Francesco, A. M., Riccardi, R., Tavecchio, M., Erba, E., Boldetti, A., Gianni, M., Fratelli, M., Pisano, C., Merlini, L., Antoccia, A., Cenciarelli, C., Terao, M. and Garattini, E. (2008). *Mol. Cancer Ther.,* **7**: 2941–2954.

Vennerstrom, J. L., Arbe-Barnes, S., Brun, R., Charman, S. A., Chiu, F. C. K., Chollet, J., Dong, Y., Dorn, A., Hunziker, D., Matile, H., McIntosh, K., Padmanilayam, M., Tomas, J. S., Scheurer, C., Scorneaux, B., Tang, Y., Urwyler, H., Wittlin, S. and Charman, W. N. (2004). *Nature*, **430**: 900–904.

Villhauer, E. B., Brinkman, J. A., Naderi, G. B., Burkey, B. F., Dunning, B. E., Prasad, K., Mangold, B. L., Russell, M. E. and Hughes, T. E. (2003). *J. Med. Chem.,* **46**: 2774–2789.

Wagner, C. E., Mohler, M. L., Kang, G. S., Miller, D. D., Geisert, E. E., Chang, Y.-A., Fleischer, E. B. and Shea, K. J. (2003). *J. Med. Chem.*, **46**: 2823–2833.

Wang, C., Takeuchi, K., Pinto, L. H. and Lamb, R. A. (1993). *J. Virol.,* **67**: 5585–5594.

Wang., J.-J. and Chern, Y.-T. (1996). *J. Biomater. Sci Polym. Ed.,* **7**: 905–915.

Waters, A. P., Higgins, D. G. and McCutchan, T. F. (1993). *Mol. Biol. Evol.,* **10**: 914–923.

Webber, W. C. and Harthoorn, P. A. (1959). Chlorinated and brominated polycyclic hydrocarbons, their preparation and insecticidal compositions containing them. British Patent GB819240.

Wei, H., Wang, C. H., Du, Q. S., *et al.* (2009). *Medicinal Chemistry*, **5**: 305–317.

Weinstock, D. M., Zuccotti, G. (2006). *JAMA,* **295**: 934–936.

WGND. (2008). Working Group TB Drug R&D Portfolio 2008. Available at StopTBDrugWG@tballiance.org. Received on 01/20/2011.

WHO. (1995). World Health Organization. Control of foodborne trematode infections. Report of a WHO study group. *WHO Tech. Rep. Series,* **849**. World Health Organization, Geneva, Switzerland, 78 p.

WHO. (2009a). World Health Organization — Recommended use of antivirals. Pandemic (H1N1) 2009 briefing note 8. Available at www.who.int/csr/disease/swineflu/notes/h1n1_use_antivirals_20090820/en/index.html. Accessed in 05/30/2011.

WHO. (2009b). World Health Organization — Tropical diseases. Available at www.who.int/topics/tropical_diseases/en/. Accessed on 02/11/2011.

WHO. (2009c). World Health Organization — Malaria Fact Sheet. Available at www.who.int/mediacentre/factsheets/fs094/en/index.html. Accessed on 02/11/2011.

Wong, S. S. Y. and Yuen, K.-Y. (2006). *Chest,* **129**: 156–168.

Xiao, S.-H., Keiser, J., Chollet, J., Utzinger, J., Dong, Y., Endriss, Y., Vennerstrom, J. L., and Tanner, M. (2007). *Antimicrob. Agents Chemother.,* **51**: 1440–1445.

Xu, G., Stoffers, D. A., Habener, J. F. and Bonner-Weir, S. (1999) *Diabetes*, **48**: 2270–2276.

Yew, W. W. and Leung, C. C. (2008). *Respirology,* **13**: 21–46.

Youmans, G. P. (1963). *Annu. Rev. Microbiol.,* **17** : 473–494.

Zdanys, K. and Tampi, R. R. (2008). *Prog. Neuropsychopharmacol. Biol. Psychiatry,* **32**: 1362–1374.

Zhang, Y. (2007). *Clin. Pharm. Therap.*, **82**: 595–600.

Zoidis, G., Fytas, C., Papanastasiou, I., Foscolos, G. B., Fytas, G., Padalko, E., De Clercq, E., Naesens, L., Neyts, J. and Kolocouris, N. (2006). *Bioorg. Med. Chem.,* **14**: 3341–3348.

Zoidis, G., Papanastasiou, I., Dotsikas, I., Sandoval, A., Dos Santos, R. G., Papadopoulou-Daifoti, Z., Vamvakides, A., Kolocouris, N. and Felix, R. (2005). *Bioorg. Med. Chem.,* **13**: 2791–2798.

Zoidis, G., Tsotinis, A., Kolocouris, N. Kelly, J. M., Prathalingam, S. R., Naesens, L. and De Clercq, E. (2008). *Org. Biomol. Chem.*, **6**: 3177–3185.

7

Diamondoids in Materials Science

7.1. Introduction

In this chapter, we focus on the modification of materials containing diamondoid molecules, such as polymers, copolymers, nanocomposites, crystal engineering, DNA-Diamondoids nanotechnology and self-assembly of diamondoids. Diamondoids, especially adamantane, its derivatives and diamantane, are already used for improvement of thermal stability and other physicochemical properties of polymers. In what follows, first, we report some advances made in improving polymer properties by incorporating adamantyl group in them, and then some general and interesting applications of diamondoids in polymers. Other important materials science applications of diamondoids which will be presented in this chapter include: (i). Their use in the production of polymer nanocomposites; (ii). Aryl-adamantanes as overcharge protection compounds for power cells; (iii). Their use as the linking units and back-bone in crystal engineering which stands for synthesis of specially designed stiff molecular structures that are in solid state at normal conditions; (iv), DNA-diamondoids nanotechnology, where, as in crystal engineering, diamondoid acts as the back-bone stiffness to produce sturdy and solid materials; and (v). Self-assembly of diamondoids and derivatives which can have tremendous applications on nanomanufacturing and design of NEMS, MEMS, thin-film nucleation, designer pharmaceuticals and sensors.

7.2. Applications of Diamondoids in Polymeric Synthesis

Adamantane, the first and most widely used diamondoid, is a molecule containing symmetric tricyclic hydrocarbon with three fused chair-form cyclohexane rings in a diamond lattice structure. Due to this feature, adamantane is thermodynamically stable, and has unusual effects on polymer behavior when it is incorporated into a polymer synthesis. Incorporation of adamantyl as a pendant group or in the backbone of polymers, usually leads to an enhanced glass-transition temperature (Tg) and thermal stability, as well as better hydrophobicity, and improved dielectric properties. Some examples of polymers in which adamantane is incorporated for such applications include polyimides (Chern, 1998a), polyamides (Chern *et al.*, 1998), polysulfones (Pixton and Paul, 1995), polyesters (Grubb and Mathias, 1997), poly(ether ketone) (Mathias *et al.*, 1997), and polymers based on acetylene-terminated adamantanes (Malik *et al.*, 1992; Burnham *et al.*, 2006). Recently, a high-refractive index polymer resulting from incorporating 1-diamantane-carboxylic acid and glycidyl methacrylate and polimerization with organic peroxide, was produced (Takano *et al.*, 2010).

In the following sections, we report some interesting examples of applications of diamondoids in polymer science and technology:

7.2.1. *Thermoset polymers*

Acetylene groups on highly hindered diamondoid cage compounds (like 1,3-diethynyladamantane) can be polymerized thermally to give thermoset polymers. The resulting thermoset polymer was shown to be stable to 475°C in air and exhibited less than 5% weight loss in air after 100 hours at 301°C. This unusually high thermal stability for an aliphatic hydrocarbon polymer obviously results from the presence of adamantyl units in the polymer backbone which due to their "diamond-like" structure, retards degradation reactions resulting from either nucleophilic or electrophilic attack, or from elimination reactions.

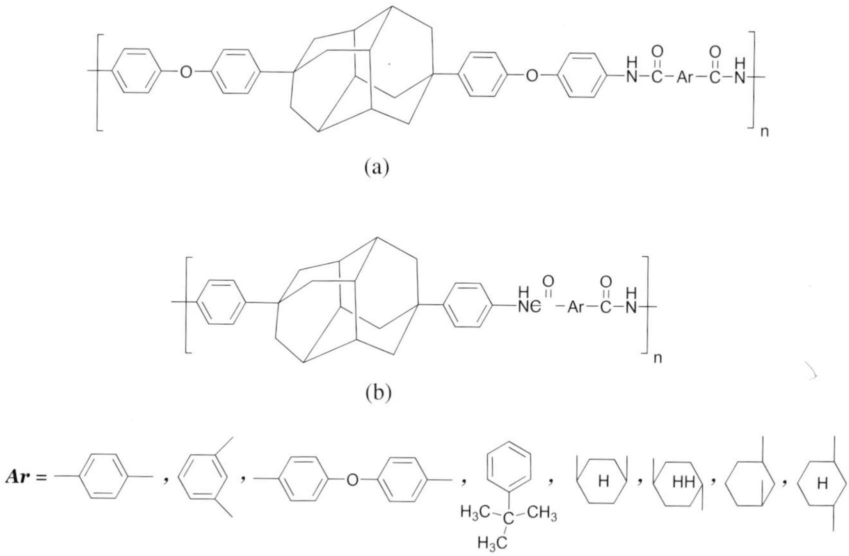

Fig. 7.1. Diamantane-based polyamides; (a) derived from 4,9-bis[4-(4-aminophenoxy) phenyl]diamantane and (b) derived from 4,9-bis(4-aminophenyl)diamantane (Chern, 1998; Chern and Wang, 1998).

7.2.2. *Incorporation of 1,6- or 4,9-diamantylene groups into polyamides*

Polyamides are high-temperature polymers and lack adequate thermal stability. The incorporating diamantyl groups in polyamides results in their thermal stability and satisfactory retention of their storage modulus (~ stress/strain) above 350°C (Chern, 1998a; Chern and Wang, 1998). These characteristics as well as improved viscosity and solubility in organic solvents, are specific to polyamides which are derived from 4,9-bis(4-aminophenyl)diamantane and 4,9-bis[4-(4-aminophenoxy)phenyl] diamantane (Fig. 7.1).

7.2.3. *Application of diamondoids in the synthesis of star polymers*

Star polymers are composed of star-shaped macromolecules containing a single branch point from which linear chains branch out. Star polymers are a class of polymers with interesting rheological and physical properties. The

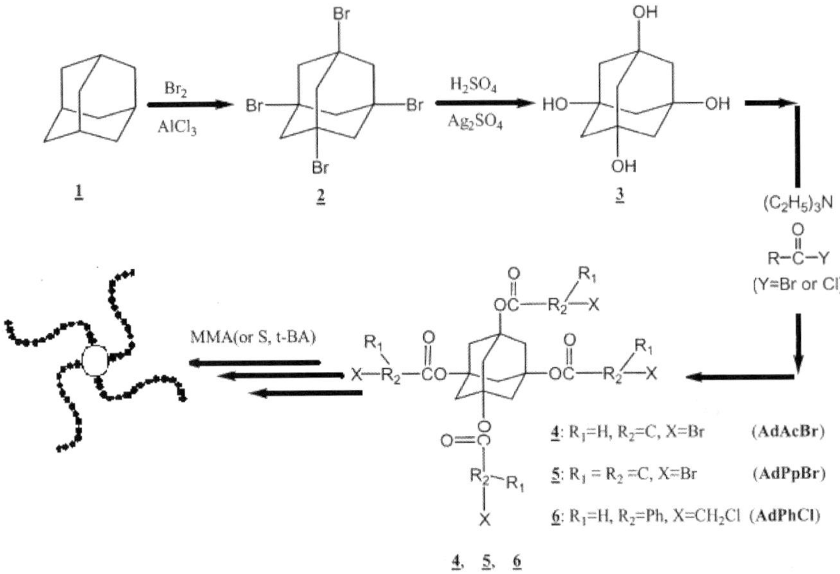

Fig. 7.2. Atom transfer radical polymerization (ATRP) synthetic route to tetrafunctional initiators of a star polymer with adamantly (adamantane core) (Huang *et al.*, 2004).

tetrafunctionalized adamantane cores (adamantyls) have been employed as initiators in the atom transfer radical polymerization (ATRP) method applied to the styrene and various acrylate monomers (see Fig. 7.2).

As a result of this process, star-like polymers have been produced with a wide range of molecular weights (Huang *et al.*, 2004).

The introduction of bulky side-chains which contain adamantyl group to the poly(p-phenylenevinylene) (PPV), a semiconducting conjugated polymer, decreases the number of interchain interactions. This action will reduce the aggregation quenching and improve the polymer photoluminescence properties (Lee *et al.*, 2002; Jeong *et al.*, 2002).

Substitution of the bulky adamantyl group on the C(10) position of the biliverdin pigment structure leads to the distortion of helical conformation thereby changing the pigment color from blue to red (Kar and Lightner, 1998).

A 3-D four-fold interpenetrating coordination diamondoid polymer framework with adjustable porous structure was synthesized recently. It was demonstrated that this polymer framework is capable of trapping gaseous molecules and may be useful for gas storage (Zhang *et al.*, 2001).

7.2.4. *Adamantyl-containing epoxy resins*

New epoxy resins as curing agents based on adamantane molecules were synthesized by Wei *et al.*, (2004). These researchers reported that an adamantyl group containing epoxy-modified epoxy polymer exhibit superior thermal properties and good dielectric performance at high frequencies, indicating good perspectives of adamantane in epoxy applications. However, the multiple steps for the synthesis of this material, as well as its poor processability may restrict its application as a curing agent. On the other hand, Su and Jing (2007) showed that epoxy compounds containing adamantyl group can be easily obtained, by a one-step method, without sacrificing other properties of the cured epoxy polymer. The authors incorporated adamantane moieties into the backbone of an epoxy resin to achieve a novel cured epoxy having better processability, increased thermal stability, lower moisture absorption, and lower dielectric constants, which are highly desirable characteristics for encapsulating materials. The synthesis of an adamantane-containing epoxy resin, diglycidyl ether of bisphenol-adamantane (DGEBAda), was carried out from 1,3-bis(4-hydroxyphen)adamantane by one-step method (Su and Jing, 2007). This new material presents high solubility in strongly polar solvents, such as dimethyl sulfoxide (DMSO), *N, N*-dimethylformamide (DMF), and *N*-methyl-2-pyrrolidone (NMP). In addition, the good solubility of DGEBAda in common solvents (e.g., acetone and ethyl acetate) suggests that such compounds can be easily processed in large, industrial, scales.

7.2.5. *Fluorinated electro-optic copolymer containing adamantane synthesized by Belardini et al. (2006)*

This copolymer presents 73% molar fraction of adamantane-bearing α-fluoroacrylate monomer, and the rest (27%) α-fluoroacrylate monomer coupled with "disperse red 1" (DR1) chromophore. Resulting Adamantane-DR1- α-fluoroacrylate copolymer (DR1AF) is an evolution of the well-known poly(methyl methacrylate) bearing the DR1 chromophore (DR1PMMA) (Singer *et al.*, 1989), where a part of the C–H groups was substituted with C–F, and with the adamantane as a supplementary side group. This copolymer has an electro-optic coefficient of 2.8 ± 0.3 pm/V (at $\lambda = 1550$ nm) and a high activation energy (105 kcal/mol) which assures high thermostability of its electro-optic properties even at a high operative temperature. These features of Adamantane-DR1AF copolymer suggest this polymer as a suitable candidate material for application in versatile integrated optical circuits.

7.2.6. *Other examples of applications of adamantyl in polymeric synthesis*

Generally, incorporation of adamantyl group into polymer materials has been reported to result in increased thermal stability and higher Tg's. For example, the introduction of an adamantyl group into the poly (etherimide) structure caused an increase in polymer glass transition temperature, Tg, and enhancements of its solubility in some solvents like chloroform and other aprotic solvents (Eastmond *et al.*, 1999). These phenomena have been associated with the "diamond-like" structure of adamantane, which inhibits degradation reactions resulting from nucleophilic attack, electrophilic attack, or elimination reactions (Fort, 1976; Fort and Schleyer, 1964; Moon *et al.*, 1970; Olah *et al.*, 1985). Thus, the adamantyl cage increases the rigidity of the polymer chain resulting in materials with higher Tg's. Moreover, aryl-adamantanes have been investigated as new overcharge protection compounds for lithium power cells due to their electrochemical properties (Watanabe *et al.*, 2006).

Adamantyl-substituted poly(m-phenylene) is synthesized starting with 1,3-dichloro-5- (1- adamantyl) benzene monomers (Fig. 7.3) and it is also shown to have a high degree of polymerization and stability, decomposing at high temperatures of around 350°C (Mathias and Tullos, 1996).

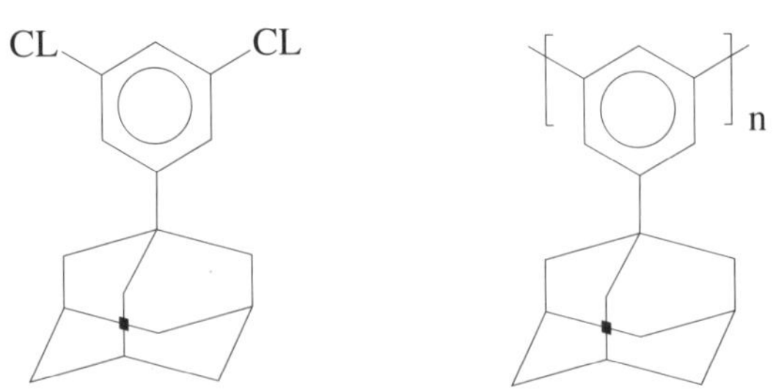

Fig. 7.3. (Left) 1,3-dichloro-5-(1-admantyl) benzene monomer and (Right) adamantly-substituted poly(m-phenylene) which is shown to have a high degree of polymerization and stability decomposing at high temperatures of around 350°C (Mathias and Tullos, 1996).

7.2.7. *Diamantyl in polymers*

The incorporation of diamantyl group into polymer materials is to take advantage of their effect on increasing stiffness, chemical and thermal stability, Tg, and solubility in organic solvents as well as the retention of polymers physical properties at high temperatures. All these special properties of an adamantylated polymer are a result of its diamantane-based molecular structure (Chern, 1998b; Chern and Wang, 1998). For example, it is demonstrated that synthesis of acetylenic derivatives of diamantane and their thermal polymerization provides greater stability to materials with thermooxidative than those of the corresponding adamantantylated polymers. The presence of diamantyl groups enhance the thermal stability of the resulting thermoset resins by nearly 50°C over the corresponding adamantane-based polymers (Malik *et al.*, 1991). Another example of the high temperature polymers of this category is polymerization of diethynyl diamantane (Yaw-Terng and Jane-Jen, 1995).

7.3. Diamondoids in Polymer Nanocomposites

Research in polymer nanocomposites (PNs) has attracted the focus of many scientists in recent years (Alexander *et al.*, 2000; Balazs, 2000; Giannelis, 1998; Soo *et al.*, 1999; Kojima *et al.*, 1993). Currently, PNs have found many different applications ranging from membranes to fuel cell (Vaia and Giannelis, 2001; Sanchez *et al.*, 2005). The enhancement of polymer properties by the addition of inorganic or organic nanoparticles is a complex function of interfacial interactions, interfacial area and the distribution of inter-nanofiller distances. It has been found that these materials have many advantages over conventional filled polymers, such as increased strength, without the degradation of other mechanical properties, decreased gas permeability, improved heat resistance, and enhanced electrical conductivity (Alexander *et al.*, 2000; Balazs, 2000; Giannelis, 1998; Soo *et al.*, 1999). Here, the fillers are dispersed particles (e.g., silicas, carbon blacks, calcium carbonate) introduced into polymer system in order to either improve physical properties or merely lower the price of the final polymer product. In comparison with inorganic fillers (e.g., silica, clays, ZnO), wholly organic diamondoids are potentially more easily dispersed in organic polymer systems. In this section, we will discuss the effect of diamondoid molecules on the thermomechanical and morphological behavior of polypropylene (PP-Dow H314–02Z, isotactic) and polycarbonate (PC-Bayer Makrolon 2205) reported by Ghosh *et al.* (2007). These specific polymers were chosen to provide insights into polar systems *versus* nonpolar systems and semicrystalline systems *versus* amorphous systems.

Diamondoids are composed of trans-fused cyclohexane rings, which form interlocking cage structures superimposable to the diamond crystal lattice (Hoft, 2000). Figure 7.4 shows Wide-Angle X-Ray Diffraction (WAXD) patterns of diamantane, triamantane, and [121]tetramantane. Adamantane and diamantane crystallize in face-centered cubic lattice structures. However, both show solid phase transitions that, in the case of adamantane, are associated

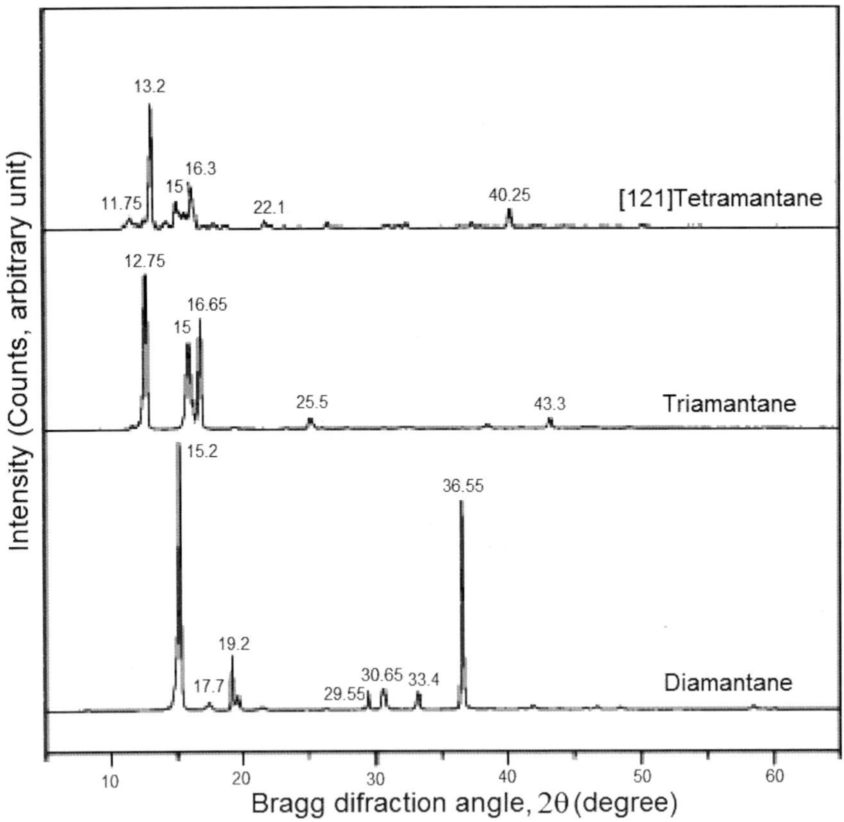

Fig. 7.4. WAXD profiles of diamantane, triamantane, and [121]tetramantane (Ghosh *et al.*, 2007).

with a change in the crystal structure (Jenkins and Lewis, 1980). This is extremely unusual for an organic compound. The molecule is expected to be completely free from both angular and torsional strains. Nevertheless, strain is now known to be present into these systems (Dheu *et al.*, 1979).

Figure 7.5 shows the WAXD profiles of an unfilled isotactic polypropylene (IPP) sample and its composites with the diamondoids: diamantane, triamantane, [121]tetramantane, and methylated tetramantanes. The Bragg reflections at 14.2°, 17°, 18.7° and 22° correspond to the indexed planes for monoclinic crystals of IPP, (α form), (110), (040), (130), and (131) + (041), and for trigonal crystals (minor β form), the Bragg angle of 21° corresponds to the index reflection of (301). The Bragg reflections at

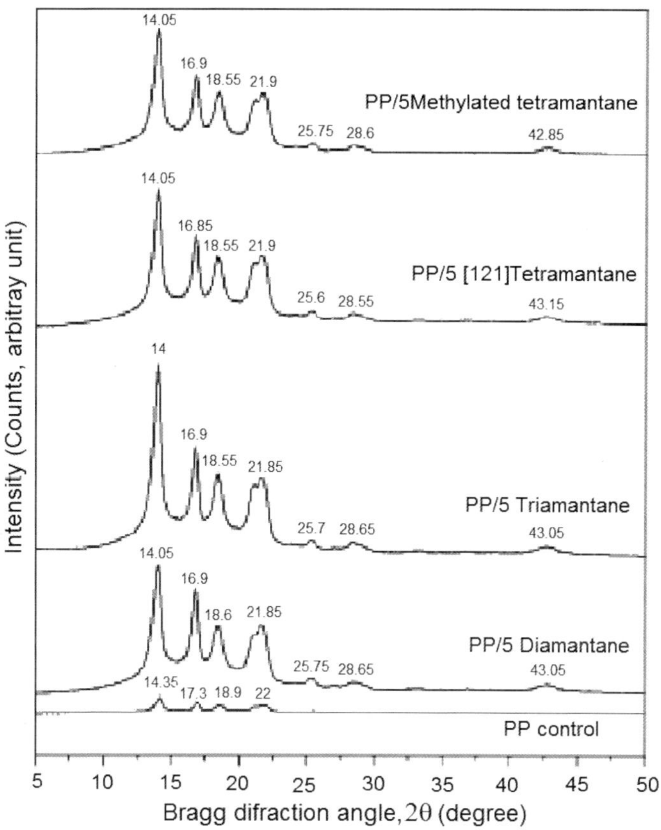

Fig. 7.5. WAXD profiles of polypropylene (PP)/diamondoid (5 wt%) composites (Ghosh *et al.*, 2007).

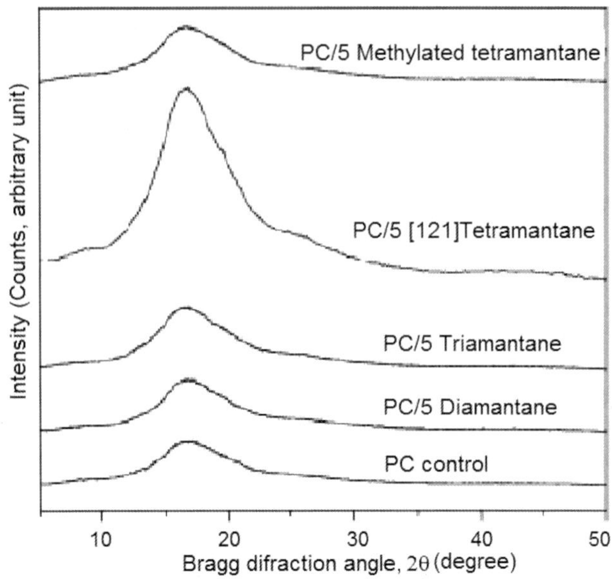

Fig. 7.6. WAXD profiles of polycarbonate/diamondoids (5 wt%) (Ghosh *et al.*, 2007).

25.7° and 29.1° correspond to (060) and (220) planes of the α phase (Weidinger and Hermans, 1961; Jehnichen *et al.*, 1998; Somani *et al.*, 2001; Supaphol and Lin, 2001). Note that polypropylene and its composites also show a minor Bragg reflection at 43°. A comparison of the WAXD profiles indicates that all peaks of IPP are retained in diamondoid/IPP composites (see Fig. 7.5). Crystalline peaks of the diamondoids are not observed in the WAXD profiles of their IPP composites. These results imply that diamondoids are homogeneously dispersed in small domains within the amorphous fractions of the IPP matrix.

Figure 7.6 shows the WAXD profiles of polycarbonate (PC) composites filled with different selected diamondoids and are quite similar to that of pure amorphous PC, with all composites showing a peak at diffraction angle $2\theta = 17.4°$. Here, the composites do not show any peaks associated with diamondoids, indicating that diamondoid molecules are homogeneously dispersed within the PC composites in small domains.

Stress-strain tests of isotactic polypropylene (IPP)/diamondoid composites showed that yield strain, fracture stress, and fracture strain increase with increasing levels of the diamondoid nanofiller (see Table 7.1). This stress-strain behavior indicates that the addition of diamondoids into IPP crystalline

Table 7.1. Stress-Strain properties of isotactic polypropylene (IPP)/diamondoid composites (Ghosh *et al.*, 2007).

Composite	Modulus at 1% strain [MPa]	Yield stress [MPa]	Yield strain [%]	Fracture stress [MPa]	Fracture strain [%]
IPP	510 ± 20	28.6 ± 0.9	17.6 ± 1.1	35 ± 0.7	1310 ± 50
IPP/5%diamantine	440 ± 20	27.2 ± 1.2	21.2 ± 0.9	39.3 ± 1.0	1440 ± 50
IPP/5%triamantane	440 ± 30	25.6 ± 1.2	20.7 ± 0.8	34.6 ± 1.3	1310 ± 30
IPP/5%tetramantane	470 ± 20	27.4 ± 0.9	21.3 ± 0.4	39.6 ± 1.0	1480 ± 40
IPP/2.5%methylated tetramantane	490 ± 20	28.1 ± 0.8	19.7 ± 0.6	39.0 ± 1.1	1460 ± 30
IPP/5%methylated tetramantane	450 ± 10	27.8 ± 0.5	21.4 ± 0.7	41.1 ± 1.1	1550 ± 40
IPP/10%methylated tetramantane	420 ± 10	25.6 ± 0.4	23.8 ± 0.7	39.1 ± 0.8	1520 ± 50

IPP = isotactic polypropylene.

network is likely to act as plasticizers, increasing the ductility of the polymer. Note that the addition of diamantane, triamantane, or [121]tetramantane to IPP marginally increases the fracture stress and fracture strain of the composites; whereas methylated tetramantane mixture behave somewhat differently than the lower diamondoid analogues in IPP-composites, significantly improving the fracture strain values. At 5–10 wt% loadings, diamondoids increase the IPP fracture stress and strain values by 10–20%, and this reflects the increased strain hardening of plasticized materials.

In contrast to the observed behaviors in IPP-composites, the stress-strain properties of polycarbonate (PC)/diamondoids composites suggest that diamondoids act as reinforcing agents. The diamondoid nanofiller increases the yield stress and reduces the yield strains of the PC matrix (see Table 7.2). The data show that the modulus at 1% strain remains unchanged, but the modulus at 10% strain increases monotonically with increased filler loadings. Thus, these stress-strain results show that nonpolar diamondoids act as plasticizers in nonpolar polypropylene matrices and as reinforcing nanofillers in moderately polar polycarbonate matrices.

Morphological changes in semicrystalline polymers induced by stress application have been previously reported by a number of authors (Flory and Yoon, 1978; Peterlin, 1971; Li *et al.*, 1999; Lin and Argon, 1974; Zhang

Table 7.2. Stress-strain properties of polycarbonate/diamondoid composites (Ghosh *et al.*, 2007).

Composite	Modulus at 1% strain [MPa]	Modulus at 10% strain [MPa]	Yield stress [MPa]	Yield strain [%]	Fracture strain [%]
PC	670 ± 10	390 ± 20	58 ± 2	17 ± 1	59 ± 26
PC/2.5%diamantane	690 ± 30	460 ± 20	61 ± 4	15 ± 0	44 ± 9
PC/5%diamantane	680 ± 30	470 ± 30	62 ± 2	14 ± 1	23 ± 6
PC/10%diamantane	670 ± 10	520 ± 20	60 ± 4	12 ± 1	14 ± 1
PC/2.5%triamantane	660 ± 20	440 ± 30	60 ± 3	16 ± 1	35 ± 12
PC/5%triamantane	700 ± 10	450 ± 20	61 ± 1	15 ± 0	38 ± 10
PC/10%triamantane	680 ± 20	490 ± 20	59 ± 1	13 ± 1	14 ± 1
PC/5%tetramantane	760 ± 20	560 ± 10	61 ± 2	14 ± 1	22 ± 12
PC/2.5%methylated tetramantane	710 ± 30	470 ± 20	59 ± 1	15 ± 1	26 ± 11
PC/5%methylated tetramantane	680 ± 20	430 ± 7	63 ± 2	15 ± 1	54 ± 26
PC/10%methylated tetramantane	700 ± 20	510 ± 15	63 ± 3	14 ± 1	30 ± 15

PC = polycarbonate.

et al., 2000; Pluta *et al.*, 2000). Crystallinity normally observed by WAXD of nonoriented semicrystalline samples can be masked through applied tensile stress, which can induce molecular orientation. The tensile stress deformation induces different effects through mechanisms such as crystalline phase gliding and mixing with amorphous phases, interlamellar shearing, and lamellar deformation. Thus, the intense shearing and other deformation mechanisms can cause lamellar fragmentation into small crystalline blocks, leading to an orientation of the chains along the stress direction. In Gosh *et al.* (2007), nonoriented IPP samples and their composites showed an average crystallinity of about 45–51%, whereas it was observed at 60% (relative) increase in crystallinity upon orientation during tensile testing.

The presence of diamondoids has an insignificant effect on the transition temperatures of IPP composites. However, the addition of 5 wt% of diamondoids reduces the storage modulus of IPP. The observed changes in the

Table 7.3. Dynamic mechanical and optical properties of polycarbonate (PC)/diamondoid composites (Ghosh *et al.*, 2007).

Composite	Storage modulus at 50°C [GPa]	Tg [°C]	Tan δ at Tg [–]	Light transmission [%]	Refractive index [–]
PC control	1.73	149	2.23	81±2	1.58
PC/5%diamantane	1.84	131	2.48	85±2	—
PC/10%diamantane	2.18	121	2.51	—	—
PC/5%triamantane	1.94	134	2.47	74±3	—
PC/10%triamantane	1.76	126	2.48	–	—
PC/5%tetramantane	1.90	136	2.43	66±1	—
PC/5%methylated tetramantane	1.93	135	2.46	81±3	1.58
PC/10%methylated tetramantane	2.02	128	2.41	78±2	1.58

dynamic mechanical behavior are indicative of the plasticization of the non-polar IPP by nonpolar diamondoids. In case of PC composites, the observed increase in the storage moduli composites are marginal to moderate due to the presence of diamondoids into polymer system. The Tgs of the PC composites shift to lower values in the presence of diamondoids (see Table 7.3). Tensile moduli of the PC composites are increased by 15–30% with 5–10% loadings of diamondoids (Table 7.2). However, these results appear to contradict the general phenomenon observed for many other filled polymer systems, in which higher moduli correlate with higher Tgs. This behavior is termed as *antiplasticization*. Ueda (2004) observed a similar behavior using *meta*-terphenyl as an antiplasticizer for polycarbonate samples in the manufacturing of optical disks. Similarly, Jackson and Caldwell (1967) reported evidences on antiplasticization of polymer composites. They defined antiplasticization as a simultaneous decrease in Tg with increased mechanical stiffening and embrittlement caused by the addition of particular substances to polycarbonate or other rigid, polar, glassy polymers.

Antiplasticization is attributed to the lowering in polymer matrix free volume and, hence, the restriction of local, noncooperative in-chain molecular motions. Note that the most important characteristics of antiplasticizers are the similarity in action with common plasticizers in the molten state and as reinforcing fillers below Tg. Nevertheless, diamondoids are not precisely antiplasticizers, as defined by Jackson and Caldwell (1967), because they are nonpolar. Diamondoid fillers decrease the Tg and moderately increase the PC stiffness. Hence they meet the functional definition of antiplasticizers. It is interesting to note that some compounds (e.g., aroclors and abietic acid esters) act as antiplasticizers for stiff polymers while serving as plasticizers for flexible polymers (Buttrey, 1960). Similar behavior was observed by adding diamondoids into IPP and PC matrices, acting as plasticizer and antiplasticizer, respectively. The mechanism of antiplasticization is perhaps a combination of several factors: Tight filling of the free volume of the polymer molecules, interaction between the polar groups of the polymer and of the antiplasticizer, and a physical stiffening action due to the presence of rigid small molecules adjacent to the polar groups of the polymer (Jackson and Caldwell, 1967). Since the most flexible portions of a rigid condensation polymer are its polar groups, e.g., carboxylate, carbonate, or sulfone groups, the interaction of these groups with stiff and polar antiplasticizer molecules would reduce flexibility. So, it is reasonable to expect that an antiplasticizer molecule containing multiple rings could stiffen a larger portion of the polymer chain. Thus, the mechanism of polymer matrix free volume tight-filling and consequent physical stiffening due to the presence of rigid diamondoid molecules may be the sole cause of the antiplasticization in this system (Ghosh *et al.*, 2007).

Optical properties of IPP/diamondoid and PC/diamondoid composites were also investigated by Ghosh *et al.* (2007). They observed that diamondoids do not promote a significant effect on the optical clarity of these composites. In particular, the PC/diamondoid composites generally exhibit transparency equivalent to that of the control PC. Again, this indicates that diamondoids are quite compatible with PC and IPP matrices and that the dispersed domains should be very small (nanoscale), giving rise to optical clarity similar to that of the control polymer. In contrast, traditional inorganic fillers (clay, silica, and ZnO) typically reduce optical clarity of polymer composites, often rendering them opaque. Refractive indices of PC/methylated tetramantanes composites can be seen in Table 7.3.

No changes in the PC composites refractive indices are observed upon the incorporation of diamondoid nanofillers. Values of nanocomposite light transmission percentage at 632.8 nm of the nanocomposites are shown in Table 7.3. Polycarbonate composites containing 5 wt% of diamondoid show almost similar values of the light transmission in comparison with those of the neat polycarbonate samples. In fact, diamondoids have specific gravities and refractive indices close to polycarbonate values (Molecular Diamond Technologies, 2007p). Therefore, diamondoids are homogeneously mixed within polycarbonate matrix, and yield composites with light transmission percentages very close to that of neat polycarbonate (Ghosh *et al.*, 2007).

7.4. Aryl-adamantanes as Overcharge Protection Compounds for Power Cells

Adamantane derivatives have been investigated as new overcharge protection compounds for 4V-class lithium cells with $LiCoO_2$ cathodes. The idea is to develop compounds to protect the cells from hazardous ignition and/or explosion when overcharged. However, these compounds must have an oxidation potential (E_{ox}) ranging from 4.6–4.8 V *versus* Li/Li^+, as well as they must not adversely influence the lithium cycling efficiency (E_{ff}). Watanabe *et al.* (2006) synthesized aryl-adamantane compounds and obtained formulations with good electrochemical properties (Fig. 7.4). These aryl-adamantanes exhibit better E_{ox} and E_{ff} values than those of biphenyl (BP), used in commercial cells as overcharge protection additive.

Watanabe *et al.* (2006) designed appropriated aryl-adamantane structures (Fig. 7.7) with features of overcharging protection additives that exhibit both an appropriate E_{ox} value (4.6–4.8 V *versus* Li/Li^+) and a high lithium charge-discharge cycling efficiency.

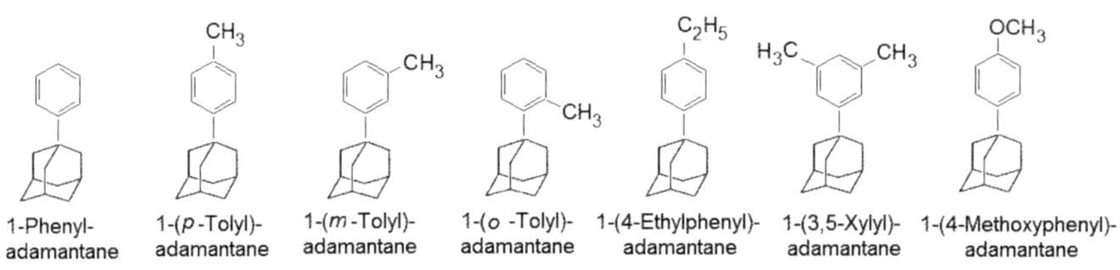

Fig. 7.7. Chemical structures of phenyladamantane derivatives (Watanabe *et al.*, 2004).

Table 7.4. E_{ox} and E_{ff} of aryladamantanes (Watanabe *et al.*, 2005).

Compound	E_{ox} **[V]** ***versus*** **Li/Li$^+$**	E_{ffAdd}/E_{ffPpC}*
Adamantane	5.17	1.080
1-phenyl-adamantane	4.88	0.936
1-(*p*-tolyl)adamantane	4.65	0.960
1-(*m*-tolyl)adamantane	4.75	0.909
1-(*o*-tolyl)adamantane	4.80	0.973
1-(4-ethylphenyl)adamantane	4.63	1.050
1-(3,5-xylyl)adamantane	4.57	0.838
1-(*p*-methoxyphenyl) adamantane	4.22	1.010

$$^{*}\ \frac{E_{ffAdd}}{E_{ffPpC}} = \frac{E_{ff}\ \text{of solution (1 M LiClO}_4\text{ / PpC) with additives}}{E_{ff}\ \text{of solution (1 M LiClO}_4\text{/PpC) without additives}}$$

These new aryl-adamantane compounds were added to the electrolyte solution ($LiClO_4$/propylene carbonate (PPC)) in order to evaluate their electrochemical properties. Table 7.4 shows four aryl-adamantanes,

1-(*p*-toly)adamantane,
1-(*m*-toly)adamantane,
1-(*o*-toly)adamantane,
1-(4-ethylphenyl)adamantane.

The above aryl-admanatanes exhibit the best performance by achieving the target E_{ox} values (4.6–4.8 V versus Li/Li$^+$) and having preferable relative E_{ff} values (0.909–1.05) among the phenyladamantane derivatives.

Thus, the new aryla-adamantane compounds designed by Watanabe *et al.* (2006) can be employed as overcharge protection additives in lithium cells with $LiCoO_2$ cathodes, since they exhibited the best performance in terms of both oxidation potential and lithium cycling efficiency.

7.5. Crystal Engineering

One of the interesting materials science applications of diamondoids is "crystal engineering". Crystal engineering refers to the design of various crystalline structures with particular morphologies and sizes starting from molecular building blocks. The approach in crystal engineering is to learn from known crystalline structures of, for example, minerals in order to design compounds with desired properties. It has been determined that a vast number of material properties including electrical and thermal conductivities, magnetic, optical, photochemical, porosity, and solubility are affected appreciably, by the crystalline structures and sizes. Crystal engineering is a new concept, through which the power of noncovalent intermolecular forces (Rafii-Tabar and Mansoori, 2004) is used in the solid-state to design new nanomaterials with desired functions. It is considered to be a key new technology with potential applications in the design of new materials with the desired properties, nanotechnology, pharmaceuticals, host–guest chemistry, supramolecular chemistry, catalysis, separations, gas storage, optical and magnetic properties, molecular sieves, gas absorption, size selective separation and more.

The structures of adamantane and other diamondoids have received considerable attention in crystal engineering due to their stability, molecular stiffness, derivatization capabilities, conformationally well defined structures, and their six or more primary, and four or more secondary, linking sites (Evans *et al.*, 1999; Evans and Lin, 2001; Reddy *et al.*, 1996; Vaidhyanathan *et al.*, 2004; Zaworotko, 1994; Zhang *et al.*, 2009; Garcia *et al.*, 2010; Tominaga *et al.*, 2009, 2011). In what follows, we present a number of examples in which diamondoids are used in crystal engineering.

1. The hydrogenbonded carboxylic acid dimer synthon is used to form hexagonal sheet structures from benzene 1,3,5-tricarboxylic acid (trimesic acid). These sheets are nonplanar and form a complex three-dimensional interpenetrating network of four separate sheets (Childs *et al.*, 2009) and three-dimensional diamond-like structures from adamantane 1,3,5,7-tetracarboxylic acid (see Fig. 7.8).

Such structures have also been widely used as synthons in crystal engineering (Johnston *et al.*, 2008; Malek *et al.*, 2005; Edwards *et al.*, 2002).

2. A molecular network known as the "disubstituted adamantane" is designed especially for application in certain supramolecular structures which are used for encapsulating and aligning guest molecules. Such systems are then used for the formation of extended polymeric networks (Edwards *et al.*, 2009). An example of such a molecular network stems

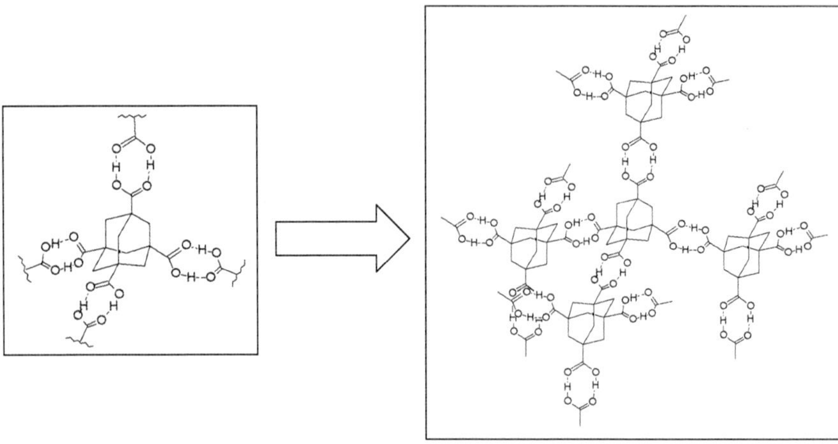

Fig. 7.8. Adamantane 1,3,5,7-tetracarboxylic acid as a synthon in crystal engineering.

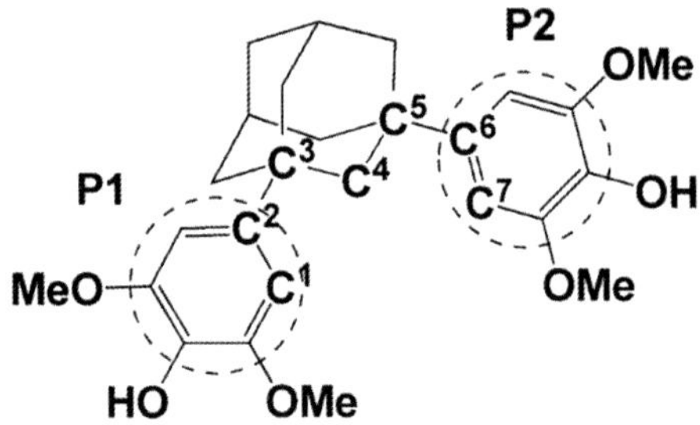

Fig. 7.9. Disubstituted adamantane (Tominaga *et al.*, 2009).

from 1,3-bis(4-hydroxy-3,5-dimethoxyphenyl)adamantane which has two dimethoxyphenols on an adamantane scaffold as shown in Fig. 7.9.

Each of the two dimethoxyphenols has three hydrogen bonding sites, and hence the resulting network can form 2D or 3D frameworks. Moreover, dimethoxyphenol is an electron donor, thus the network will offer kinetically labile complexes with an electron acceptor compound due to donor-acceptor interactions. Depending on the kind of solvent (chloroform, benzene, or *n*-hexane/chloroform) used for recrystallization in the presence of 1,3,5-trinitrobenzene, a variety of structures will be formed such as 1D infinite zigzag chains (Fig. 7.10 A), a continuous

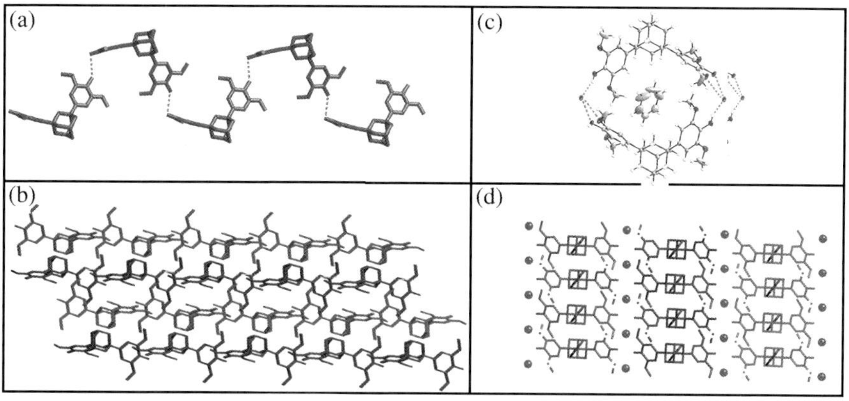

Fig. 7.10. Crystal structures of disubstituted adamantane crystals (Tominaga *et al.*, 2009).

2D layered structure through multiple hydrogen bonding, C-H· · · O (Fig. 7.10 b), cyclic networks (Fig. 7.9 c), and tubes (Fig. 7.9 d — Side view shown).

3. As a third example, we report here the design of an engineered crystal (Fig. 7.11 d) which mimics the crystal structure of platinum oxide (PtS) (Fig. 7.11 c).

This engineered structure (Fig. 7.11 d) is made of $Cu_2(OCO)_4$ cluster looking like a paddle-wheel (Fig. 7.11 a) and adamantane (Fig. 7.11 b). These building blocks assemble to adopt the platinum oxide (PtS) network (Fig. 7.11 c) with its vertices occupied by the clusters to a decorated PtS framework. The product will adopt the platinum sulfide (PtS) structure topology (Fig. 7.11 c) (the simplest, and most symmetrical linkage of tetrahedra and squares), in which each square Pt atom and tetrahedral S atom in PtS would be replaced by a square and a tetrahedral cluster, respectively, to give a decorated form of PtS (Fig. 7.11 d).

7.6. Diamondoids-DNA Nanoarchitecture

Recently a route for the DNA-directed assembly as a method of choice to assemble diamondoid-made nanostructures was introduced by Ramezani *et al.* (Ramezani *et al.*, 2007a; Ramezani *et al.*, 2007b; Ramezani and

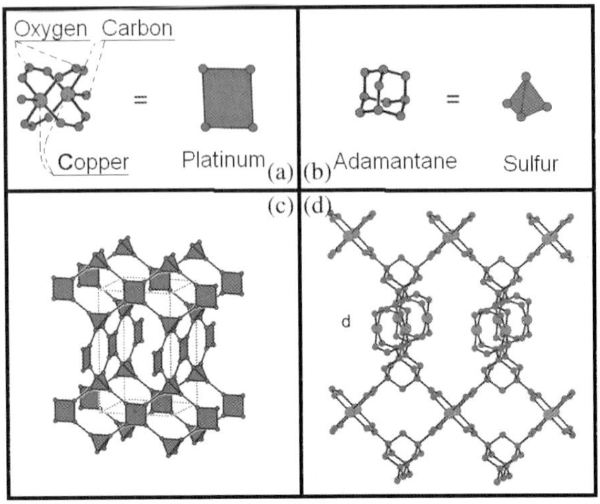

Fig. 7.11. Design of an engineered crystal based on adamantane plus $Cu_2(OCO)_4$ cluster which mimics the crystal structure of platinum oxide (PtS) (Chen *et al.*, 2000).

Mansoori, 2007). In this section the concept of diamondoid-DNA nanoarchitecture (DDN) will be presented.

DNA is a supramolecule (a nanoscale entity) that has entered into nanotechnology in many different ways, like in host-guest chemistry for gene delivery, in DNA nanoscale sequencing and assembly, in DNA-based nanoelectronics, just to name a few (Seeman and Kallenbach, 1983; Nikakhtar *et al.*, 2007). Utilization of DNA in the Diamondoid-DNA nanoarchitecture (DDN) takes advantage of the DNA properties based upon which DNA-nanotechnology (Seeman and Kallenbach, 1983; Pinto *et al.*, 2005) has been underpinned. While supramolecular DNA is a carrier of biological information, in DNA-nanotechnology, the molecules that constitute the DNA are used as structural materials to create two- and three-dimensional nanostructures.

The main idea in DDN is to incorporate DNA strands into a diamondoid molecule to act as linkers. DNA hybridization is a valuable tool to accomplish nanostructure self-assembly. It would enable us to control the orientation of a nanostructure's spatial extension through specific molecular recognition sites existing between two complementary DNA sequences. The diamondoids-DNA nanoarchitectures add stiffness to the DNA-directed assembly which adds to its applicability for solid state devices.

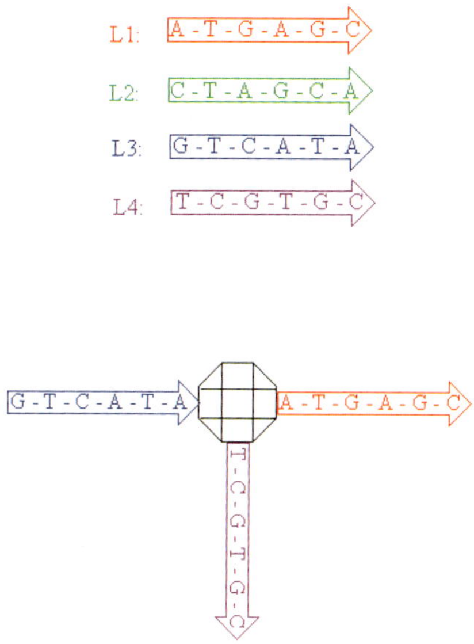

Fig. 7.12. The hexanucleotide linkers L1–L4 used in constructing a DDN and schematic representation of a DDN: Three exemplary linker (L1, L3, L4) sequences and a diamondoid-made nucleus (Insert symbol) (Ramezani *et al.*, 2007b).

7.6.1. *Design of Diamondoids–DNA Nanostructure (DDN)*

A DDN consists of a diamondoid-derived nucleus and three or four DNA sequences as linkers. Ramezani *et al.* (2007a,b) reported how to construct a cube using DDNs. The hexanucleotide linkers L1–L4 reported in Fig. 7.12 are used in constructing of a DDN.

The letters "A", "T", "G", and "C" are representative of adenosine, Thymidine, Guanosine, and Cytidine nucleotides, respectively. Also reported in Fig. 7.12 are the schematic representation of a DDN consisting of three linkers and a diamondoid as chosen by Ramezani *et al.* (2007a) to construct a cube.

The three-dimensional structure of DDN using adamantane and tetramantane nucleus will look like what is shown in Fig. 7. 13.

Eight such DDNs (with three DNA single strands as linkers) are necessary to form a cube. We also need to design a bridging DNA sequence in such a way that its first half is complementary to one of the linkers of

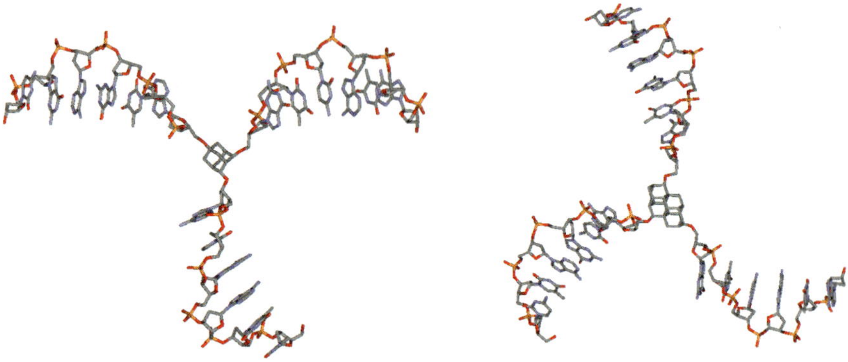

Fig. 7.13. The chemical structure of DDNs (i) and (ii) consisting of adamantane and tetramantane nucleus, respectively, and their three hexanucleotide linkers. Hydrogen atoms are removed for clarity (Ramezani *et al.*, 2007b).

Fig. 7.14. The DNA Bridge-based Self-assembly: Two successive nanostructures, (a) and (b), are attached together via their related DNA bridge (Ramezani *et al.*, 2007a,b).

DDN "a" and its second half is complementary to one of the linkers of the neighbor DDN "b", as represented in Fig. 7.14.

In view of the fact that diamondoids are rigid molecules and the DNA's length is about 50 nm, combination of DNA double strands about 4 to 7 nm in length (less than two turns) with diamondoids is very likely to lead to the rigid nanostructures with well-founded and predictable geometries.

The hybridization of DNA single stranded linkers of each set of two DDNs with their complementary single stranded bridge was investigated using the Autodock Tools 3.0 software working under the Linux operating system. The two hexanucleotide linkers were defined for the software as if they had been a single strand consisting of 12 nucleotides. However, the phosphate-sugar backbone of the strand was cut between residues 6 and 7 to simulate the presence of two separate linkers but they were not displaced with respect to each other. These sequences were introduced to Autodock Tools as a macromolecule and the bridge sequence was designated to be the ligament. The software calculated the docking parameters and yielded the docked double helix structures.

The DNA-sequence bridge comprises ten nucleotides to bridge the two linkers and attach them together by hybridization. Each linker consists of six nucleotides and thus the two successive linkers ranged in row create a gap of twelve nucleotides between two neighboring nucleuses. Only five out of six nucleotides of each linker would participate in the hybridization. The first nucleotide of each linker, immediately attached to the diamondoid-nucleus, would not interact with the bridge. It is supposed to play the role of a spacer separating the double strand structure and nucleus in order to avoid steric congestions.

To complete the cube, the next assignment was to select the appropriate linkers for each DDN. The following special considerations are made regarding the selection rules:

A. The linkers should not be complementary on their own. If there are two complementary sequences on neighboring nanostructures, they would interact and might form improper local structures. The resulted false pattern of interaction would prevent the linkers from hybridization with the bridging strand.

B. The linkers are supposed to be adopted in such a way that only one interacting sequence exists for attachment to the nearby DDN. In other words, the bridging sequences used to attach a DDN to its neighbors should not be identical, otherwise, it is not possible to manage the three dimensional extension of the nanosystem to come up with a cube.

C. There are $(4)^3 = 64$ possible combinations for choosing three linkers out of the four given sequences provided that the orders of linkers are considered. However, many repetitive combinations are regarded to be the same. For instance, the L1-L2-L3, L1-L3-L2, L2-L1-L3, L2-L3-L1, L3-L2-L1, and L3-L1-L2 are all identical as a single combination (Table 7.5).

Table 7.5. The eight DDNs chosen to build the cube shown in Figs. 7.16 and 7.17.

(a): L1, L1, L1	(e): L1, L2, L4
(b): L2, L2, L2	(f): L1, L4, L4
(c): L3, L3, L3	(g): L3, L3, L4
(d): L4, L4, L4	(h): L1, L1, L3

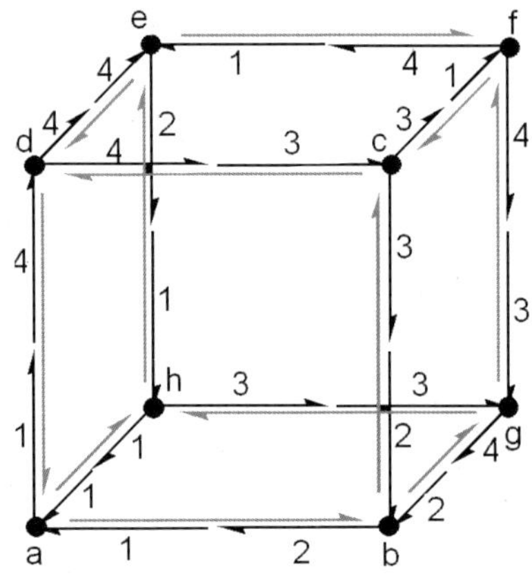

Fig. 7.15. The molecular blueprint used for designing a cubic nanostructure based on a bottom-up strategy. The solid spheres at the corners correspond to the positions of the diamondoid nucleuses. The letters stand for the DDNs given in Table 7.5 (Ramezani *et al.*, 2007a,b).

As a result, there are only 20 possible independent states to be considered.

To construct the cube, eight states out of the 20 combinations for the linkers (Ls) were selected as is depicted below (Ramezani *et al.*, 2007a,b):

The blueprint showing the detailed description of the cube-like nanostructure is given in Fig. 7.15.

The chemical structures of the cubic nanostructure shown in Figs. 7.16 and 7.17 were built by Ramezani *et al.* (2007b) using the above mentioned procedure through the Hyperchem release 7.0.

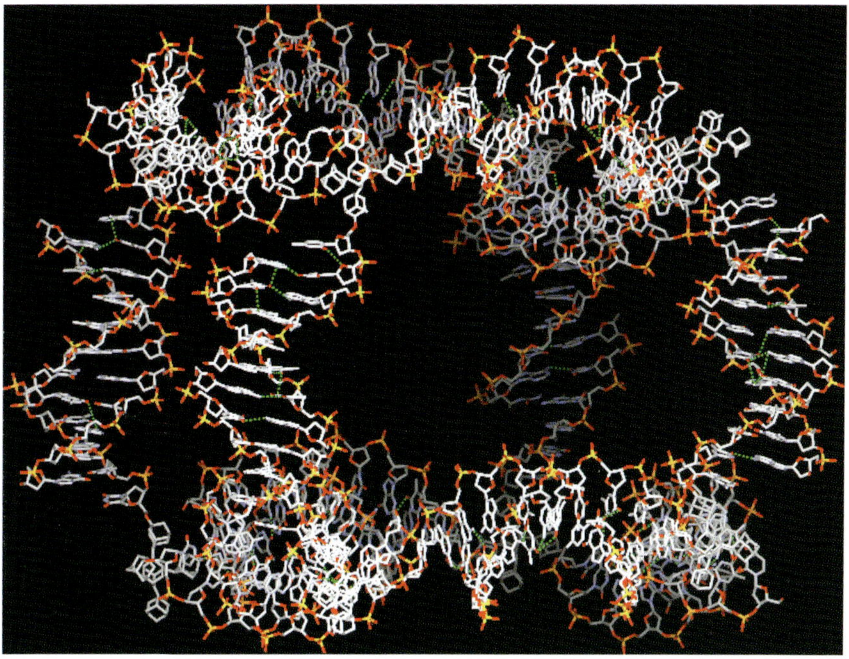

Fig. 7.16. The cube-shaped nanostructure made of eight DDNs. The green dotted lines show hydrogen bonding between complementary sequences (Ramezani *et al.*, 2007a,b).

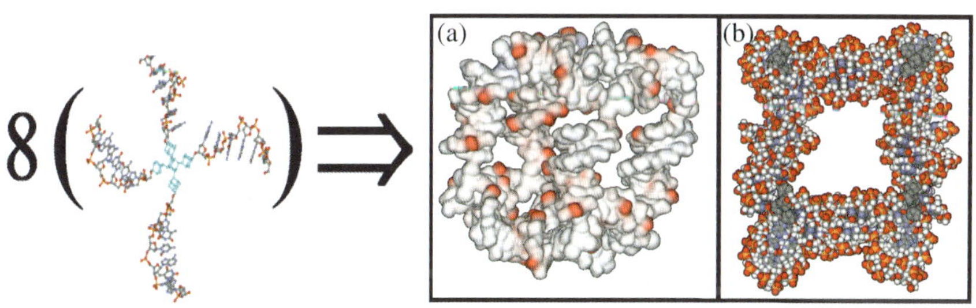

Fig. 7.17. The diamondoid-DNA cube-shaped nanostructure, A simulated cube-shaped nanostructure assembled by connection of eight nanomodules (Fig. 7.13 Right) using DNA bridge-based self-assembly: (a) The electrostatic potential surface of the cube-shaped nanostructure. The red regions bear negative partial charges whereas the positive sites are colored in blue. (b) The CPK (Corey–Pauling–Koltun) model of the cube-shaped nanostructure reflects the square shape of each face (Ramezani *et al.*, 2007a, b).

To achieve a realistic cube-shaped nanostructure the following procedure was chosen (Ramezani *et al.*, 2007b):

A. The geometries of diamondoids and their derivatives were initially optimized using the MM+ force field of molecular mechanics in Hyperchem.
B. The deoxyribonucleic acid residues (nucleotides) were taken from the nucleic acid database of Hyperchem. The sequences were chosen to be in the B-form secondary conformation (right-hand helices) as that is the most prevalent form of DNA under physiological conditions.
C. The pucker of the furanose ring was selected to be 2′ endo consistent with the B-form conformation.
D. In importing nucleotides from the Hyperchem's database Ramezani *et al.* (2007b) took advantage of working with the well-defined and parameterized nucleotide templates which possessed structural properties and information in accordance with their real molecular structures in nature.
E. After building the initial models, they were all optimized with different force fields of molecular mechanics. Those force fields were MM+ and Optimized Potentials for Liquid Simulations (OPLS). For MM+, the electrostatics was set to be based on bond dipoles while for AMBER the effective dielectric constant was defined to be distance-dependent with the scale factor of 1. For the OPLS and BIO+ the effective dielectric constants were set to be constant. No cutoff distance for denying the nonbonded interactions was specified.
F. Three different optimization algorithms were used to minimize the nanostructure energies. The Polak–Ribiere and the Fletcher–Reeves optimizers are among the conjugate gradient methods used for this minimization. The optimization was continued until the root mean square (*RMS*) gradient reached the specified value of 0.1 kCal.mol^{-1}.Å^{-1}. The *RMS* gradient is calculated according to formula (1):

$$RMSG = (3N)^{-1}\left(\sum_{A}\left(\frac{\partial E}{\partial X_A}\right)^2 + \left(\frac{\partial E}{\partial Y_A}\right)^2 + \left(\frac{\partial E}{\partial Z_A}\right)^2\right)^{1/2}, \qquad (7.1)$$

where "*N*" is the number of atoms, "*X*", "*Y*", and "*Z*" are the Cartesian coordinates of atom "*A*", and "*E*" is the energy. The minimization history was used to obtain new directions with optimized coordinates. The optimized coordinates were expected to satisfy the following constrain:

$$\frac{\partial V}{\partial r_i} = 0, \tag{7.2}$$

where "V" is the potential energy and "r_i" stands for the related Cartesian coordinates. The third optimizer used was the block-diagonal Newton–Raphson algorithm. The results of energy minimization by the aforementioned optimizers were recorded for further comparison and studied for monitoring hydrogen bonds. The electrostatic potential surface and the Corey–Pauling–Koltun model of the cube-shaped nanostructure is shown in Fig. 7.17, which were produced by using the Accelrys ViewerLite 4.2 software. The visual graphics of figures representing DDNs and nanostructures were produced using the same software. The Polak–Ribiere optimization algorithm was used among the conjugate gradient methods (Ramezani *et al.*, 2007a,b).

G. The whole computations were run *In Vacuo* without adding solvent molecules to the system explicitly. However, the solvent effects are accounted to some extent by setting the effective dielectric constants. Different energies including bond, angel, dihedral, bending, and electrostatic were recorded for comparison and data analysis.

The proposed hybrid diamondoid-DNA nanoarchitecture is capable of forming highly stable networks and could serve as a useful building block for DNA-based nanoconstructions. It may have a wide variety of applications which range from self-assembly of nanoelectronic components to acting as the 3-D scaffolds for crystallography of biomolecules, molecular cages for drug delivery purposes, and as elements in molecular manufacturing of nanorobotics machinery parts (Pathak and Marx, 2011).

7.7. Self-assembly of Diamondoid Molecules and Derivatives

Diamondoid molecules and their derivatives have been recognized as molecular building blocks in nanotechnology (Drexler, 1999; Mansoori,

2005, 2007; Mansoori *et al.*, 2007). They have been drawing more and more research attention due to their highly symmetrical and strain free structures, controllable nanostructural characteristics, nontoxicity and their applications in producing a variety of nanostructural shapes in molecular manufacturing, in nanotechnology and in NEMS and MEMS (Ramezani and Mansoori, 2007c; Xue and Mansoori, 2008). A feasible way to produce nanostructures for such applications may be through the self-assembly of appropriate diamondoids and derivatives.

Self-assemblies of two lower diamondoids (adamantane and diamantane), three adamantane derivatives (amantadine, memantine, and rimantadine) and two organometallic molecules (substituting one hydrogen atom in adamantane and diamantine with one sodium atom) have been recently studied by Xue and Mansoori (2010a,b). For further details see Sec. 5.3.6. The molecular structures of these seven molecules are classified in groups (see Table 5.2). *Group* 1: Adamantane and diamantane, the lowest two diamondoids. Due to their six or more primary and four or more secondary linking sites, they have found major applications as templates and as molecular building blocks in nanotechnology, polymer synthesis, drug delivery, drug targeting, DNA-directed assembly, DNA-amino acid nanostructure formation, and host-guest chemistry (Mansoori, 2007). However these diamondoids do not have good electronic properties which are necessary for building molecular electronics, but some of their derivatives do. *Group* 2: Memantine, rimantadine, and amantadine, the three derivatives of adamantane, which have medical applications as antiviral agents and due to their amino groups, they could be treated as molecular semiconductors (Xue and Mansoori, 2008, Marsusi *et al.*, 2009). *Group* 3: ADM•Na, DIM•Na, the two artificial molecules, substituting one hydrogen ion in adamantane and diamantane with a sodium ion which could have potential applications in NEMS and MEMS (Xue and Mansoori, 2008; Mansoori *et al.*, 2009). In the following sections we report the self-assembly and phase transition properties of these seven molecules.

Xue and Mansoori (2010a,b) performed density functional theory (DFT) calculations to optimize theses seven molecules' initial structures and obtained atomic electronic charges. Then they performed nanoscale molecular dynamic (MD) simulation (Esfarjani and Mansoori, 2005) in the study of their self-assembly and phase transition behaviors. They chose Optimized Potentials for Liquid Simulations All Atom (OPLS-AA) force field in their MD simulations since it is a reliable force field for hydrocarbons. In the OPLS-AA force field, the nonbonded interactions are represented by the general Lennard–Jones plus Columbic form as shown by Eq. 7.3, where E_{ij} is the interaction energy of two sites for both intermolecular and intramolecular nonbonded cases,

$$E_{ij} = \sum_{i,j} \left\{ 4\varepsilon_{ij} \left[\left(\frac{\sigma_{ij}}{r_{ij}} \right)^{12} - \left(\frac{\sigma_{ij}}{r_{ij}} \right)^{6} \right] + \frac{q_i q_j e^2}{r_{ij}} \right\} f_{ij}. \tag{7.3}$$

The combining rules used along with this equation are $\sigma_{ij} = (\sigma_{ii}\sigma_{jj})^{1/2}$ and $\varepsilon_{ij} = (\varepsilon_{ii}\varepsilon_{jj})^{1/2}$. For intermolecular interactions (when i, j corresponds to different molecules), $f_{ij} = 1.0$. While for intramolecular nonbonded interactions between all pairs of atoms ($i < j$) separated by three or more bonds, the 1, 4-interactions are scaled down by $f_{ij} = 0.5$. Furthermore, in order to obtain the geometries and atomic charges for the MD simulations of all the seven molecules Xue and Mansoori (2010a,b) applied the DFT calculation using the B3YLP exchange-correlation functional method with cc-pVDZ basis, for all the molecules except for adamantane+Na and diamantane+Na for which 6–311+G(d, p) basis was used. All the MD simulations involved simulated-annealing procedure on 125 molecules in the seven different types of diamondoids and their derivatives.

The snapshots of the gas-to-solid MD simulations for 125 molecules of each of the seven molecules (*Adamantane, Diamantane, Amantadine, Rimantadine, Memantine, Adamantane+Na., Diamantane+Na*) are reported in Fig. 7.18.

From these snapshots one can directly observe clear phase transitions for each kind of molecule, from the gaseous state a highly condensed solid (self-assembled) state. According to Fig. 7.18, adamantane, diamantane, adamantane+Na and diamantane+Na form ordered condensed (crystalline) states. However, amantadine, rimantadine and memantine, while they self-assemble, they do not seem to form clear ordered condensed (crystalline) states. Xue and Mansoori (2010a,b) further produced the hydrogen bonds locations of the same self-assembled snapshots of amantadine, memantine and rimantadine as shown in Fig. 7.19.

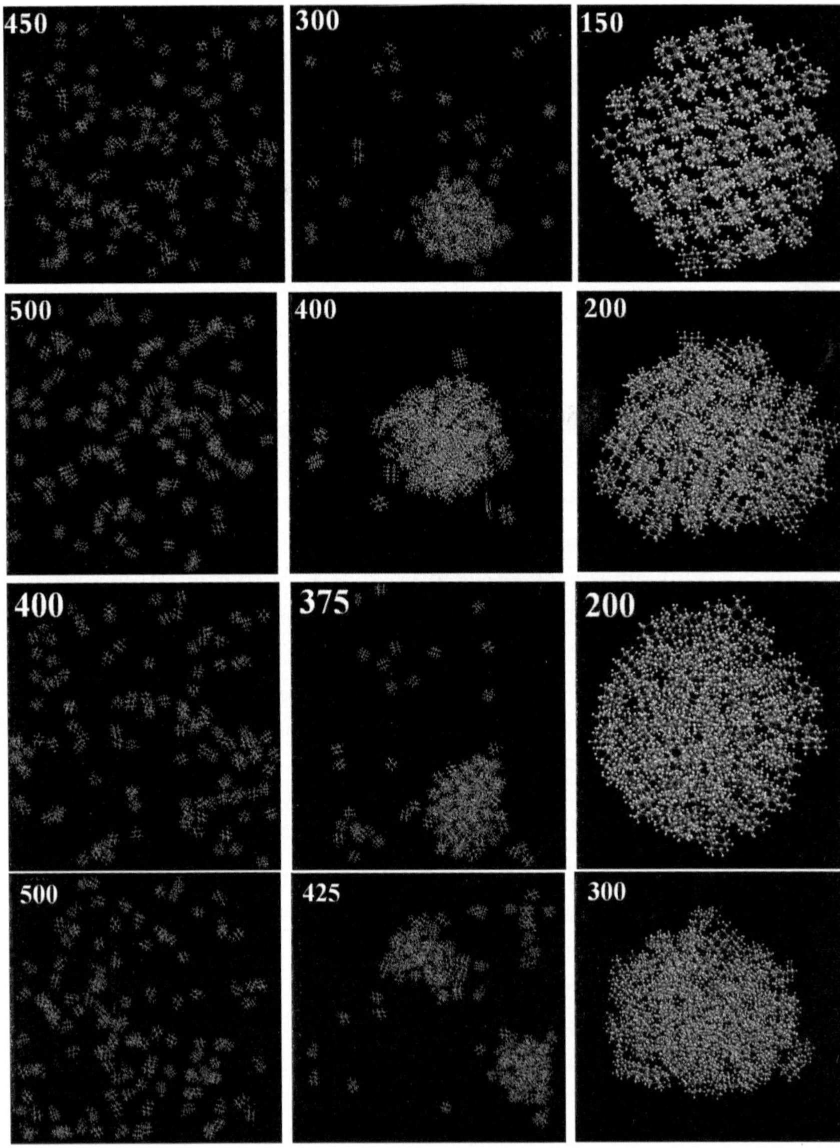

Fig. 7.18. Self-assembly snapshots of 125 molecules of the seven compounds (from top: Adamantane, Diamantane, Amantadine, Rimantadine, Memantine, Adamantane+Na., Diamantane+Na) as temperature [K] is decreased (Xue and Mansoori, 2010).

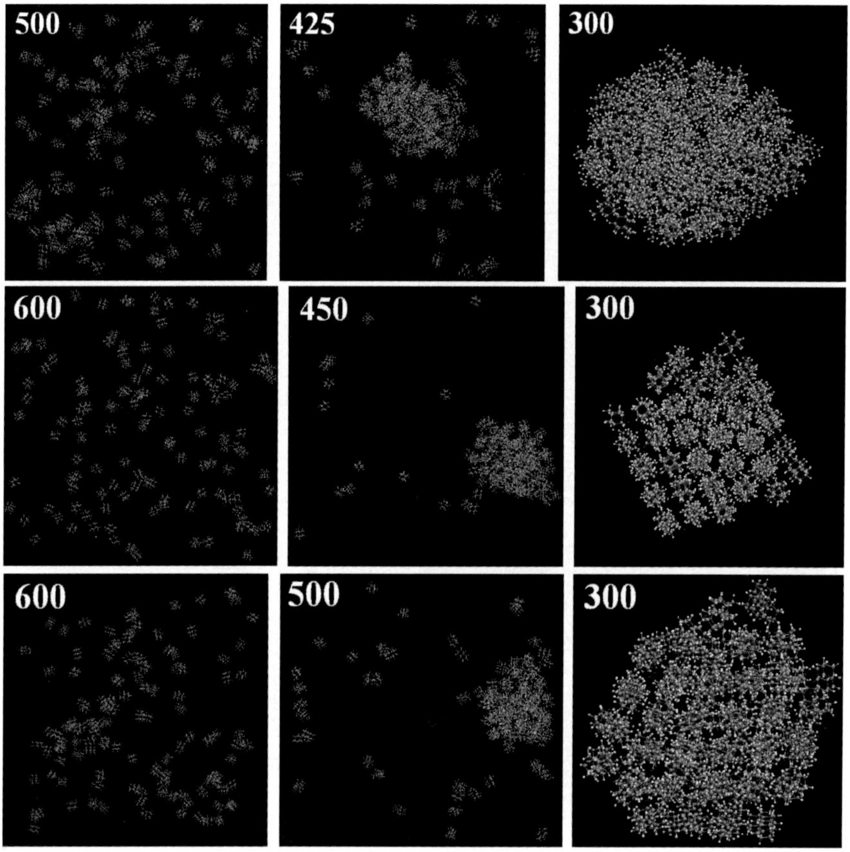

Fig. 7.18. (*Continued*)

According to this figure no ordered format for the location of hydrogen-bonds is observed. This is an indication of the noncrystalline self-assembled states of these three kinds of molecules. The presence of partial hydrogen-bonds between these molecules is responsible for the lack of a clear crystalline state in their self-assemblies.

Xue and Mansoori (2010a,b) also performed studies in the hydrogen-bonds saturation as a function of temperature as well as hydrogen-bond distribution of the three molecules (amantadine, rimantadine, and memantine) as shown in Figs. 7.20 and 7.21.

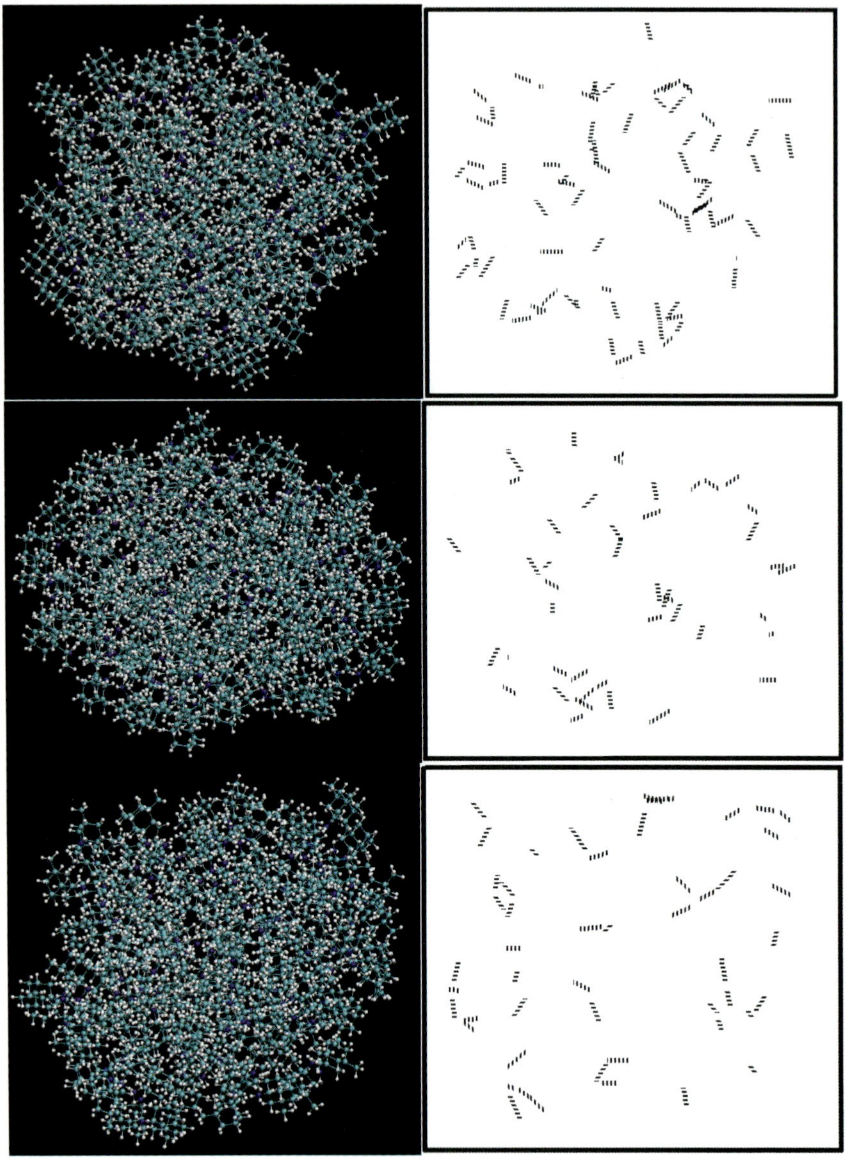

Fig. 7.19. MD snapshots of (from top to bottom) amantadine, rimantadine and memantine and at 50K hydrogen bonds locations (Xue and Mansoori, 2010).

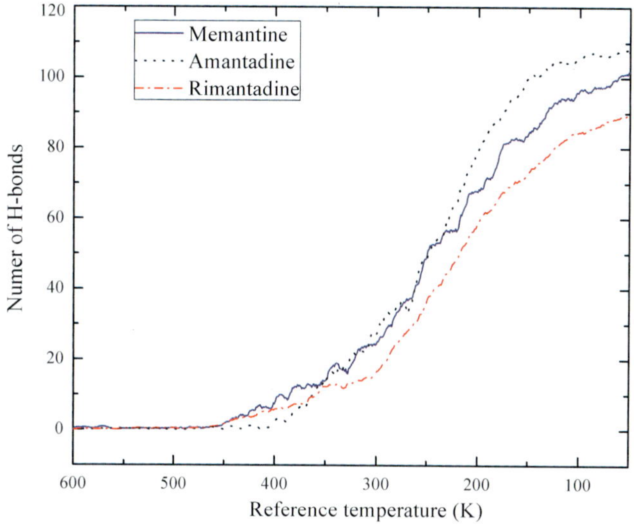

Fig. 7.20. The number of hydrogen bonds for amantadine, memantine and rimantadine versus reference temperature respectively. As temperature decreases, the numbers increase, and tend to maximum numbers for the three derivatives (Xue and Mansoori, 2010).

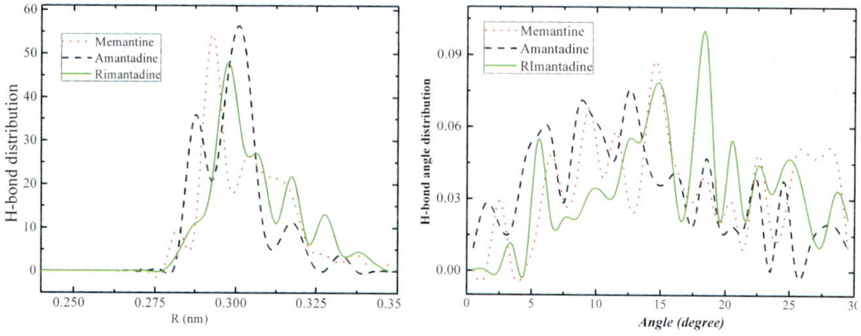

Fig. 7.21. Hydrogen-bond distance distribution (Left) and hydrogen-bond angle distribution (Right) at 60 K for Amantadine, Rimantadine and Memantine. The most possible hydrogen-bond lengths are around 0.3 nm for the three molecules. The hydrogen-bond angles seem randomly distributed (Xue and Mansoori, 2010).

According to Fig. 7.20, during the process of phase transition and self-assembly of these three molecules, the number of their hydrogen-bonds saturate at relatively low temperatures (below 100 K). From the hydrogen-bond distance-distribution graph of these three molecules at

60 K (Fig. 7.21-Left), one can observe three peaks at about 0.3 nm, which indicate that the structures are close to uniform crystal structures. However, the angle distributions (Fig. 7.21-Right) show that the hydrogen-bond angles are rather randomly distributed for all the three molecules. Both Figs. 7.19 and 7.21 indicate that these structures are not neat crystal structures contrary to adamantane self-assembled state. These three hydrogen-bonded molecules just self-assemble at particular temperatures, but may not form well-organized self-assembled crystal structures.

Xue and Mansoori (2010a,b) also performed the self-assembled (solid-state) radial distribution functions and structure factors calculations of the seven molecules as are reported in Figs. 7.22 and 7.23.

According to these figures, the investigators concluded that adamantane and adamantane+Na have more sharp peaks than other five molecules, which indicate compact solid state structures. While other five

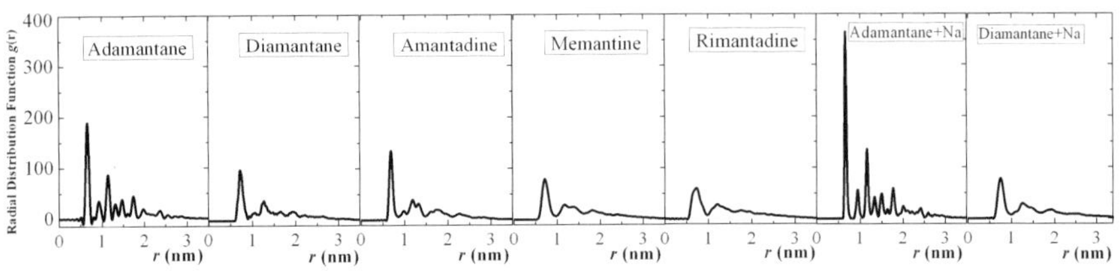

Fig. 7.22. Radial distribution functions of the seven compounds (from left: adamantane, diamantane, amantadine, rimantadine, memantine, adamantane+na., diamantane+Na) in the self-assembled (solid) state (Xue and Mansoori, 2010).

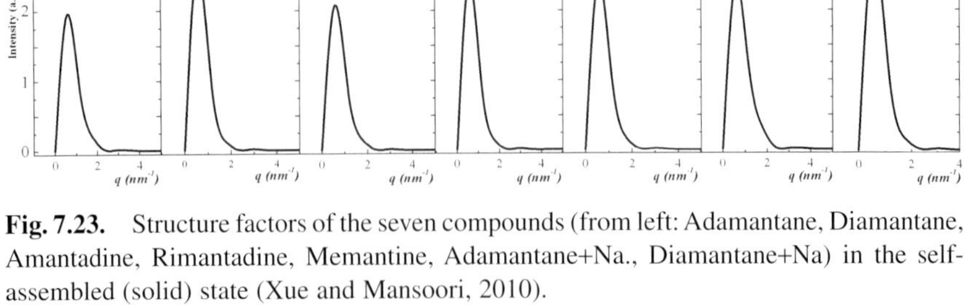

Fig. 7.23. Structure factors of the seven compounds (from left: Adamantane, Diamantane, Amantadine, Rimantadine, Memantine, Adamantane+Na., Diamantane+Na) in the self-assembled (solid) state (Xue and Mansoori, 2010).

molecules also show solid characteristic peaks, they have less number of such sharp peaks than adamantane and adamantane+Na. These results match the image observation of simulations (Fig. 7.18), i.e. adamantane and adamantane+Na have neater structures in solid state due to their molecular symmetry, which thus prove that self-assembly of those molecules are structure-dependent.

7.8. Concluding Remarks

Applications of diamondoids in materials science are quite diverse. The major reasons for their applications are their effect on increasing stiffness, chemical and thermal stability, Tg, controllable and strain free as well as retention of materials physical properties at high temperatures and under shear. In this chapter we presented modification of materials containing diamondoid molecules such as polymers, copolymers, nanocomposites, power cells overcharge protecting compounds, crystal engineering, DNA-diamondoids nanotechnology and self-assembly of diamondoids.

In applications of diamondoids in polymer synthesis the general purpose is to increase thermal stability and higher Tgs of the synthesis results. We discussed some interesting examples of applications of diamondoids in polymer science and technology which included production of thermoset polymers, incorporation of 1,6- or 4,9-diamantylene groups into polyamides, application of diamondoids in synthesis of star polymers, production of adamantyl-containing epoxy resins, synthesis of fluorinated electro-optic copolymer containing adamantane and introduction of diamantyle group in polymer structures. Diamondoids and derivatives are also being used in production of polymer nanocomposites.

Some examples in which diamondoids are used in crystal engineering include: Adamantane 1,3,5,7-tetracarboxylic acid as a synthon to form hexagonal sheet structures; the "disubstituted adamantane" molecular network for application in supramolecular nanostructures; and design of engineered crystals to mimic inorganic crystal structures.

The diamondoids-DNA nanoarchitectures are considered a route to add stiffness to the DNA-directed assembly which is lacking in the DNA-based nanostructures.

Self-assembly of diamondoid molecules and derivatives is shown to be a feasible way to produce highly symmetrical, nontoxic, controllable and strain free nanostructures for producing a variety of shapes for applications in molecular manufacturing, in nanotechnology and in NEMS and MEMS.

As we have shown in this chapter, diamondoids have been used as templates for crystallization, in the synthesis of high-temperature polymers, in crystal engineering, as molecular building blocks in nanotechnology, in DNA directed assembly, in host-guest chemistry and dynamic combinatorial chemistry, the latter being the basic science behind molecular design. Their self-assembly is being now investigated as a means to develop components suitable for MEMS and NEMS. *Sui generis* properties of diamondoids have provoked an extensive range of applications in different fields of science and technology and recently in nanoscience and nanotechnology.

Bibliography

Alexander, M. and Dubois, P. (2000). *Mater. Sci. Eng. Rev.*, **28**: 1–63.

Balazas, A. C. (2000). *Curr. Opin. Colloid Interface Sci.*, **4**: 443–448.

Belardini, A., Dominici, L., Larciprete, M. C. and Michelotti, F. (2006). *Appl. Phys. Lett.*, **89**: 231110.

Burham, K. S., Roth, R., Zhou, F., Fan, W., Brouk, E., and Stifanos, M. (2006). *J. Polym. Sci. Part A: Polym. Chem.*, **44**: 6909–6909.

Buttrey, D. N. (1960). Plasticizers, 2nd ed; Franklin: Palisades, NJ, p. 89 and 110.

Canters, G. W. and de Boer, E. (1967). *Mol. Phys.*, **13**: 395.

Canters, G. W., Hendriks, B. M. P. and de Boer, E. (1970). *J. Chem. Phys.*, **53**: 445–446.

Chen, B., Eddaoudi, M., Reineke, T. M., Kampf, J. W., O'Keeffe, M. and Yaghi, O. M. (2000). *J. Am. Chem. Soc.*, **122**: 11559–11560.

Chern, Y. T. (1998). *Macromoleucles*. **31**: 5837–5844.

Chern, Y. T. (1998). *Polymer*, **39**: 4123–4127.

Chern, Y. T., Shiue, H. C., Kao, S. C. (1998). *J. Polym Sci Part A: Polym. Chem.*, **36**: 785–792.

Chern, Y. T. and Wang W. L. (1998). *Polymer*, **39**: 5501–5506.

Childs, S. L., Wood, P. A., Rodrı'guez-Hornedo, N., Reddy, L. S. and Hardcastle, K. I. (2009). *Crystal Growth & Design*, **9**: 1869–1888.

Dheu, M. L., Gagnaire, D., Duddeck, H., Hollowood, F. and McKervey, M. A. (1979) *J. Chem Soc. Perkin Trans.*, **2**: 357–359.

Drexler, K. E. (1999). *TIBTECH.*, **17**: 5–7.

Eastmond, G. C., Gibas, M. and Paprotny, J. (1999). *European Polymer J.*, **35**: 2097–2106.

Edwards, M. R., Jones, W. and Motherwell, W. D. S. (2002). *Cryst. Eng.*, **5**: 25–36.

Esfarjani, K. and Mansoori, G. A. (2005). Chap. 16, in the *Handbook of Theoretical and computational Nanotechnology*, 2005.

Evans, O. R., Xiong, R. G., Wang, Z., Wong, G. K. and Lin, W. (1999). *Angew. Chem. Int. Ed. Engl.*, **38**: 536–538.

Evans, O. R., and Lin, W. B. (2001). *Chem. of Materials*, **13**: 2705–2712.

Flory, P. J. and Yoon, D. Y. (1978). *Nature*, **272**: 226–229.

Fort, R. C., Jr. Adamantane. (1976). The Chemistry of Diamond Molecules, Dekker, New York.

Fort, R. C. Jr. and Schleyer, P. V. R. (1964). *Chem. Rev.*, **64**: 277–300.

Garcia, J. C., Justo, J. F., Machado, W. V. M. and Assali, L. V. C. (2009). *Phys. Rev. B.*, **80**: 125421 (2009).

Garcia, J. C., Assali, L. V. C., Machado, W. V. M. and Justo, J. F. (2010). *J. Phys.: Condens. Matter*, **22**: 315303.

Giannelis, E. P. (1998). *Appl Organomet Chem.*, **12**: 675–680.

Ghosh, A., Sciamanna, S. F., Dhal, J. E., Liu, S., Carlson, R. M. K. and Schiraldi, D. A. (2007). *J. Polym. Sci. Part B: Polym. Phys.*, **45**: 1077–1089.

Grubb, T. L., Mathias, L. J. (1997). *J. Polym. Sci. Part A: Polym. Chem.*, **35**: 1743–1751.

Hoft, H. (2000). Classes in Hydrocarbon Chemistry: Syntheses, Concepts, Perspectives; Wiley-VCH: Weinheim, p. 547.

Huang, C. F., Lee, H. F., Kuo, S. W., Xu, H. and Chang, F. C. (2004). *Polymer*, **45**: 2261–2269.

Jackson, W. J., Jr. and Caldwell, J. R. (1967). *J. Appl Polym Sci.*, **11**: 211–226.

Jehnichen, D., Friedel, P., Bergmann, J., Taut, T., Tobish, J., and Pospiech, D. (1998). *Polymer*, **39**: 1095–1102.

Jenkins, T. E. and Lewis, J. (1980). *Spectrochim Acta A.*, **36**: 259–264.

Jeong, H. Y., Lee, Y. K., Talaie, A., Kim, K. M., Kwon, Y. D., Jang, Y. R., Yoo, K. H., Choo, D. J. and Jang, J. (2002). *Thin Solid Films,* **417**: 171–174.

Johnston, A., Johnston, B. F., Kennedy, A. R. and Florence, A. J. (2008). *Cryst. Eng. Comm.*, **10**: 23–25.

Kar, A. K., Lightner, D. A. (1998). *Tetrahedron, Asymmetry*, **9**: 3863–3880.

Kojima, Y. Usuki, A., Kawasumi, M., Okada, O., Fukushima, Y., Kurachi, T. and Kamigato, O. (1993). *J. Mater. Res.*, **8**: 1185–1189.

Lee, Y. K., Jeong, H. Y., Kim, K. M., Kim, J. C., Choi, H. Y., Kwon, Y. D., Choo, D. J., Jang, Y. R., Yoo, K., Jang, H. and Talaie, J. A. (2002). *Current Appl. Physics*, **2**: 241–244.

Li, J. X., Cheung, W. L. and Chan, C. M. (1999). *Polymer*, **40**: 2089–2102.

Lin, C. H., Huang, J. M., Wang, C. S. (2002). *Polymer*, **43**: 2959–2967.

Lin, L. and Argon, A. S. (1974). *J. Mater Sci.*, **9**: 2034–2051.

Malek, N., Maris, T., Simard M. and Wuest, J. D. (2005). *Acta Crystallogr. Sect. E: Struct. Rep. Online*, **61**: o518–o520.

Malik, A. A., Archibald, T. G., Baum, K. and Unroe, M. R. (1991). *Macromolecules*, **24**: 5266–5268.

Malik, A. A., Archibald, T. G., and Unroe, M. R. (1992). *J. Polym. Sci. Part A: Polym. Chem.*, **30**: 1747–1754.

Mansoori, G. A. (2005). *Principles of Nanotechnology- Molecular-Based Study of Condensed Matter in Small Systems*; World Scientific Pub. Co., Hackensack, NJ.

Mansoori, G. A. (2007). Diamondoid Molecules, in Advances in Chemical Physics **136**: 207–258.

Mansoori, G. A., George, T. F., Zhang G. P. and Assoufid L. (Ed's). (2007). Molecular Building Blocks for Nanotechnology: From Diamondoids to Nanoscale Materials and Applications, Springer, New York, Topics in Applied Physics **109**: 44–71.

Mansoori, G. A., George, T. F., Zhang G. P. and Assoufid, L. (2009). Structure and Opto-Electronic Behavior of Diamondoids, with Applications as MEMs and at the Nanoscale Level, *Progress in Nanotechnology Research*, Nova Pub's., Chapter 1, pp. 1–19.

Mathias, L. J., Lewis, C. M., Wiegel, K. N. (1997). *Macromolecules* **30**: 5970–5975.

Mathias, L. J. and Tullos, G. L. (1996). *Polymer*, **37**: 3771–3774.

Molecular Diamond Technologies. www. chevron. com/moleculardiamond Accessed Feb 19, 20011.

Moon, S., Schwartz, A. L., and Hecht, J. K. (1970). J. *Polym. Sci.*, A-1 **8**: 3665.

Nikakhtar, A., Nasehzadeh, A., Mansoori, G. A. (2007). *J. Comput'l & Theor'l Nanoscience*, **4**: 1–8.

Olah, G. A., Prakash, K. S., Shih, J. G., Krishnamurthy, V. V., Mateescu, G. D., Liang, G., Sipos, G., Buss, V., Gund; T. M., and Schleyer; P. V. R. (1985). *J. Am. Chem. Soc.*, **107**: 2764–2772.

Pathak, R. and Marx, A. (2011). *Chemistry — An Asian Journal* **6**: 1450–1455.

Peterlin, A. (1971). *J. Mater. Sci.*, **6**: 490–508.

Pinto, Y. Y., Le, J. D., Seeman, N. C., Forsyth, K. M., Taton, T. A. and Kiehl, R. A. (2005). *Nano Lett.*, **5**: 2399–2402.

Pixton, M. R., Paul, D. R. (1995). *Polymer*, **36**: 3165–3172.

Pluta, M., Bartczak, Z. and Galeski, A. (2000). *Polymer*, **41**: 2271–2288.

Rafii-Tabar, H. and Mansoori, G. A. (2004). Interatomic Potential Models for Nanostructures. *Encyclopedia of Nanoscience & Nanotechnology*. E. D. Carlson (Ed.), Nova Sci Publ's, Hauppauge, NY, **4**: 231–248.

Ramezani, H. and. Mansoori, G. A. (2007c). Diamondoids as Molecular Building Blocks for Nanotechnology, in Molecular Building Blocks for Nanotechnology: From Diamondoids to Nanoscale Materials and Applications", (Springer, New York), Topics in Applied Physics **109**: 44–71.

Ramezani, H., Mansoori, G. A. and. Saberi, M. R. (2007b). *J. Comput'l & Theor'l Nanoscience*, **4**: 96–106.

Ramezani, H., Saberi, M. R. and Mansoori, G. A. (2007a). *International Journal of on Nanoscience and Nanotechnology*, **3**: 21–36.

Reddy, D. S., Craig, D. C. and Desiraju, G. R. (1996). *J. Amer. Chem. Soc.*, **118**: 4090–4093.

Sanchez, C., Julian B., Belleville, P. and Popall, M. (2005). *J. Matter. Chem.*, **15**: 3559–3592.

Seeman, N. C. and Kallenbach, N. R. (1983). *Biophys. J.*, **44**: 201–209.

Singer, K. D., Sohn, J. E., King, L. A., Gordon, J. M., Katz, H. E., and Dirk, C. W. (1989). *J. Opt. Soc. Am. B.*, **6**: 1339–1350.

Somani, H. R., Hsiao, S. B. and Nogales, A. (2001). *Macromolecules*, **34**: 5902–5909.

Soo, P. P., Huang, B. Y., Jang, Y. I., Chiang, Y. M., Sadoway, D. R. and Mayers, A. M. (1999). *J. Electrochem. Soc.*, **146**: 32–37.

Su, X. and Jing, X. (2007). *J. Appl. Polym. Sci.*, **106**: 737–743.

Supaphol, P. and Lin, J. S. (2001). *Polymer*, **42**: 9617–9626.

Takano, T., Lin, Y. C., Shi, F. G., Carlson, B. and Sciamanna, S. (2010). *Obtical Mat*, **32**: 648–651.

Terng, C. Y. and Jen, W. J. (1995). *Tetrahedron Lett.*, **36**: 5805–5806.

Tobishima, S., Ogino, Y. and Watanabe, Y. (2003). *J. Appl. Electrochem.*, **33**: 143–150.

Tobishima, S., Takei, K., Sakurai, Y. and Yamaki, J. (2000). *J. Power Sources*, **90**: 185.

Tobishima, S. and Yamaki, J. (1999). *J. Power Sources*, **81–82**, 882–886.

Tominaga, M., Katagiri, K. and Azumaya, I. (2009). *Crystal Growth & Design*, **9**: 3692–3696.

Tominaga, M., Masu, H. and Azumaya, I. (2011). *Cryst. Growth Des.*, **11**: 542–546.

Ueda, M. (2004). *Polym. Eng. Sci.*, **44**: 1877–1884.

Vaia, R. A. and Giannelis, E. P. (2001). *Mater. Res. Soc. Bull.*, **26**: 394–401.

Vaidhyanathan, R., Natarajan, S. and Rao, C. N. R. (2004). *Encyclopedia of Supramolecular Chemistry*, 1–12.

Xu, K., Chen, M. C., Zhang, K., Hu, J. W. (2004). *Polymer.*, **45**: 1133–1140.

Xue Y. and Mansoori, G. A. (2008). *International Journal of Nanoscience*, **7**: 63–72.

Xue, Y. and Mansoori, G. A. (2010a). *Int'l J Molecular Sciences*, **11**: 288–303.

Xue, Y. and Mansoori, G. A. (2010b). Phase Transition and Self-assembly of Lower Diamondoids and Derivatives. In Proceedings of Fall 2010 MRS Meeting, Boston, MA, Nov. 30-Dec. 2, 2010.

Zaworotko, M. J. (1994). *Chem. Soc. Rev.* **23**: 283–288.

Zhang, H. X., Yao, Q. X., Jin, X. H., *et al.* (2009). *Cryst Eng Comm.*, **11**: 1807–1810.

Zhang, J., Lin, W. B., Chen, Z. F., Xiong, R. G., Abrahams, B. F. and Fun, H. K. (2001). *J. Chemical Soc. -Dalton Trans.*, **12**: 1806–1808.

Zhang, X. C., Butler, M. F. and Cameron, R. E. (2000). *Polymer*, **41**: 3797–3807.

Watanabe, Y., Morimoto, H. and Tobishima, S. (2006) *J. Power Sources*, **154**: 246–254.

Wei, Q., Lazzeri, A., Di Cuia, F., Scalari, M., Galoppini, E. (2004). *Macromol. Chem. Phys.*, **205**: 2089–2096.

Weidinger, A. and Hermans, P. (1961). *Makromol Chem.* (*Basel*). **50**: 1733.

Glossary

1,3-Diethynyladamantane

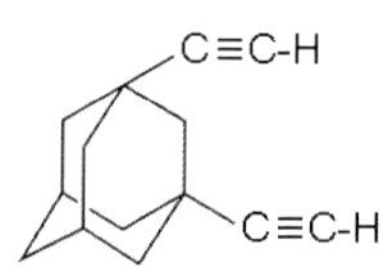

5-HT3 receptor = 5-Hydroxytryptamine receptor

A receptor for serotonin (a neurotransmitter), which activates a variety of second messenger signaling systems and through them indirectly regulates the function of ion channels.

6-31G*, 6-31G, D95(d,p), cc-pVDZ

Types of basis sets used in DFT calculations of physical properties of nanosystems.

ab initio = from the beginning (a Latin word)

Ab initio methods are based entirely on theoretical principles, with no inclusion of experimental data as a quantum mechanical approximation theory to solve the Schrödinger equation. The approximations are generally of mathematical nature.

acetylcholinesterase

An enzyme that degrades the acetylcholine (a neurotransmitter).

Ada = adamantane

adamantyl 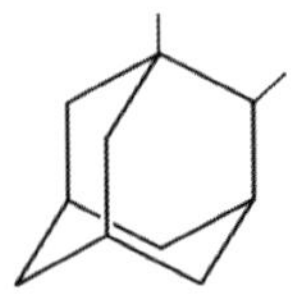

Adamantane-based chemical group or radical.

ADM = adamantane

AFM = atomic force microscope

One of the scanning probe microscopy (SPM) techniques. It allows the register of topographic images, the mapping of adhesive and elastic properties of a surface and measurement of intermolecular forces between pairs of molecules.

agonist

A compound that triggers a response by a cell when binds to its receptor.

AMPA = **a**lfa-amino-3-hydroxy-5-**m**ethyl-4-isoxazole **p**ropionic **a**cid

AMPAR = AMPA receptor

A non-NMDA-type ionotropic transmembrane receptor for glutamate that mediates fast synaptic transmission in the central nervous system.

antagonist

A compound which is able to bind to a drug or endogenous chemical's receptors and block them. Hence, it competes with the drug/endogenous chemical to occupy the receptors and inhibits the pharmacological effects of a drug/endogenous chemical to appear.

antisense RNA

Single-stranded RNA that is complementary to a mRNA strand transcribed within a cell. It may be introduced into a cell to inhibit translation of a complementary mRNA by base pairing to it and physically obstructing the translation machinery.

apoptosis

Programmed cell death.

antinociceptive

Analgesic; painlessness.

aprotic solvent

A solvent that has no H bonded to a highly electronegative atom and therefore cannot form a hydrogen bond.

aryladamantane 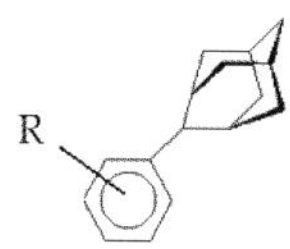

artemisinin

A drug used to treat multi-drug resistant strains of falciparum (a parasite causing malaria).

asialoglycoprotein receptor

Receptors located on liver cells, made of lectins (sugar-binding proteins) which bind asialoglycoprotein (glycoproteins from which a sialic acid has been removed) to expose galactose residues.

asphaltene

A high-molecular weight, polycyclic and polydispersed substance found in fossil fuels. It has C, H, O, N, S in its diverse chemical structure If present in crude oil it causes major fouling in petroleum flow conduits. Its fractions are deposited from petroleum in the laboratory by addition of normal alkanes.

association constant (K)

In host–guest chemistry between host (H) and guest (G) molecules is defined as $K=[H\text{–}G][H]^{-1}[G]^{-1}$.

ATRP = atom transfer radical polymerization

B3LYP

Level of the density functional theory (DFT) corresponding to the three-parameter hybrid functionals of Becke and the correlation functional of Lee, Yang, and Parr.

base pairing rule

Molecular structure constraints of DNA and RNA in hydrogen bonds formation among the four purine and pyrimidine bases.

basin

Low-lying or depressed area that collects sediments.

BBB = blood-brain barrier

bcr/abl

An abnormal gene which causes the body to make too many abnormal white blood cells (chronic myeloid leukemia).

BE = binding energy

Biliverdin pigment

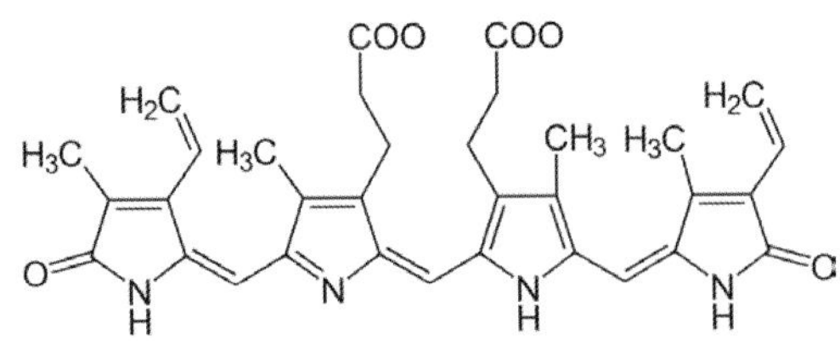

Biomarker

Organic compound whose carbon structure is synthesized by formerly living organisms and is sufficiently stable to be recognized in crude oil or in the organic matter of ancient sediments and rocks.

birefringence

Splitting of light into two unequally reflected waves by an optically anisotropic medium like quartz.

bitumen

Native substance of variable color, hardness and volatility, composed predominantly by carbon and hydrogen and sometimes associated with minerals, the nonmineral constituents are asphaltenes, polycyclic aromatic hydrocarbons and resins and it is largely soluble in carbon disulfide.

BPEI = branched polyethylenimine

CAIC = cyclodextrin-adamantane inclusion complexation

carbocation

An ion with a positively charged carbon atom.

carbocation rearrangement

Transformation from an unstable to a more stable state of an ion with a positively charged carbon ion, through the use of various shifts within the molecule. It is quite commonly used in organic chemicals reactions.

carbonium ion = carbocation

carbon cycle

Geochemical cycle by which carbon atom of carbon dioxide molecule is fixed to form organic matter by photosynthesis or chemosynthesis to further return to their original state through biological or chemical oxidation.

carbonium mechanism = carbocation rearrangement

CDM = Conduction Band Minimum

Minimum range of electron energies, higher than valence band, sufficient enough to free an electron from binding with individual atom and release it to move within the lattice of a solid crystal.

CDx = cyclodextrin

chloroquine

A drug commonly used in the treatment and prevention of malaria (4-aminoquinoline).

Cmpd = compound

condensate

Also known as NGL standing for "natural gas liquid". Hydrocarbon mixture that is gaseous in the underground high pressure reservoirs but turns into liquid and separates from natural gas when reaches at atmospheric pressure.

CNS = central nervous system

CPK = Corey–Pauling–Koltun model

crystal engineering

Utilization of non-covalent intermolecular forces in the solid-state to design new nanomaterials with desired functions.

curing agent

Hardening agent — For example addition of adamantly group to polymers cause their curing.

CVD = chemical vapor deposition

cyclodextrin = CDx

A cyclic oligosaccharide compound containing commonly six to eight glucose units.

cytokines

Nonspecific water-soluble glycoproteins with a short half-life produced and secreted abruptly by white blood cells in response to an external stimulus, and which act as chemical messengers between cells.

DDN = diamondoid-DNA nanoarchitecture

DDS = 4,4′-diaminodiphenyl sulfone

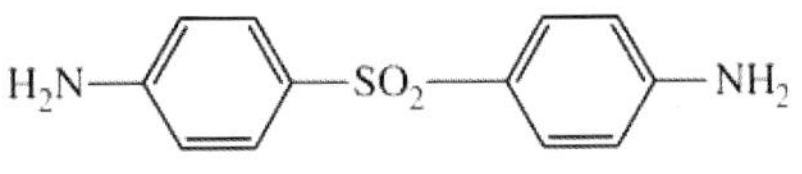

DFT = density functional theory

A first principles technique to solve the Schrödinger equation. A general approach to the *ab initio* description of quantum many-particle systems, in which the original many-body problem is rigorously recast in the form of an auxiliary single-particle problem It was developed by Walter Kohn and co-workers in the 1960's based on the original ideas of Thomas and Fermi.

DGEBAda = diglycidyl ether of bisphenol-adamantane

An adamantane-containing epoxy resin.

DICY = dicyandiamide (NH_2–CNH–NH–CN)

diamondoid

Cage-like saturated hydrocarbons with strong stiff structures resembling diamond. Contain three dimensional networks of covalent bonds. Considered the favorite molecular building blocks for nanotechnology.

DOS = density of states

DSC = differential scanning calorimeter

EC_{50} = A measure of a drug's potency — It is the drug concentration that will induce a response halfway between the baseline and maximum after some specified exposure time.

EDOS = electronic density of states

electron affinity = EA

An energy distance between the bottom of the conduction band of a semiconductor or isolator and the energy level in vacuum.

electrophile = electron-lover

A reagent which is attracted to electrons participating in a chemical reaction by accepting an electron pair to bond to a nucleophile.

enantiomeric

Optical antipode, enantiomers are a pair of optical isomers.

esterases

Enzymes that catalyze the hydrolysis of an ester bond.

exergonic

Capable of releasing energy in the form of work.

facies

A rock or stratified body that is formed under certain conditions of sedimentation, reflecting a particular process or environment and can be distinguished from others by its appearance or composition.

FCC = face-centered cubic = cubic closest packing (ccp)

A cubic close-packed structure with an extra point in the centre of each cube face.

Fibrinogen-GPIIB/IIIA

A complex responsible for aggregation of platelets in the bloodstream

FISH = fluorescent-*in-situ* hybridization

The method for utilizing fluorescently labeled DNA probes to detect or confirm gene or chromosome abnormalities that are generally beyond the resolution of routine cytogenetics.

FTIR = Fourier-transform infrared spectroscopy

germa-adamantane = $Ge_{10}H_{16}$

GABA receptors

Receptors for γ-aminobutyric acid. GABA is an amino acid that acts as an inhibitory neurotransmitter in the CNS.

GC/MS/SIM = gas chromatography–mass spectrometry with selected ion monitoring

glycans

Refers to polysaccharides or oligosaccharides.

glycosylation

The enzymatic process that produces one of the fundamental biopolymers found in cells (along with DNA, RNA, and proteins). This process links saccharides and produces glycans attached to proteins, lipids, or other organic molecules.

GOR = gas-to-oil ratio

A terminology used in petroleum engineering defining the gas-to-oil ratio of a (reservoir fluid) crude oil brought to standard state condition.

GOX = glucose oxidase

Gram-positive bacteria

Bacteria which are stained dark blue or violet by Gram staining.

Gram staining

Bacteria differentiation into Gram-positive and Gram-negative groups based on the physicochemical properties of their cell walls.

H_2N_2

A subtype of *influenza* A viruses, presenting hemagglutinin type 2 and neuraminidase type 2 proteins.

H_3N_2

A mutated strain of H_2N_2, containing hemagglutinin type 3, instead of type 2 protein.

Hartree–Fock (HF) approximation

It is a self-consistent field interactive procedure to calculate the "best possible" single determinant solution to the time-independent Schrödinger equation. This approximation considers that in a many-electron system, the interaction of electrons can be represented by a potential. Each component particle of the interaction process is governed by an effective potential due to all other particles of the system.

HbA1c

Glycosylated hemoglobin A1c. A form of hemoglobin used primarily to identify the average plasma glucose concentration over prolonged periods of time A marker of two to three months blood glucose levels for diabetes testing.

hemagglutinin

An influenza virus surface glycoprotein which is responsible for viral binding and entry into host epithelial cells. Sixteen hemagglutinin serotypes (H1 to H16) are present in animal species.

hepatoma

The most common type of liver-originating cancer.

hexogen

Also known as RDX, cyclonite and with IUPAC name: 1,3,5-Trinitroperhydro-1,3,5-triazine.

histamine

A chemical present in cells throughout the body and released during an allergic reaction.

HOMO = highest occupied molecular orbital

HOMO level for organic semiconductors is like valence band to inorganic semiconductors.

host–guest chemistry

Construction of molecular receptors by self-assembly so that such receptors could, to some extent, gain molecular recognition capability.

Humic acid

The major organic constituents of sediments. It generally consist of a complex and heterogeneous mixture of polydispersed materials formed by biochemical and chemical reactions during the decay and transformation of plant and microbial remains Plant lignin and its transformation products, as well as polysaccharides, melanin, cutin, proteins, lipids, nucleic acids and fine char particles are important components of humic acid.

hybridization = hybridization

Forming of new orbitals by mixing atomic orbitals. Used to qualitatively describe properties of atomic bonding.

hydrocracking

A catalytic cracking followed by hydrogenation process at high pressure and temperature to lower the molecular weight of, for example, heavy hydrocarbons to lighter ones.

Hydrogen Index = HI ≡ $[100 \times S_2]$/TOC)

HI is a parameter used to characterize the origin of organic matter in geological samples like kerogen S_2 stands for hydrocarbons produced during cracking of kerogen HI typically ranges from ~100 to 600 in geological samples.

IC_{50}

Half maximal inhibitory concentration. An effectiveness measure of a compound in inhibiting a bio function.

imatinib

An anticancer drug

impetigo

Contagious skin infection that usually produces blisters or sores on the face, neck, hands, and diaper area. It is generally caused by one of two bacteria: *Staphylococcus aureus* or group A *streptococcus*.

incretin biology

Biology of the gastrointestinal hormones causing increase in the amount of insulin in blood.

inertinite

Inert maceral that have angular shapes, typically showing cellular structure. Its phytoclasts exhibit high reflectance, no fluorescence and are opaque in transmitted light.

IP = Ionization Potential

IPDT = integral procedure decomposition temperature

IPP = isotactic polypropylene 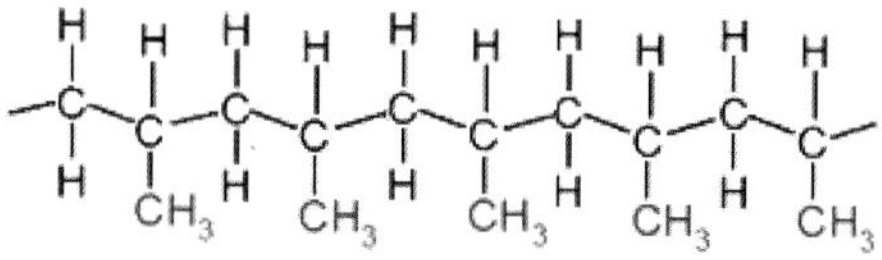

kerogen

Organic matter in sedimentary rocks that is insoluble in organic solvents and resists to acid digestion of the mineral matrix.

kerogen types

Classification of kerogen based on composition and oil-forming characteristics: Type I kerogen presents highly oil-prone organic matter showing Rock-Eval pyrolysis hydrogen index (HI) over 600 mg. hydrocarbons/g TOC when thermally immature. Type II kerogen is oil-prone organic matter showing Rock-Eval pyrolysis HI in the range of 300–600 mg hydrocarbons/g TOC when thermally immature. Type IIS kerogen is similar to type II, but presents high organic sulfur contents (8–14% wt %) and appears to generate oil at lower thermal exposure. Type III kerogen is gas-prone organic matter showing Rock-Eval pyrolysis HI in the range of 50–150 mg hydrocarbons/g TOC when thermally immature. Type IV kerogen is inert organic matter showing Rock-Eval pyrolysis HI below 50 mg hydrocarbons/g TOC in thermally immature rocks. In this type of kerogen, organic matter has been extensively reworked and oxidized.

LDA = local density approximation

A method of calculation of the electronic band structure of solids. Approximations to "exchange-correlation energy functional" in DFT. They purely depend upon the value of the electronic density at each point in space.

liptinite

Oil-prone maceral that shows low reflectance, high transmittance and intense fluorescence at low levels of maturity.

Log P $\equiv \log_{10}$ {[Compound]octanol phase/[Compound]water phase}

The partition coefficient used to describe lipophilicity for neutral compounds, or for compounds that exist in a single form. It is measured through the differential solubility of a compound in two immiscible solvents. The most commonly used solvent system is octanol/water.

LPEI = linear polyethylenimine

LPG = liquefied petroleum gas

Also called GPL, LP Gas and autogas. It is the gas separated from liquid petroleum upon expansion of reservoir crude oil to atmospheric condition during petroleum production. It is then compressed and liquefied and used as a fuel.

LUMO = lowest unoccupied molecular orbital

LUMO level for organic semiconductors is like conduction band to inorganic semiconductors.

maceral

Particulate organic component of kerogen that can be distinguished by its microscopic morphology and undergoes changing in physicochemical properties with thermal maturity. It is divided in three main groups: liptinites, vitrinites, and inertinites.

macrophage

A type of white blood cell that ingests (takes in) foreign material. Macrophages are key players in the immune response to foreign invaders such as infectious microorganisms.

MBB = molecular building blocks

A term used in nanotechnology to certain molecules like diamondoids to define the building blocks of nanostructures.

McMurry coupling reaction

A reaction of two ketones or aldehydes producing an alkene in presence of a titanium chloride and a reducing agent.

mechanosynthesis

Chemical synthesis controlled by mechanical means operating on the atomic and molecular scale and performing direct positional selection of chemical reaction sites. Provided such mechanical manipulation means with atomic- and molecular-precision could be developed in the future mechanosynthesis may become possible.

mefloquine = Lariam

A preventive oral malaria drug

MIC = minimal inhibitory concentration

The lowest concentration at which no bacteria are observed to grow as determined by standardized OD_{595} (optical density at 595 nm).

minus-sense RNA

A single-strained RNA genome of the opposite polarity as mRNA.

***M. tuberculosis* H37Rv strain**

The most commonly studied laboratory TB strain, which has been subject to *in vitro* passage.

MPECVD = microwave plasma enhanced chemical vapor deposition

mRNA = messenger RNA

The main intermediary in gene expression, translating the DNA's genetic code into the amino acids that make up proteins.

MRSA = methicillin-resistant *Staphylococcus aureus*

A bacterial infection which is quite resistant to some antibiotics.

nanometer

A billionth of the meter or 10^{-9} m.

NCD = nanocrystalline diamond

ND = nanodiamond

NEA = negative electron affinity

NEMS = nanoscale electromechanical systems

neuraminidase

An influenza virus surface glycoprotein involved in the budding (release) of new virions from infected cells. Nine neuraminidase serotypes (N1 to N9) are present in animal species.

NEXAFS = XAS = near edge x-ray absorption fine structure spectroscopy

Developed for elucidating structure of molecules bonded to surfaces, in particular low-Z molecules.

NLE = nonlinear ellipsometry

Also called "Teng and Man technique", investigates the electro-optical (EO) susceptibility of organic polymer films.

NMDAR = *n*-methyl-*d*-aspartate receptor

Ionotropic receptor for glutamate. It plays a critical role in synaptic plasticity mechanisms and thus is necessary for several types of learning and memory.

NMR = nuclear magnetic resonance

NNI = National Nanotechnology Initiative

Program established in USA in fiscal year 2001 to coordinate US nanotechnology research and development activities.

norbornene

A bicyclic olefin.

nucleotide

The group of molecules that, when linked together, form the structural of DNA and RNA. A nucleotide is composed of a nucleobase (nitrogenous base), a five-carbon sugar (either ribose or 2′-deoxyribose), and one to three phosphate groups.

OI = oxygen index

oil deadline

The depth at which most liquid hydrocarbons in petroleum are cracked to gas.

oil shale

Compact rock of sedimentary origin presenting ash content of more than 33% and bearing organic matter that yields oil when pyrolyzed, but not appreciable when extracted with solvents.

orphan disease

Designation for a disorder affecting fewer than 200,000 people. It gives little incentive for pharmaceutical industry to produce medicines to treat it.

orphan drug

Designation given to a drug for treatment of an orphan disease.

Oxygen Index = OI ≡ [100 × S3]/TOC

A parameter used to characterize the ratio of oxygen to carbon and the type of organic matter in geological samples like kerogen. OI values range from 0–150.

PAH = polyaromatic hydrocarbon

pandemic

An epidemic that occurs over a large geographical area, including multiple countries.

PANI = polyaniline

A conducting polymer, which its electrical behavior depends on oxidation and doping states.

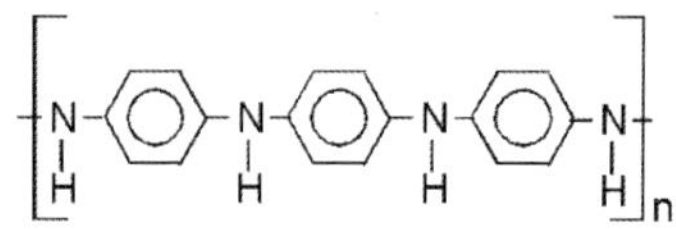

PAP = poly-adamantane-pyrrole

PBS = phosphate-buffered saline

PC = polycarbonate

PEGylation

Refers to attachment of polyethylene glycol to another molecule or coating of a nanostructure with it.

PES = photoelectron spectra

pharmacophore

A part within a drug molecule which plays a major role in interaction of that drug with its target. There may be more than one pharmacophoric site within the chemical structure of a drug molecule.

PES = photoemission spectroscopy = photoelectron spectroscopy

For bonding energy measurement of electrons emitted from a substance by the photoelectric effect.

phytoclast

An identifiable particle or maceral in kerogen.

plasmid DNA

A circular DNA separated from chromosomal DNA.

PN

polymer nanocomposite

polyplex

A molecule created to protect DNA from undesirable degradation during the transfection process and improve its delivery into the cell.

polycation

A chemical entity (molecule or complex) with positive charges at several parts.

polymantane = diamondoid

PP = polypropylene

prion

A modified form of a normal protein. Prion protein changes into a pathogenic (harmful) form and damages the central nervous system of cattle and humans.

prodrug

Inactive precursor of a drug, which is converted into its active form in the body by normal metabolic processes.

proteolytic enzyme

An enzyme that catalyzes the breakdown of proteins into amino acids.

pyrobitumen

Thermally altered bitumen that is infusible and relatively insoluble in carbon disulfide and has a maximum atomic hydrogen/carbon ratio of 05.

pyrrole

QC = quantum confinement

QCE = quantum confinement effect

QMC = quantum Monte Carlo calculation

A stochastic computation/simulation technique based on random number generation of the variable to estimate statistically the function.

quantum confinement = QC

The phenomenon of nonzero lowest energy and quantization of the allowed energy levels in low-dimensional structures, arising from the confinement of electrons within a limited space.

$R_0 = R_{0\ random} = R_m$

Random vitrinite reflectance measured with non-polarized light.

R_{max}

Mean maximum vitrinite reflectance measured with polarized light.

R_{min}

Mean minimum vitrinite reflectance measured with polarized light.

racemates = racemic mixture

A mixture containing equal amounts of left- and right-handed enantiomers of a chiral molecule.

reassortment

May occur when one animal is infected by two different strains of the same virus. These two strains may mix in a cell, exchange one or more of their eight genetic segments and result in a new virus that were not observed in either donor strain.

RNA = Ribonucleic acid

A molecule present in the cell of all living beings and essential for the synthesis of proteins.

Rock-Eval

A commercial pyrolysis instrument used to identify the type and maturity of organic matter and to detect petroleum potential in sediments through four basic parameters (S_1, S_2, S_3 and T_{max}), as explained below.

S_1

The amount of free hydrocarbons (gas and oil) in the sedimentary sample (in milligrams of hydrocarbon per gram of rock).

S_2

The amount of hydrocarbons generated through thermal cracking of non-volatile organic matter in sedimentary rocks. S_2 is an indication of the quantity of hydrocarbons that the rock has the potential of producing should burial and maturation continue.

S_3

The amount of CO_2 (in milligrams CO_2 per gram of rock) produced during pyrolysis of kerogen. S_3 is an indication of the amount of oxygen in the kerogen and is used to calculate the oxygen index.

SAM = self-assembled monolayer

self assembly

A bottom-up process in nanotechnology. The spontaneous assembly of molecules into structured, stable, non-covalently joined aggregates.

sila-adamantane = $Si_{10}H_{16}$

source rock

A fine-grained organic-rich sediment that in its natural setting could generate or has already generated hydrocarbons.

***sp3* hybridized**

A type of hybridization of 3 P and 1 S orbitals forming a new orbital having properties of both S and P orbitals.

SXE = soft-X-ray emission

A spectroscopy method used for determination of electronic material structure. For example it probes the highest occupied sp^3 hybridized states in diamondoid molecules, by fluorescent photons observation emitted as valence states decay into the carbon 1*s* core-hole.

synaptic transmission

Transmission through the junction across which a nerve impulse passes from an axon terminal to a neuron, muscle cell, or gland cell.

T_d group symmetry

Tetrahedral rotational symmetry with reflection symmetry (like methane molecule).

TEY = total electron yield

T_g

Glass transition temperature. Associated with cessation (or "freezing in") of molecular motion of polymer chains.

TGA = thermogravimetric analysis

thermal maturity

An indicator of the maximum temperature a rock has experienced.

tight-binding (TB) approximation

The method of assuming valence electrons in an atom are completely delocalized when a solid is formed.

TMBP = tetramethyl biphenol

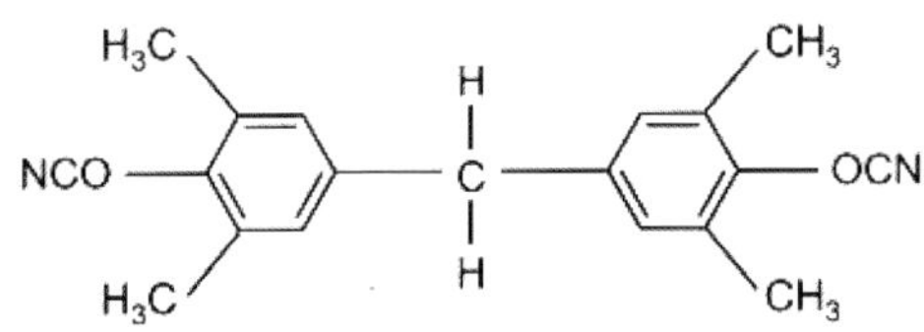

TNF = tumor necrosis factor

One of the important cytokines that plays a key role in activation and induction of some immune system cells and cellular immunity processes responsible for proinflammatory and inflammatory response reactions.

TOC = total organic carbon

The sum of pyrolyzed carbon plus residual organic carbon.

Tp

Poling temperature

transfection

Deliberate introduction of nucleic acids into a cell which may cause unexpected morphologies and abnormalities in the cell.

tropism

Response/reaction of an organism to an external stimulus shown as movement.

TSR = thermochemical sulfate reduction

UDD = ultradispersed diamond

Nanoparticles obtained by a detonation process.

UDOS = Unoccupied local electronic density of states

UNCD = ultrananocrystalline diamonds

van Hove singularity

A discontinuity in solids' density of states. A general application of it is in optical absorption spectra analysis.

vitrinite

A gas-prone maceral having angular shape and derived mostly from lignin, cellulose and tannins of vascular plants. It shows intermediate reflectance (low-grey) and transmittance and no fluorescence unless contaminated by liptinite.

vitrinite reflectance

The major maturity parameter used to assess thermal history of dispersed organic matter of source rocks. Its measurements are done on polished minerals under an oil-immersion microscope objective, using filtered (546 nm) incident light with or without polarizer. About 25–100 randomly oriented vitrinite grains (known as phytoclasts) are present.

vulcanization

To design crosslinking: formation of polymer networks by interchain chemical bonds. Originally developed by Charles Goodyear in 1839 to vulcanize sulfur for rubber crosslinking.

WAXD = wide-angle X-ray diffraction

To determine polymers crystalline structure.

wet gas

Natural gas consisting of methane and vapors of heavier hydrocarbons that are in liquid state in pure normal conditions.

XAS = X-ray absorption spectroscopy = NEXAFS

Uses intense and tunable X-ray beam produced by a synchrotron radiation source. Measures electronic and local geometric structure of materials in gas-, liquid(dissolved)-, or solid- state.

XPS = X-ray photoelectron spectroscopy

Measures elemental composition, empirical formula, chemical state and electronic state of elements in a material.

zero-shear viscosity

Viscosity at rest and undisturbed, measured at the limit of low shear rate.

zinc-finger protein

A DNA-binding protein that contains a zinc atom.

Index